Springer

Berlin
Heidelberg
New York
Barcelona
Budapest
Hong Kong
London
Milan
Paris
Santa Clara
Singapore
Tokyo

192

Topics in Curren

Organofluorine Chemistry

Fluorinated Alkenes and Reactive Intermediates

Volume Editor: R. D. Chambers

With contributions by
B. Améduri, V. V. Bardin, B. Boutevin,
R. D. Chambers, W. R. Dolbier, Jr.,
U. A. Petrov, J. F. S. Vaughan

Springer

This series presents critical reviews of the present position and future trends in modern chemical research. It is addressed to all research and industrial chemists who wish to keep abreast of advances in the topics covered.

As a rule, contributions are specially commissioned. The editors and publishers will, however, always be pleased to receive suggestions and supplementary information. Papers are accepted for "Topics in Current Chemistry" in English.

In references Topics in Current Chemistry is abbreviated Top. Curr. Chem. and is cited as a journal.

Springer WWW home page: http://www.springer.de
Visit the TCC home page at http://www.springer.de/

ISSN 0340-1022
ISBN 3-540-63171-2
Springer-Verlag Berlin Heidelberg New York

Library of Congress Catalog Card Number 74-644622

Printed in Germany

Cover design: Friedhelm Steinen-Broo, Barcelona
Typesetting: Fotosatz-Service Köhler OHG, 97084 Würzburg
SPIN: 10556003 66/3020 – 5 4 3 2 1 0 – Printed on acid-free paper

Volume Editor

Prof. Richard D. Chambers, F. R. S.

Department of Chemistry
University of Durham
South Road
DH1 3LE Durham, Great Britain
E-mail: R. D. Chambers@durham.ac.uk

Editorial Board

Preface

Fluorine is unique, in that it is possible to replace hydrogen in an organic molecule by fluorine either singly or multiply and, in so-doing, create a potentially infinite extension to organic chemistry that is entirely synthetic. The excitement of the chemistry of these systems stems from the unique reactions that ensue and the 'special-effects' that introduction of fluorine impart. Indeed, these effects are now exploited in a remarkable array of applications across the whole of the chemical, pharmaceutical, and plant-protection industries, although this is not widely appreciated. In this book, we have gathered authors with immense experience in various aspects of their field and each is a world-authority on the important topics they have described. Some topics, like the use of elemental fluorine, and enzymes in synthesis, are relatively new areas that are rapidly growing.

We dedicate the book to a long standing friend, Professor George Olah, in the year of his 70th birthday, in recognition of his massive achievements.

Durham, May 1997 Dick Chambers

Contents

Contents of Volume 193
Organofluorine Chemistry: Techniques and Synthous

Volume Editor: R. D. Chambers

Nucleophilic Reactions of Fluorinated Alkenes

R. D. Chambers and J. F. S. Vaughan

Department of Chemistry, University of Durham, South Road, Durham, DH1 3LE, U.K.
E-mail: r.d.chambers @ durham. ac. uk

Fluorinated -alkenes and -cycloalkenes have a special relationship with their hydrocarbon analogues, usually exhibiting a chemistry that is complementary. For example, the fluorinated systems are frequently susceptible to nucleophilic attack, in some cases dramatically so, and therefore reactions of nucleophiles with fluorinated alkenes often reveal unique new chemistry. This chapter covers electrochemical reduction, principles governing orientation and reactivity of fluorinated alkenes towards nucleophiles, fluoride ion as a nucleophile and the 'mirror-image' relationship of this chemistry with that of proton-induced reactions, reactions with nitrogen-, oxygen-, carbon- centred nucleophiles etc., and, finally, chemistry of some oligomers of fluorinated -alkenes and -cycloalkenes.

Keywords: Fluorinated alkenes; nucleophilic attack; fluoride ion; oligomerisation; rearrangements; displacement; carbanions.

Topics in Current Chemistry, Vol. 192

1
Introduction

Highly fluorinated alkenes are generally very electron-deficient species and therefore susceptible to attack by a wide range of nucleophiles. In this chapter we will attempt to give an overview of this chemistry, with emphasis of more recent work. The subject was reviewed many years ago [1] and other publications have included discussion of this subject [2–9]. We will also include some chemistry of fluorinated dienes, where appropriate.

We can summarise the sequence of steps in the reaction of a nucleophile with a fluorinated alkene, as outlined in Scheme 1, where charge-transfer interaction (1) would occur, without being rate-determining, and then either single-electron transfer (SET) giving (2) or two-electron transfer, giving carbanionic intermediates (3) could occur, probably on the same reaction path. A halophilic process, leading to (4), is also possible in some cases. For example, nucleophilic attack on bromine occurs with bromotrifluoroethene [10–12], in competition with attack on carbon, to give mixtures of products (Scheme 2).

Clearly, factors influencing electron-affinity of the fluorinated alkene, i.e. the influence of F or perfluoroalkyl (R_F) as a substituent at the double-bond, are important to consider, as are the effects of these substituents on carbanion (3) stabilities.

$$\text{Nuc} + \text{C=CX} \longrightarrow \left[\text{Nuc}\right]^{\delta+}\left[\text{C=CX}\right]^{\delta-} \longrightarrow \text{Nuc—C—CX}^-$$
$$\text{(1)} \qquad \qquad \text{(3)}$$

$$\text{NucX} + \text{C=C}^- \qquad \left[\text{Nuc}\right]^{+\bullet}\left[\text{C=CX}\right]^{\bullet-} \longrightarrow \text{Product Formation}$$
$$\text{(4)} \qquad \qquad \text{(2)}$$

Scheme 1

$$t\text{-BuO}^- + \text{F}_2\text{C=CFBr} \longrightarrow t\text{-BuOCF}_2\text{CFBr}^- \xrightarrow{\text{F}_2\text{C=CFBr}} t\text{-BuOCF}_2\text{CFBr}_2 \quad \text{etc.}$$
$$\text{(30\%)}$$

$$\text{F}_2\text{C=CFBr} + \underset{\text{N}}{\overset{\text{F}}{\bigcirc}} \xrightarrow{\text{i}} \text{CF}_3\text{CFBr}_2 + \text{N} \overset{\text{CF}_3}{\underset{\text{F}}{\bigcirc}} \text{N}$$
$$\text{(25\%)} \qquad \qquad 15\%$$

i, CsF, 90 °C, sulfolan

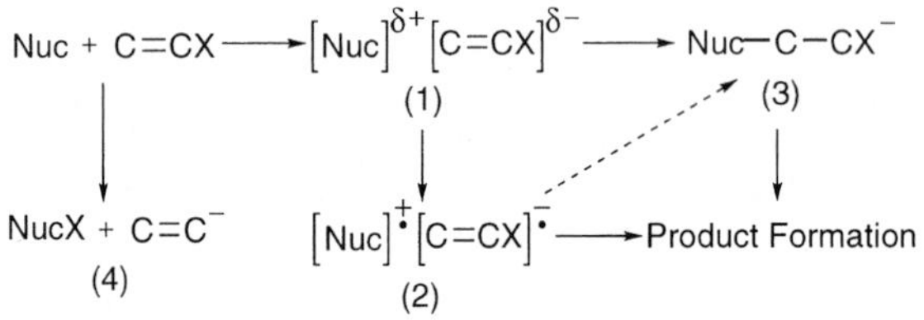

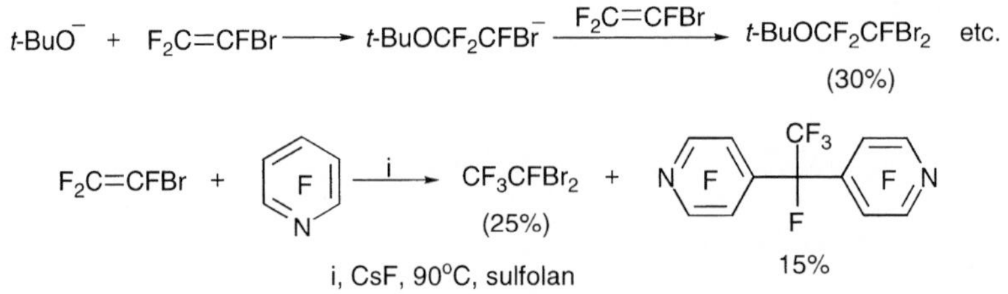

Scheme 2

2
Substituent Effects of Fluorine and Perfluoroalkyl Groups

Perhaps the most revealing information regarding the influence of fluorine and fluorocarbon groups on π-bonds comes from photo-electron spectroscopy [13, 14] where it has been clearly established that a perfluoroalkyl group *lowers* π-orbital energies, as we would anticipate for an electron-withdrawing substituent, but a fluorine atom directly attached to the carbon of the double bond has an effect that is not very different from that of a hydrogen atom. Qualitatively, these data suggest that, for the ground-state of an alkene, we can describe simply the effects as illustrated, (5) and (6), where the electron-withdrawing effect of a fluorine atom (6a) may be offset by p-π interaction, which increases π-electron density (6b) [15–17] (Scheme 3).

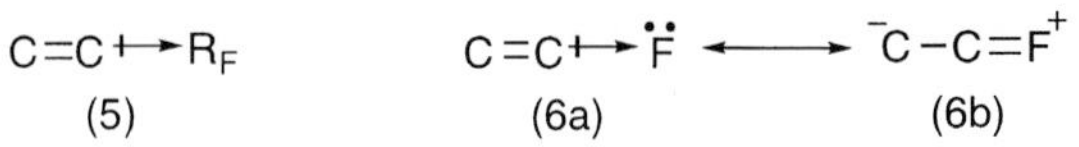

$$C=C \longleftrightarrow R_F \qquad C=C \longleftrightarrow \ddot{F} \longleftrightarrow {}^-C-C=F^+$$

$$\qquad (5) \qquad\qquad (6a) \qquad\qquad (6b)$$

Scheme 3

(7) Strongly stabilising   (8a) Stabilising   (8b) Slightly destabilising

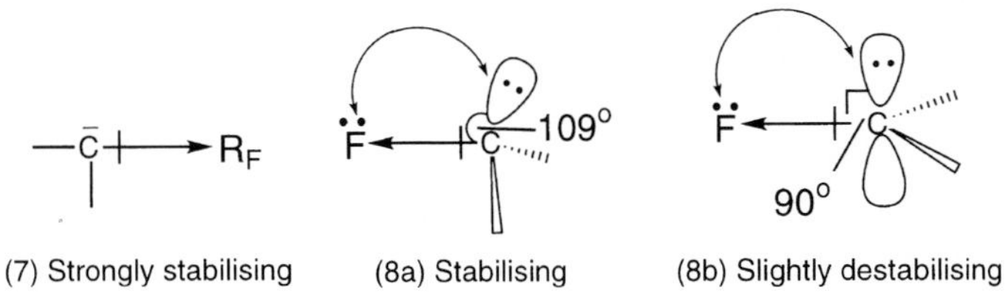

Scheme 4

A similar situation applies for the analogous effects on carbanion stabilities (Scheme 4) [3, 18, 19], where perfluoroalkyl is always strongly carbanion stabilising (7) but the effect of fluorine is more complex. For a tetrahedral structure (8a), fluorine is significantly *stabilising* (NB., CF_3H is 10^{40} times more acidic than CH_4 [3, 19]), but the situation changes substantially with the shape of the carbanionic site because, for a planar system (8b), fluorine is slightly *destabilising*. These simple generalisations allow us to rationalise a great deal of the chemistry described below. Negative hyperconjugation has been a much debated subject [3, 19] and the effect, relating to perfluoroalkyl groups, could be represented as in (9a, b), where charge from the carbanionic centre is transferred to the σ^* orbital of the C-F bond (Scheme 5).

$$F_3C-\bar{C}- \quad \longleftrightarrow \quad F^- \ F_2C=C\big\langle$$

$$(9a) \qquad\qquad (9b)$$

Scheme 5

There is little doubt that this is a real effect, well supported by calculation [20, 21], and there is structural evidence arising from bond lengths in the salt $CF_3O^-(Me_2N)_3S^+$ [22]. Nevertheless, it is not clear how much the effect contributes to carbanion stabilities because most kinetic data are ambiguous in the need to invoke the effect [3, 23], but here this is not considered to be an important issue.

3
Electrochemistry and Related Electron-Transfer Processes

In principle, the simplest form of nucleophile is, of course, a cathode and reduction potentials are particularly revealing about structure and reactivity [24]. Furthermore, in some cases, electrochemical reduction is synthetically useful. Much of the numerical data currently available comes from systems where the reduction process is non-reversible, although not all [25].

Table 1. Half-wave reduction potentials of fluorinated alkenes ($-E_{1/2}$, V vs SCE) [24]

Alkene	Reduction potential/V
$CF_2=CF_2$	3.20
$CF_2=CFR_F$ ($R_F=CF_3$, n-C_3F_7, i-C_3F_7)	2.60–2.80
$CF_2=C(CF_3)_2$	2.20
$CF_3CF=CFR_F$ ($R_F=CF_3$, C_2F_5, n-C_3F_7, i-C_3F_7, C_5F_{11})	2.12–2.24
$(CF_3)_2C=CFR_F$ ($R_F=CF_3$, C_2F_5)	1.40–1.53
$CFH=CFC_2F_5$	2.46
$CF_3CH=CHCF_3$	2.34

$$C=C-CF \xrightarrow[\text{Slow}]{+e^-} \left[\overset{\bullet}{C}-\overset{-}{C}-CF \right] \xrightarrow{-F^-} \left[\overset{\bullet}{C}-C=C \right] \xrightarrow{\text{etc}}$$
$$(10) \qquad\qquad\qquad (11) \qquad\qquad (12)$$

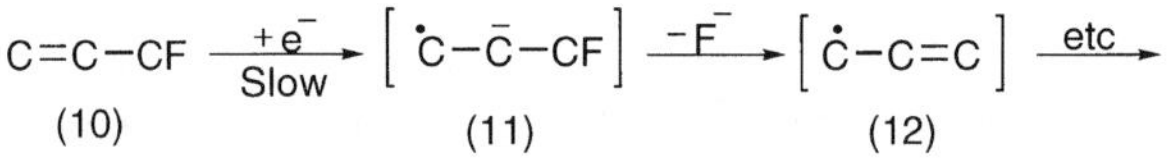

Scheme 6

Consequently, if we consider the type of process involved, e.g. (10) to (11) to (12) (Scheme 6), then loss of fluoride ion could, in principle, occur in concert with addition of an electron to (10). However, the low influence of added proton-donors [26, 27] indicates that addition of the electron to (10) is the slow step and, therefore, the reduction potential data may be meaningfully related to orbital energies [24, 25]. A selection of data is contained in Table 1 and these are reasonably interpreted on the basis that a perfluoroalkyl group leads to a significant lowering of π^* energies and thus we have electron accepting properties diminishing in the order:

$$(R_F)_2C=C(R_F)_2 > (R_F)_2C=CFR_F > (R_F)_2C=CF_2 \sim R_FCF=CFR_F$$
$$> R_FCF=CF_2 > CF_2=CF_2$$

i.e., tetrafluoroethene has the highest reduction potential in the series.

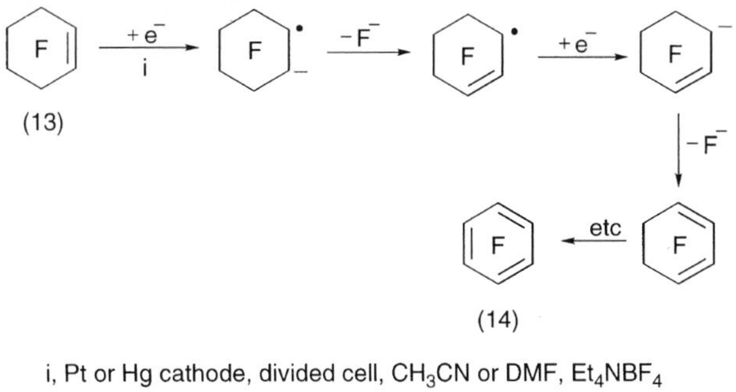

Scheme 7

Reductive defluorination occurs with perfluorocyclohexene (**13**), forming hexafluorobenzene (**14**) [26–28] by a series of one-electron transfer steps (Scheme 7).

However, perfluoro-cyclobutene (**15**) and -cyclopentene (**18**) behave quite differently [28] because electropolymerisation occurs at the cathode, to give interesting conducting materials (**17**), (**20**) respectively, whose properties have not been explored. It seems most likely that the propagation process in each case involves nucleophilic attack on the perfluorocycloalkene by an intermediate carbanion, (**16**) or (**19**). The likely propagating processes are outlined in Scheme 8. In contrast, cathodic reductive dimerisation of the trifluoromethyl derivative (**21**) occurs [29].

In other cases, e.g. **22**, electro-reduction involving replacement of fluorine by hydrogen (**23**) [24, 30] or dimerisations (**24, 25**) [31], have been observed [30] (Scheme 9).

Interestingly, some of these processes are mimicked by reactions with nucleophiles and it is clear that one-electron transfer from the nucleophile is involved. Indeed, a remarkable process [32], Scheme 10, begins with perfluorodecalin (**26**) and must proceed via intermediate polyfluorocycloalkene derivatives, e.g. (**27**), in which successive electron transfers occur, and the final product is a naphthalene derivative (**28**). So far, this is the only case in which a *saturated* perfluorocarbon has been reported to react in this way, to give meaningful products.

Tertiary phosphines lead to a variety of reductive defluorinations, e.g. (**29**) to (**31**) and coupling reactions, e.g. (**29**) to (**35**) and (**32**) to (**34**) (Scheme 11) [29, 30]; the intermediate species in these cases are, however, ylides, i.e. (**33**) and (**30**), which then react further giving (**34**) and (**35**), respectively. Other examples of ylide formation are described later (Sect. 7).

Reaction of perfluoroisobutene with phosphines also gives coupling reactions (Scheme 12) [33].

In contrast, reaction of triphenylphosphine with perfluorocyclobutene gave a product which Burton and co-workers established as the ylide (**36**), by X-ray crystallography [34], with significant double-bond character (Scheme 13). An analogous arsine derivative has also been described [35].

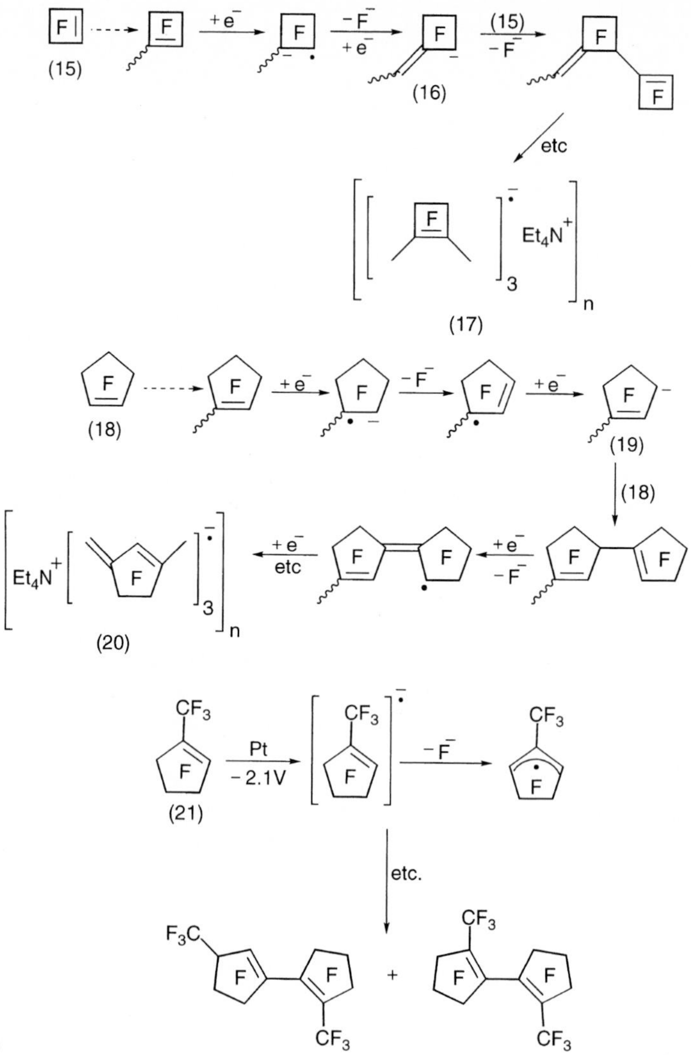

Scheme 8

Sodium amalgam is effective for promoting defluorination (Scheme 14), to produce an interesting series of dienes (**38**), (**40**) and (**42**), especially (**38**) [7, 8, 36], but tetrakis(dimethylamino)ethene (TDAE) (**43**) (Scheme 15) is a much safer system to use and, consequently, with the latter reagent, the process may be scaled up [36]. TDAE (**43**) is successful because its donor capacity is high, i.e. it has been variously described as being similar to that of zinc [37] or to alkali-metals [38] and will react, therefore, with most perfluorinated alkenes or -cycloalkenes. Furthermore, the fact that the dienes (**38**), (**40**) and (**42**) may be isolated derives from the fact that the dienes have more CF= sites than the starting alkenes and consequently their respective reduction potentials vary by

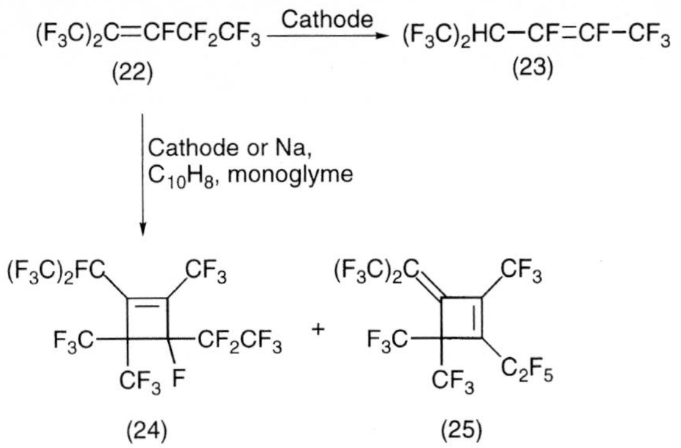

$$(F_3C)_2C=CFCF_2CF_3 \xrightarrow{\text{Cathode}} (F_3C)_2HC-CF=CF-CF_3$$

(22)   (23)

Cathode or Na,
$C_{10}H_8$, monoglyme

(24)   +   (25)

Scheme 9

(26)   (27)   (28)

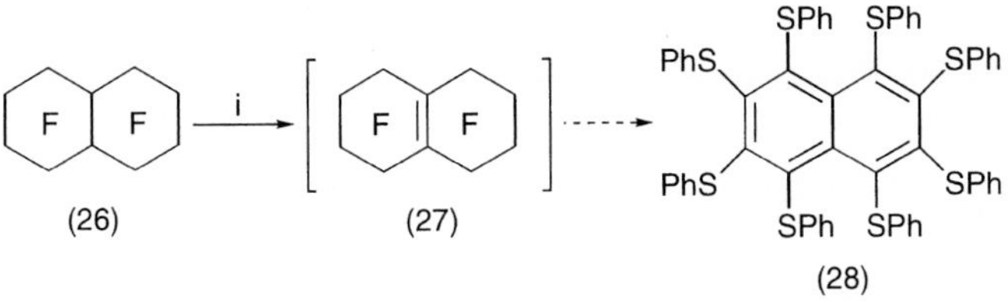

i, PhSNa, 1,3-Dimethylimidazolidin-2-one, 60-70°C, 10 d (65%)

Scheme 10

$$(F_3C)_2C=CFC_2F_5 \longrightarrow (F_3C)_2FC-\overset{-}{C}-C_2F_5 \longrightarrow (F_3C)_2C=C-C_2F_5$$

(29)   (30) $\overset{|}{PR_3}$ ... PR_3 / F

$-R_3PF_2$ etc

$F_2C=C\overset{CF_3}{\underset{C_3F_7}{}}$

(31)

(32)   (33)

(R = NMe$_2$)   (32)

CF$_3$ (34)   i, R$_3$P, 15°C $-R_3PF_2$

Scheme 11

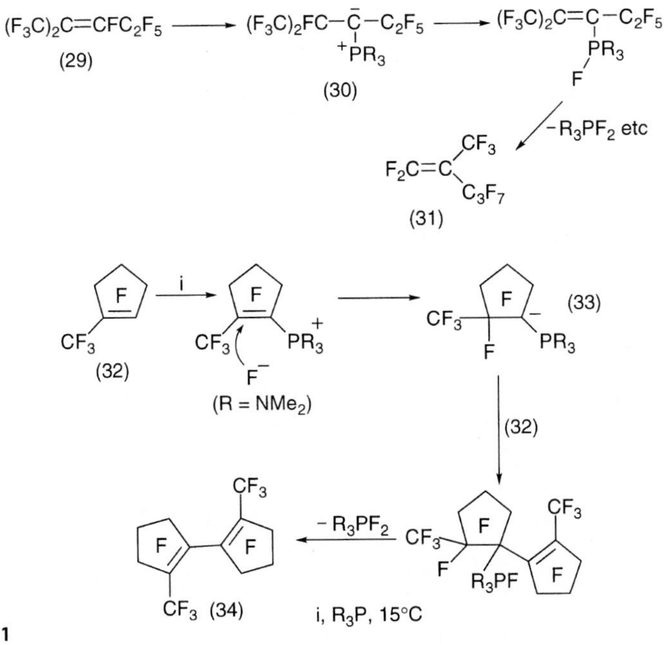

$$(30) + (31) \longrightarrow (F_3C)_2FC-\overset{\overset{\displaystyle C_2F_5}{|}}{\underset{\underset{\displaystyle PR_3F}{|}}{C}}-CF=C\overset{\displaystyle CF_3}{\underset{\displaystyle C_3F_7}{<}} \longrightarrow \overset{\displaystyle F_3C}{\underset{\displaystyle F_3C}{>}}C=C\overset{\displaystyle C_2F_5}{\underset{\displaystyle CF=C<_{C_3F_7}^{CF_3}}{<}}$$

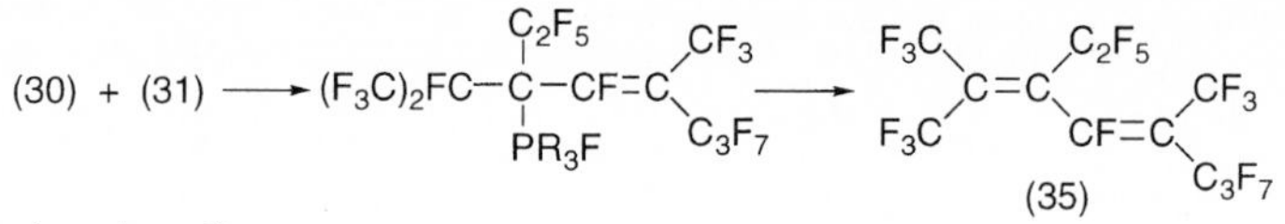

(35)

Scheme 11 (continued)

$$2\ \overset{\displaystyle F_3C}{\underset{\displaystyle F_3C}{>}}C=CF_2 \ + \ PBu_3 \ \xrightarrow[-Bu_3PF_2]{i} \ \overset{\displaystyle F_3C}{\underset{\displaystyle F_3C}{>}}C=C\overset{\displaystyle CF:C(CF_3)_2}{\underset{\displaystyle CF:C(CF_3)_2}{<}}$$

+

$$\overset{\displaystyle F_3C}{\underset{\displaystyle F_3C}{>}}C=C\overset{\displaystyle F}{\underset{\displaystyle CF:C(CF_3)_2}{<}}$$

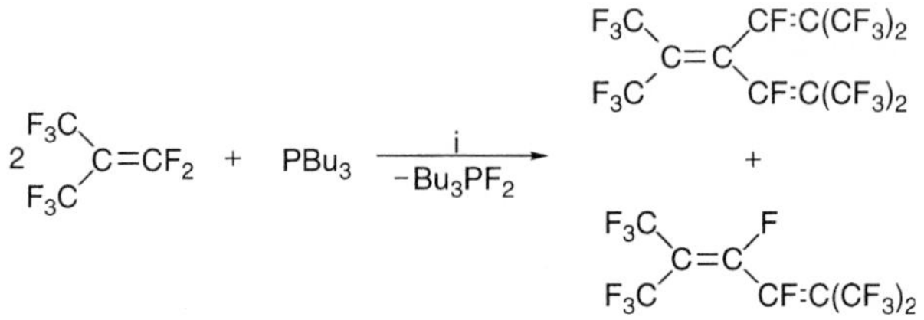

i, CH$_3$CN, -30°C

Scheme 12

$$\boxed{F\,|} \ + \ Ph_3P \ \xrightarrow{i} \ \langle\ \rangle{F}=PPh_3$$

(36)

i, Et$_2$O, 25°C

Scheme 13

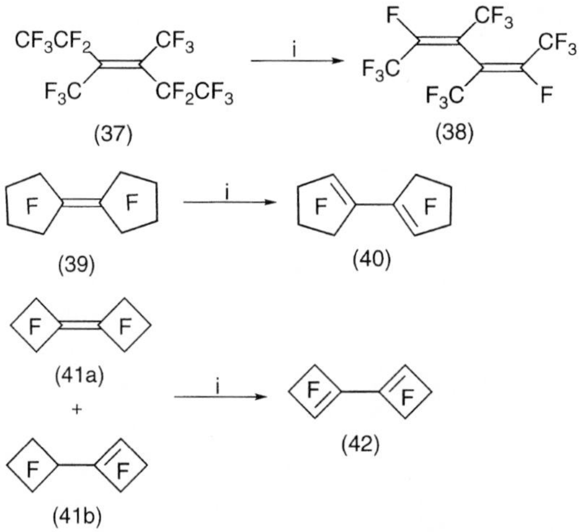

$$\overset{\displaystyle CF_3CF_2}{\underset{\displaystyle F_3C}{>}}C=C\overset{\displaystyle CF_3}{\underset{\displaystyle CF_2CF_3}{<}} \ \xrightarrow{i} \ \text{(38)}$$

(37)

(39) $\xrightarrow{i}$ (40)

(41a)

+

(41b)

$\xrightarrow{i}$ (42)

i, Na, Hg (0.5% w/w), water cooling

Scheme 14

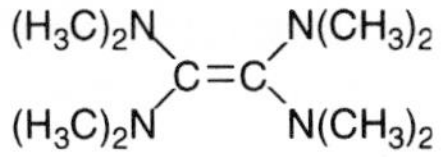

(43)

Scheme 15

Table 2. Reduction potentials of some perfluorinated alkenes and dienes (vs S.C.E.) [36]

Alkene	Reduction potential/V	Diene potential/V	Reduction
(37)	−1.62 ± 0.05	(38)	−2.35 ± 0.10
(39)	−1.06 ± 0.03	(40)	−2.03 ± 0.01
(14a)	−1.10 ± 0.03	(42)	−2.25 ± 0.01
(41b)	−1.23 ± 0.03	(42)	−2.25 ± 0.01

almost 1 volt, thus making the dienes isolable in preference to further reaction with (**43**) (Table 2).

4 Reactivity and Regiochemistry of Nucleophilic Attack

A great deal of chemistry involving nucleophilic attack on fluorinated alkenes (**44**) may be rationalised on the basis of some simple ground-rules and assumptions.

i. There is a significant ion-dipole interaction [15] that contributes to the much greater reactivity of alkenes bearing fluorine (**46**) vs chlorine (**47**) at comparable sites [39], and a terminal difluoromethylene (**44**) is especially reactive (Schemes 16, 17) [39, 40].

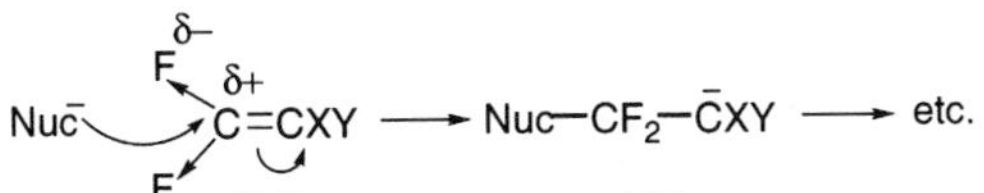

(44) (45)

Scheme 16

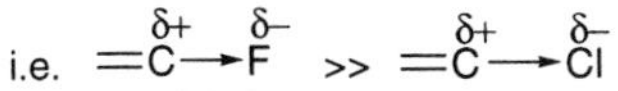

(46) (47)

Scheme 17

ii. Fluorine attached to carbon, which is itself adjacent to the carbanionic site (**45**) is carbanion stabilising and therefore strongly activating, e.g. when X or Y in (**45**) is CF_3.
iii. When fluorine is directly attached to the carbanionic site, e.g. X or Y = F in (**45**), the result is usually activating but much less so than in (ii). Thus, we

have an increase in reactivity in the series (**48**)–(**50**) [16, 24, 41, 42], and we can see that this corresponds to an increase in stabilities of the derived intermediate carbanions (Scheme 18) (**48a**)–(**50a**):

$$\underset{(48)}{CF_2{=}CF_2} < \underset{(49)}{CF_2{=}CFCF_3} < \underset{(50)}{CF_2{=}C(CF_3)_2}$$

$$\underset{(48a)}{Nuc{-}CF_2{-}\bar{C}F_2} < \underset{(49a)}{Nuc{-}CF_2{-}\bar{C}FCF_3} < \underset{(50a)}{Nuc{-}CF_2{-}\bar{C}(CF_3)_2}$$

Scheme 18

However, this type of argument is insufficient to account for the greater reactivity of perfluoropropene (**49**) than perfluoro-2-butene (**51**) (Scheme 19) because the corresponding intermediate (**51a**) could only have marginally different stability from that of (**49a**). There is also the greater reactivity of (**52**) than (**53**) (Scheme 20) to account for, where any difference in stability of intermediate carbanions would be marginal.

$$Nu\bar{c} + \underset{(51)}{CF_3CF{=}CFCF_3} \longrightarrow \underset{(51a)}{Nuc{-}CF(CF_3){-}\bar{C}FCF_3} \longrightarrow \text{Products}$$

Scheme 19

$$\underset{(52)}{(R_F)CF{=}C(R_F)_2} > \underset{(53)}{(R_F)_2C{=}C(R_F)_2}$$

Scheme 20

Consequently, a Frontier-Orbital approach has also been used to account for reactivity and orientation of attack [43]. This approach recognises that HOMO-LUMO interaction, between nucleophile and fluorinated alkene respectively, will be important and that replacing fluorine in a fluorinated alkene by trifluoromethyl reduces LUMO energy. This increases reactivity, providing that the trifluoromethyl groups are on the same carbon atom of the double bond, i.e. (**48**)–(**50**). However, coefficients also appear to be important and introduction of trifluoromethyl increases the coefficient in the LUMO at the adjacent carbon, i.e. (**54**) (Scheme 21). When two trifluoromethyl groups are attached to adjacent carbon atoms (**55**) then their effect on coefficients, and hence reactivity, is opposing and the reactivity order $CF_2{=}CF_2 < CF_2{=}CFCF_3 > CF_3CF{=}CFCF_3$ is observed.

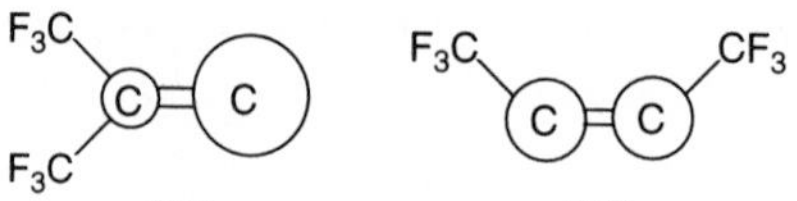

(54) (55)

Scheme 21

Strain is a very important factor affecting reactivity and this is probably best illustrated by the relative reactivities of the dienes (**38**), (**40**) and (**42**) [44], towards methanol giving the methoxy derivatives indicated (Scheme 22). Diene (**42**) reacts vigorously with neutral methanol, (**40**) reacts only over several days, while base is required to induce reaction with (**38**).

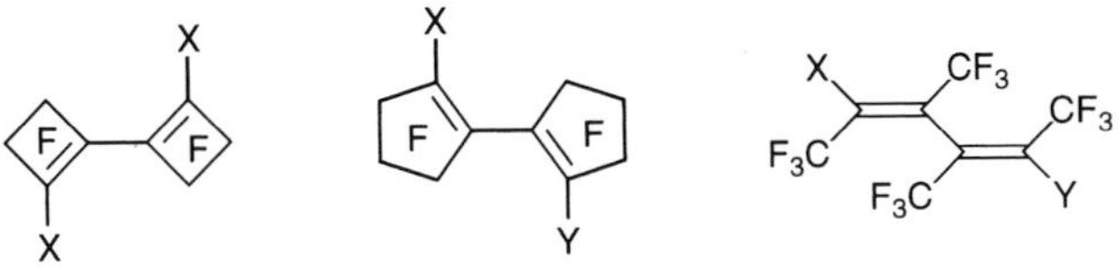

X, Y = F (42)
X = OCH_3

X, Y = F (40)
X = F, Y = OCH_3;
X, Y = OCH_3

X, Y = F (38)
X = F, Y = OCH_3
X, Y = OCH_3

Scheme 22

Electronic effects in the dienes (**38**), (**40**) and (**42**) are essentially equivalent and, therefore, these considerable differences may be taken, generally, as illustrative of the contributions that angle-strain may introduce.

5
Reactions Involving Fluoride Ion [3, 9]

There is, essentially, a mirror-image relationship between the chemistry of alkenes and highly fluorinated alkenes, in that the latter are intrinsically susceptible to nucleophilic attack and this analogy is very well illustrated by reactions that involve addition of fluoride ion. We are familiar with the massively important developments of Professor Olah and his co-workers [45] in generating observable cations (**56**) by using super-acids and there is an interesting analogy, therefore, in the fact that fluoride ion may be added to a variety of unsaturated fluorocarbons to give observable anions (**57**) (Scheme 23).

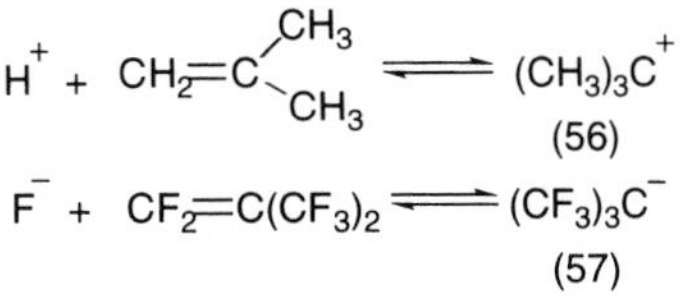

Scheme 23

The nature of the counter-ion and the solvent are extremely important in determining the stability of these anions [46] and cesium and the "TAS" cation, i.e. $(Me_2N)_3S^+$, are currently the most effective [9, 46, 47]; the latter is particularly useful since the starting fluoride, $TAS^+Me_2SiF_2^-$, is soluble in various organic media. Examples of stable anions are shown in Scheme 24 [46].

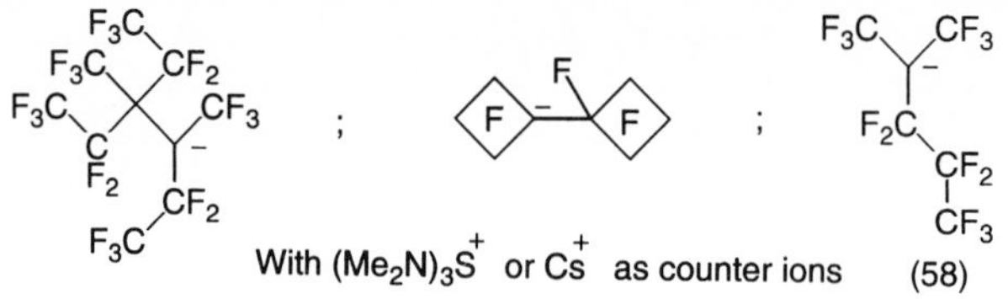

(58)

Scheme 24

It is claimed that addition of catalytic quantities of a quaternary ammonium salt greatly enhances both reactivity and yield [48] in the formation and subsequent reactions of (**58**). Tertiary fluorinated carbanions, generated from corresponding unsaturated precursors, react with a variety of electrophiles [46, 49, 50] (Scheme 25), the most surprising being with fluorocarbon iodides, to give remarkable fluoride-bridged products, e.g. (**60**) [47].

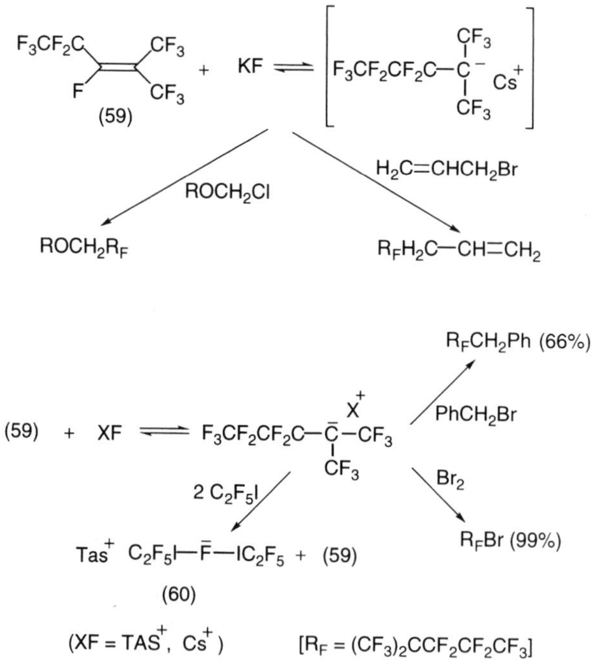

Scheme 25

Reactions with hexafluoropropene oxide (**61**) provide a good route to perfluorinated ketones [51, 52] and trapping with acid fluorides directly also gives ketones [53] (Scheme 26).

The inverse relationship that exists between fluoride ion-induced reactions of fluorinated alkenes [3] and proton-induced reactions of alkenes may be adduced from the processes described in the following sections.

i.e. F_3C—[ring with F, F, O]—F + F^- $\rightleftharpoons$ CF_3CF_2COF

(61)

$CF_3CF_2COF + R_F^- \longrightarrow CF_3CF_2COR_F$

e.g. (61) + $F_2C{=}CFCF_3$ $\xrightarrow{KF}$ $C_2F_5COCF(CF_3)_2$ (92%)

$RCOF + F_2C{=}CFCF_3 \xrightarrow{F^-} RCOCF(CF_3)_2$

(e.g. R = CH₃, CH₃CH₂, etc.)

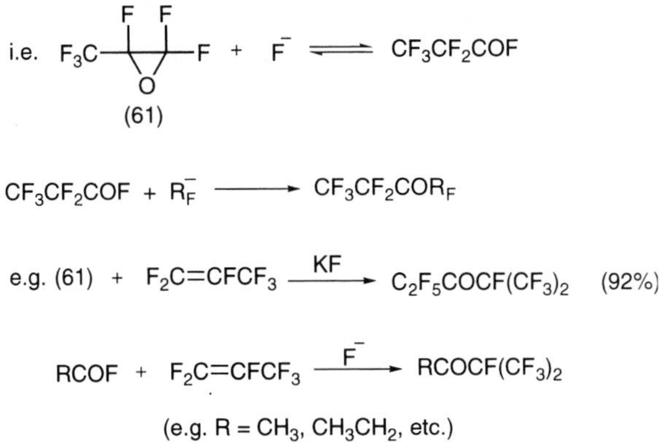

Scheme 26

5.1
Additions [3]

Even short-lived fluorinated carbanions, generated by addition of fluoride ion, have been trapped by various electrophiles (Scheme 27).

$F^- + CF_2{=}C\big\langle \rightleftharpoons CF_3{-}C\big\langle$

I_2 / 1. CO_2 2. H_2SO_4 \ $HgCl_2$

$CF_3{-}\overset{|}{\underset{|}{C}}{-}I$ $CF_3{-}\overset{|}{\underset{|}{C}}{-}COOH$ $F_3C{-}\overset{|}{\underset{|}{C}}{-}HgCl$

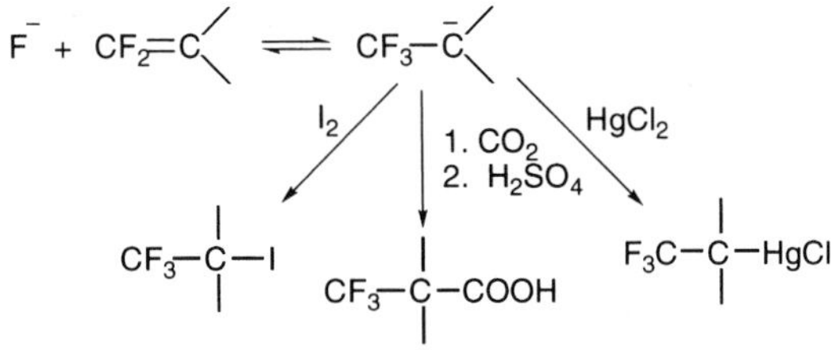

Scheme 27

5.2
Nucleophilic Analogues of Friedel Crafts Reactions [3, 54]

Trapping of carbanions by reactive fluorinated aromatic compounds gives a very simple method of introducing bulky perfluoro substituents (Scheme 28), and the more recent use of TDAE (43) to initiate these processes is discussed in the next section (see Scheme 32).

[pyridine with F] + $CF_2{=}CFCF_3$ $\xrightarrow{i}$ [R_F-substituted pyridine with F] + [R_F-substituted pyridine with F]

i, CsF, 70°C, Sulpholan $R_F = (CF_3)_2CF$

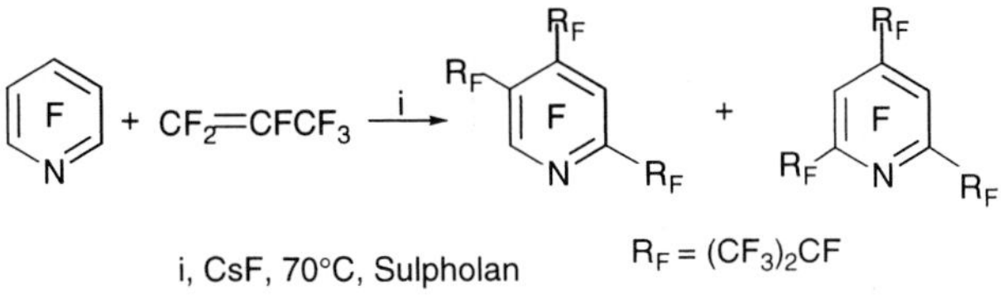

Scheme 28

5.3 Oligomerisation

Oligomerisation of tetrafluoroethene [55–57] leads to the interesting "internal" perfluoroalkene (**37**), a tetramer [58] which, together with the pentamer (**66**) are the principal products (Scheme 29). Remarkably, the hexamer (**67**), i.e. with a terminal difluoromethylene group, is also formed rather than an internal isomer and this is probably attributable to the destabilising influence of steric effects on the alternative structures.

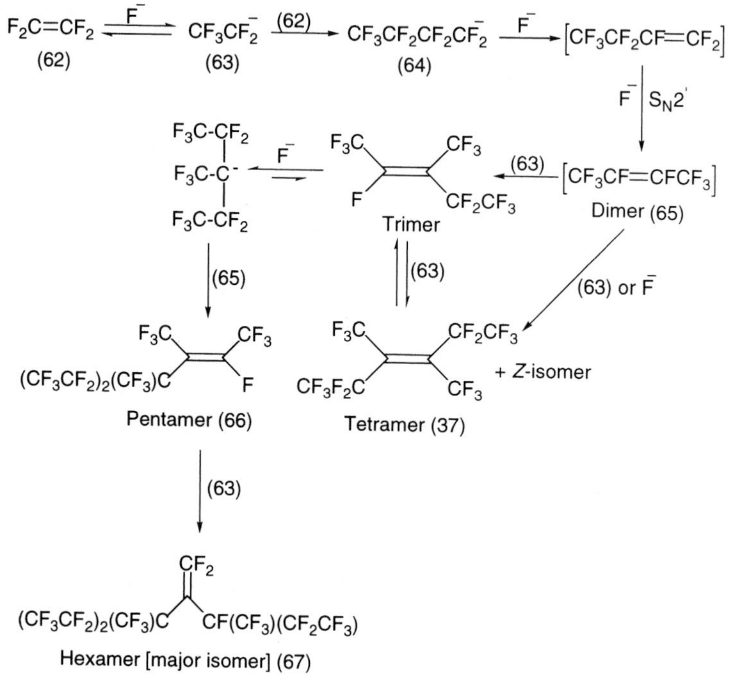

Scheme 29

Other systems, e.g. perfluoro-propene [54, 59–61] and -cycloalkenes [62, 63] (Scheme 30), may also be oligomerised by fluoride ion. A number of co-oligomerisations have also been successfully carried out [54, 64–66].

A particularly useful reaction of this type involves the direct formation of hexakis(trifluoromethyl)cyclopentadiene (**71**) (Scheme 31), or the corresponding cyclopentadienide (**72**), from the diene (**38**) by a fluoride ion induced reaction with pentafluoropropene [67–69]. Recent work [54] has shown that very active sources of fluoride ion can be generated by direct reaction of amines, especially TDAE (**43**), with perfluorinated alkenes or perfluorinated aromatic compounds and these essentially solventless systems promote both oligomerisations (see above) and polyfluoroalkylations. The absence of solvent makes recovery of product very easy, e.g. in high-yielding formation of (**73**), (**74**) or (**75**) (Scheme 32).

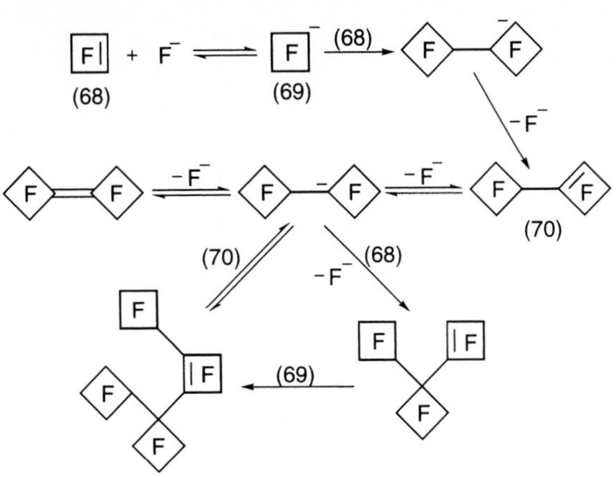

Scheme 30

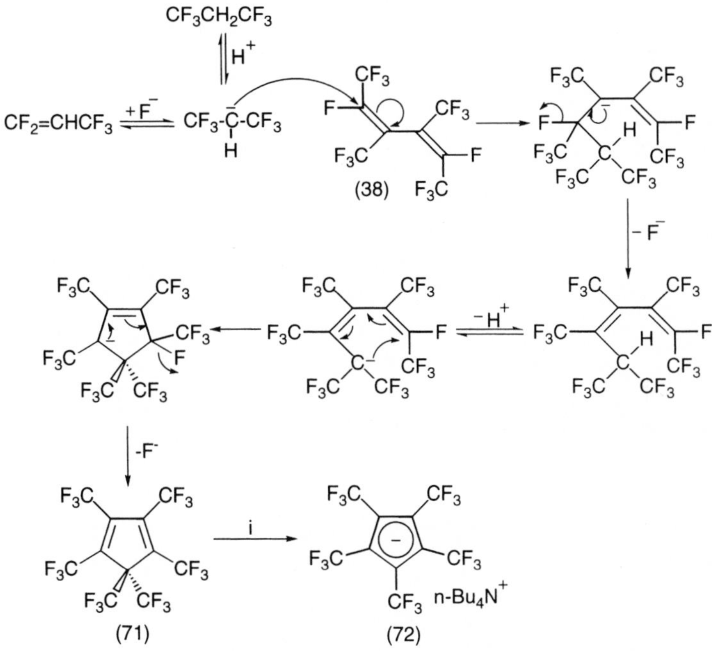

i, n-$Bu_4N^+ I^-$, CH_3CN, reflux

Scheme 31

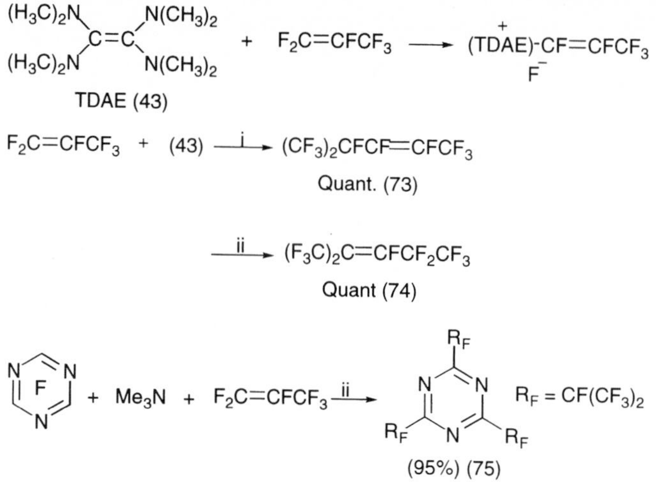

i, No solvent, room temp.; ii. No solvent, 60°C

Scheme 32

5.4 Rearrangements

Fluoride-induced rearrangements of terminal to "internal" isomers of perfluoroalkenes are well established [70–72], for example the rapid isomerisation of perfluoro-1-pentene to the thermodynamically more stable isomer occurs in the presence of cesium fluoride (Scheme 33) [73]. Allylic displacement reactions occur readily and with a stereochemistry suggesting steric control (Scheme 34) [74].

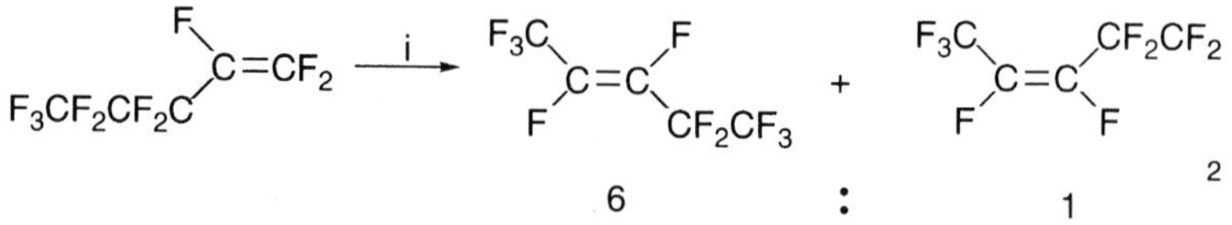

i, CsF, diglyme, ca. 20°C, 20 min

Scheme 33

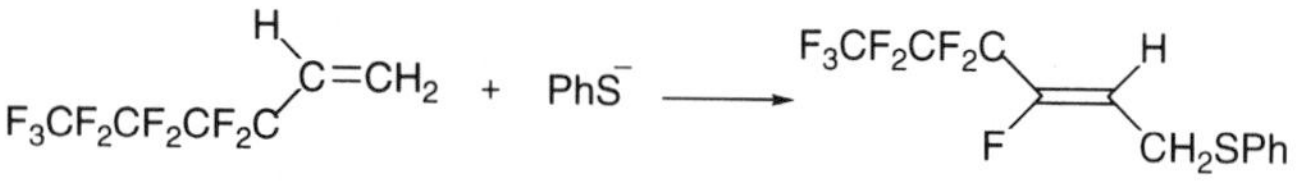

Scheme 34

A remarkable rearrangement of the strained system (**70**)–(**76**) has been observed, where three of the carbon atoms of a four-membered ring appear as trifluoromethyl groups [75] (Scheme 35).

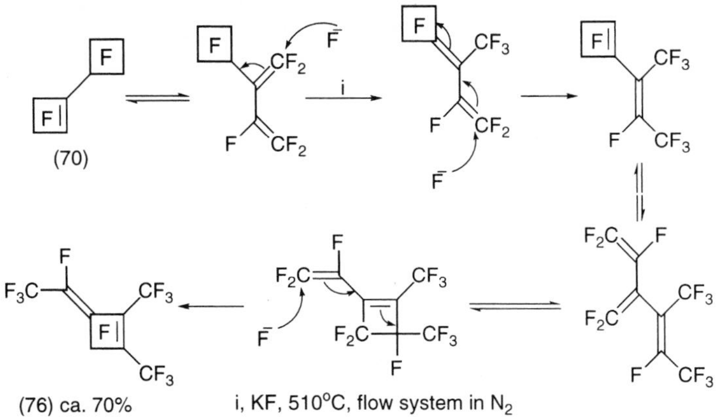

Scheme 35

Intermolecular transfer of trifluoromethyl has been demonstrated in the fluoride ion induced rearrangement of the perfluorinated alkene (**77**) to the isomer (**78**). Again, the driving force in this process is to produce an isomer (**78**) with fewer vinylic fluorine sites than in the starting isomer (**77**). The intermediate trifluoromethyl anion has been trapped with perfluoropyrimidine (**79**) [76] (Scheme 36), demonstrating the intermolecular nature of the process.

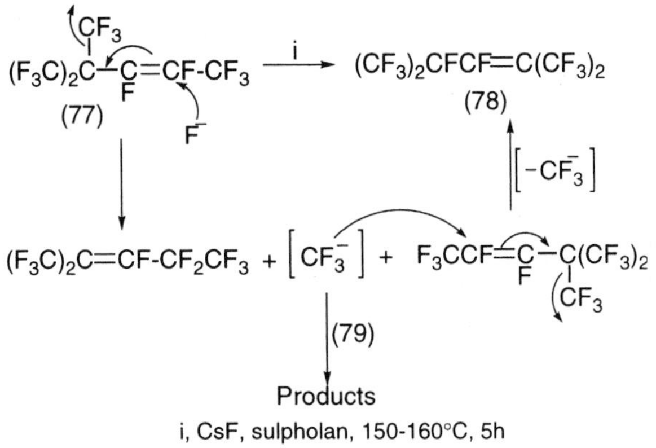

Scheme 36

Conversion of the trimer (**80**) to the seven-membered ring system (**81**) occurs readily with cesium fluoride [3, 75], whereas in the presence of TAS fluoride, the di-anion (**82**) is trapped. These observations lead to the most likely mechanism for rearrangement as that in Scheme 37 [3]. The cyclisation step shown in Scheme 37 is made more easily accepted by the fact that the diene (**83**) (Scheme 38), undergoes rapid rearrangement in the presence of fluoride ion, giving the cyclic system (**86**) [77]. The ready cyclisation of a crowded anion (**84**) to give what appears to be a sterically unfavourable intermediate (**85**) would not be easily predictable!

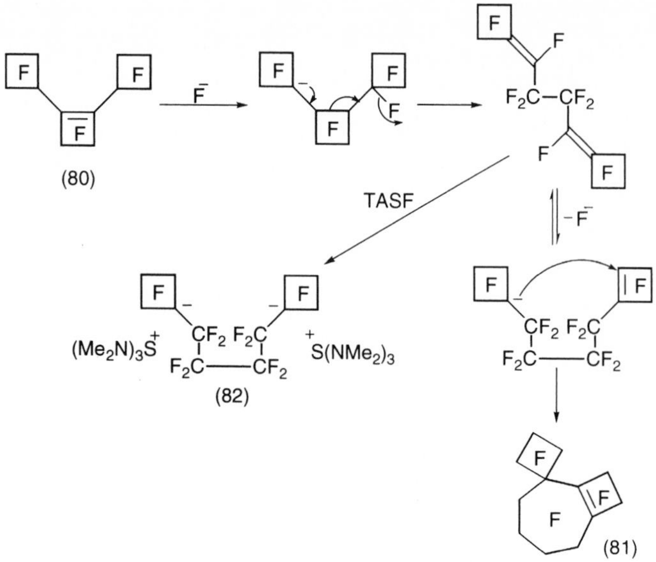

Scheme 37

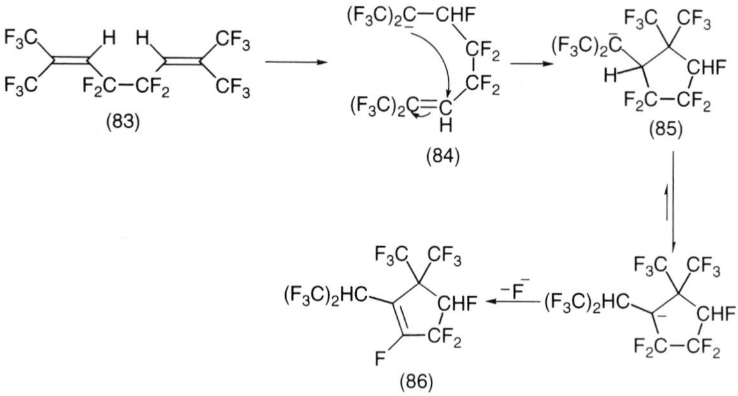

Scheme 38

6 Attack by O, S and N Centred Nucleophiles

The subject has been covered in various discussions [1, 78–84] previously but a striking example in this group of reactions is the occurrence of nucleophilic epoxidation. These reactions emphasise the frequently stressed "mirror-image" relationship of the chemistry of alkenes and their perfluorinated analogues that we have emphasised earlier. Bleaching powder has proved to be a very effective reagent for various epoxidations, except when the fluoroalkene is very reactive and then chloride ions, present in high concentration, begin to compete effectively (Scheme 39). A number of variations in technique have been described [52, 85–90].

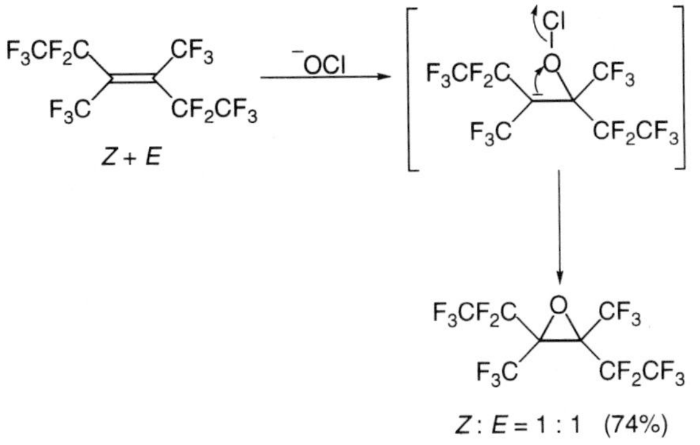

Scheme 39

Lithium *tert*-butylperoxide has been used effectively as an epoxidising agent with electron deficient alkenes (Scheme 40) [91, 92]. However, application of this methodology to systems containing fluorine has only recently been explored, and it is now established that this can be a very successful procedure with fluoroalkenes. Indeed, the Lithium *tert*-butylperoxide system worked in some cases where the calcium hypochlorite reaction was ineffective [93, 94].

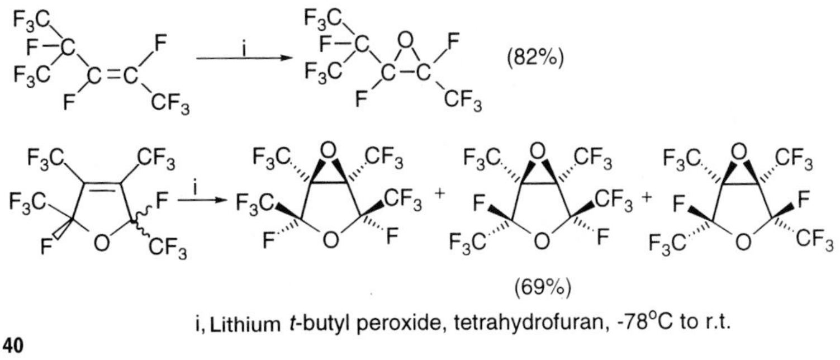

Scheme 40

The nature of the products formed during nucleophilic attack may be regarded as dependent on the fate of an intermediate carbanion (**87**); this can then lead to the various processes shown in Scheme 41.

In some cases, the intermediate carbanion can be trapped, e.g. with dimethylcarbonate or esters etc. This approach has been developed elegantly by Krespan to synthesise di-functional derivatives derived from tetrafluoroethene [95] (Scheme 42).

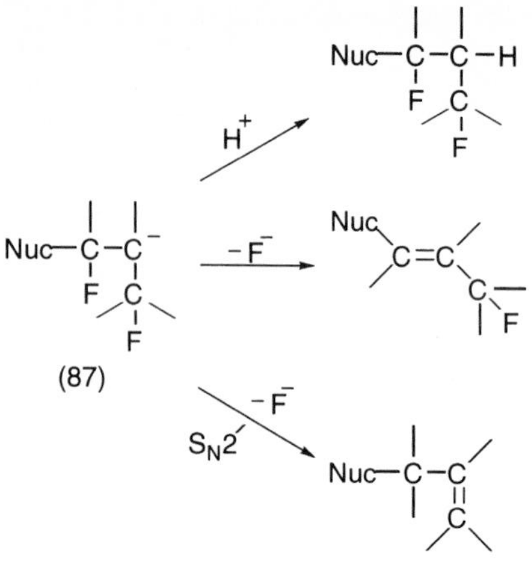

Scheme 41

However, the fate of the intermediate carbanion can depend on the nucleophile, the presence of appropriate electrophiles as trapping agents and the solvent, e.g. Scheme 42 [96]. Examples to illustrate these processes are drawn from more recent literature in Table 3.

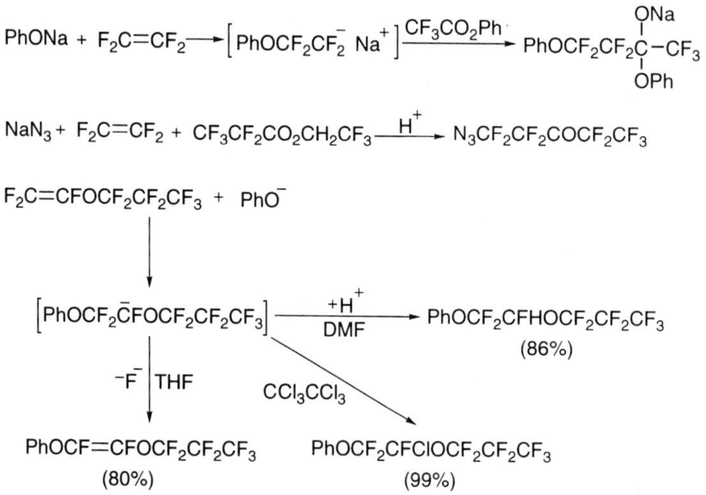

Scheme 42

Reactions of diethylamine and of alcohols with dienes gives products that depend critically on the reaction conditions [98, 99] (Scheme 43).

Neither the regiochemistry nor the stereochemistry is clear cut in reactions of aryl derivatives of hexafluoropropene [100–102] (Scheme 44).

Formate ion shows high nucleophilicity towards fluorinated alkenes and provides a route to fluorinated carboxylic acids [103] (Scheme 45).

Table 3. Some nucleophilic reactions of unsaturated fluorocarbons

Alkene or diene	Nucleophile	Products	References
$F_2C{=}CF{-}CF_2C_4F_9$	CH_3ONa	$CH_3OCF_2CFHCF_2C_4F_9$ 45% $CH_3OCF_2{-}CF{=}CF{-}C_4F_9$ 40%	[79]
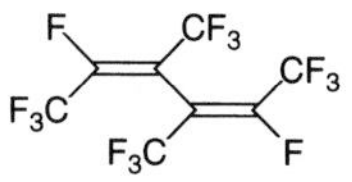	K_2S	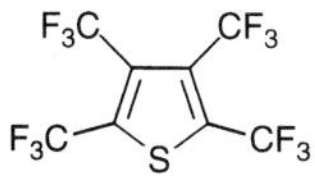	[44]
	CH_3CH_2SH	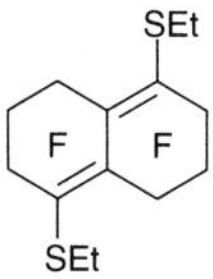	[97]
$F_2C{=}C{=}CF_2$	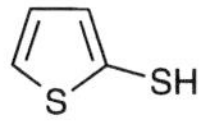	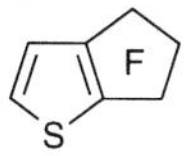	[82, 83]
$CF{=}C(F)CF_2OSO_2F$	N_3^-	$N_3CF_2CF{=}CF_2$	[84]

$$F_2C{=}CFCF_2CFClCF{=}CF_2 \xrightarrow{\text{i}} F_2C{=}CFCF_2CF{=}CFCONEt_2 \ (97\%)$$

$$F_2C{=}CFCF_2CFClCF{=}CF_2 \xrightarrow{\text{ii}} EtOF_2CFC{=}CFCF_2CHFCF_2OEt \ (100\%)$$

Scheme 43 i, Et_2N, Et_2O, -20°C to r.t.; ii, EtOH, 20°C

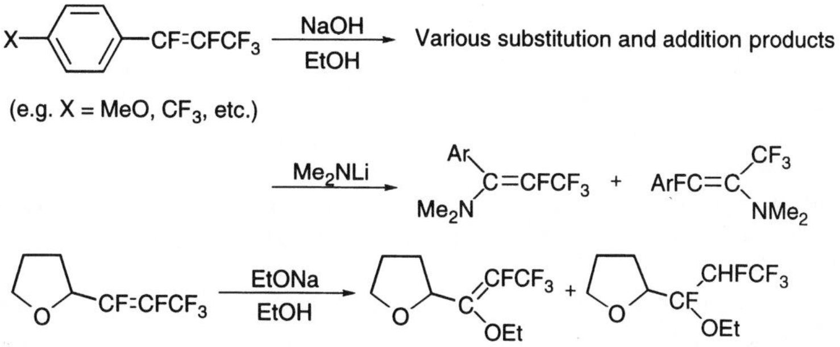

Scheme 44

$$R_FCF_2CF{=}CF_2 + HCOO^- \xrightarrow[\text{HCOOH}]{\text{DMF}} R_FCF_2CFHCOOH + R_FFC{=}CFCOOH$$

(with NaOH arrow over the products)

e.g. $R_F{=}(CF_2)_5F$; $(CF_2)_3H$ etc.

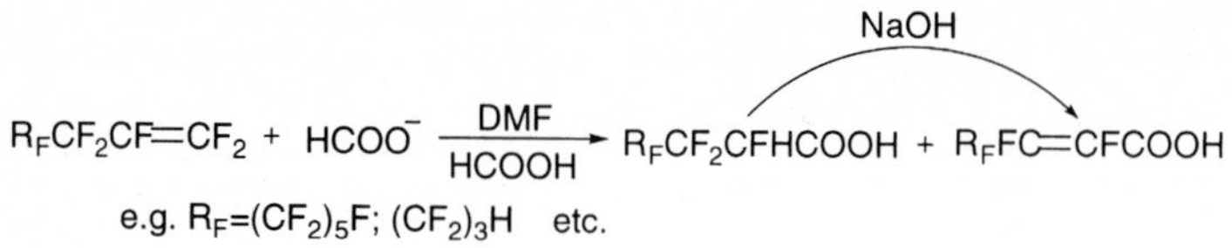

Scheme 45

Heterocycles are formed with difunctional alcohols [104]. Anions generated by de-silylation procedures have been particularly successful, especially for reactions involving difunctional nucleophiles (Scheme 46), and the subject has been reviewed by Farnham [105].

i, CsF, glyme

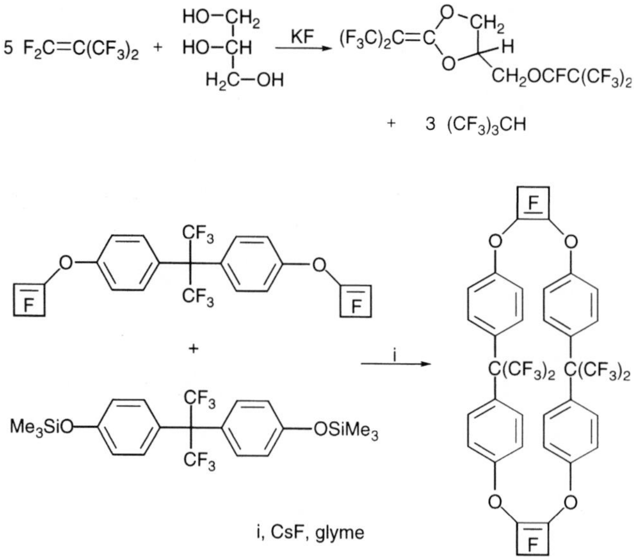

Scheme 46

Heterocycles can be formed in a number of systems and perhaps the most surprising involves reaction of the diene (**38**) (Scheme 47) in moist ether! Obviously, the process depends on the high reactivity of the vinylic fluorine atoms in (**38**) and the subsequent step is most likely an electrocyclisation, to form (**88**). Also, pyrroles, e.g. (**89**) and, more remarkably, pyrrolopyridines, e.g. (**90**) may be synthesised from (**38**) using aniline derivatives [44, 106].

A surprising series of reactions occur with heptafluorobut-2-ene (**91**) and 1,8-diazabicyclo[5,4,0]undec-7-ene (DBU) (Scheme 48), previously thought to be a non-nucleophilic base [107], giving the heterocycle (**92**) as the ultimate

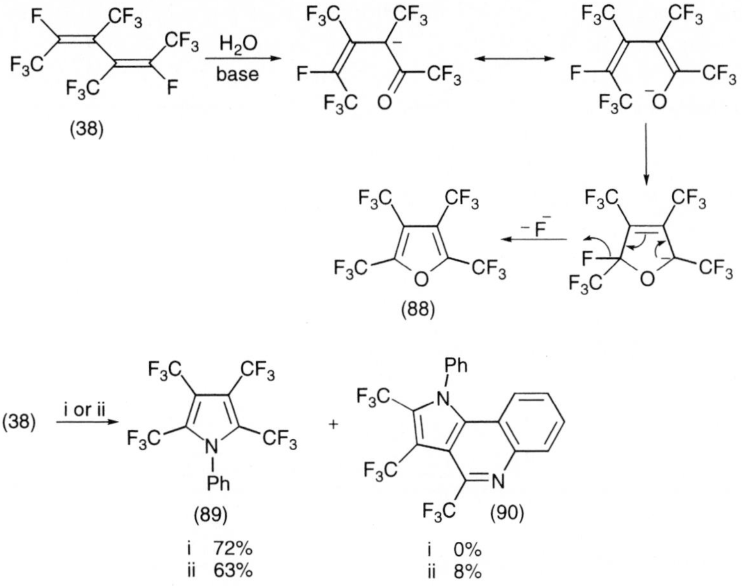

i, $PhNH_2$, CsF, CH_3CN, r.t.; ii, $PhNH_2$, KF, CH_3CN, r.t.

Scheme 47

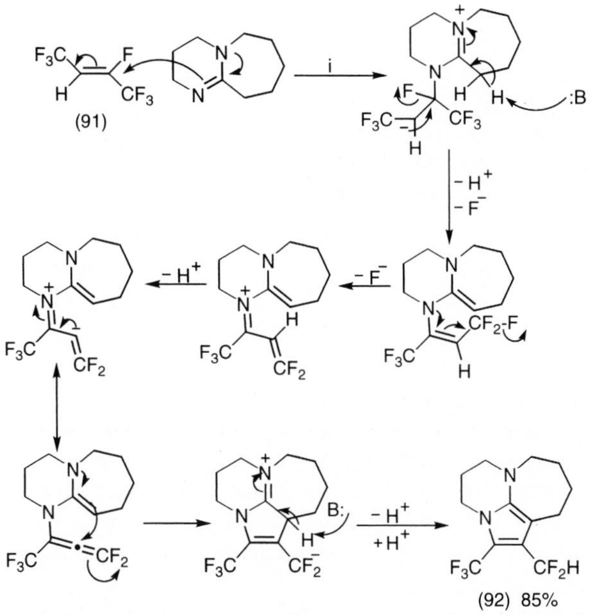

i, DBU : $CF_3CH{=}CFCF_3$ = 4:1, hexane, sealed tube, room temp, 2 d

Scheme 48

product. Indeed, DBU reacts with other fluorinated alkenes, but the products are difficult to separate and identify.

Displacement of fluorine from *gem*-difluoroalkenes has been developed as methodology for *"fluorine-free"* synthesis [108] (Scheme 49).

$$PhCH{=}O \xrightarrow{\ i\ } PhCH{=}CF_2 \xrightarrow{c.\ H_2SO_4} PhCH_2COOH \quad (93\%)$$

$$n\text{-}C_6H_{13}CH{=}O \xrightarrow{\ i\ } n\text{-}C_6H_{13}CH{=}CF_2 \xrightarrow{ii,\ iii} n\text{-}C_6H_{13}CH_2COOH \quad (69\%)$$

i, $Ph_3P{=}CF_2$; ii, $Hg(OAc)_2$, CF_3CO_2H; iii, aq. $NaHCO_3$, H_2S

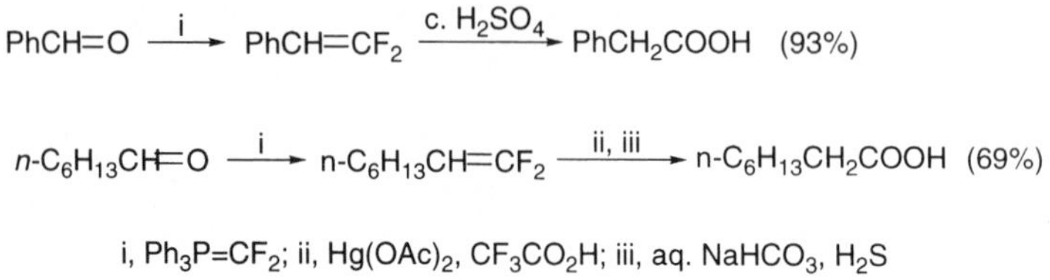

Scheme 49

Sodium sulfite acts as a strong nucleophile [109] (Scheme 50).

$$F_3C(CF_2)_4FC{=}CF_2 + NaSO_2OH \longrightarrow F_3C(CF_2)_4\bar{C}FCF_2\overset{+}{S}O(OH)ONa$$

$$\downarrow -HF$$

$$CF_3(CF_2)_4CF{=}CFCF_2SO_2ONa$$

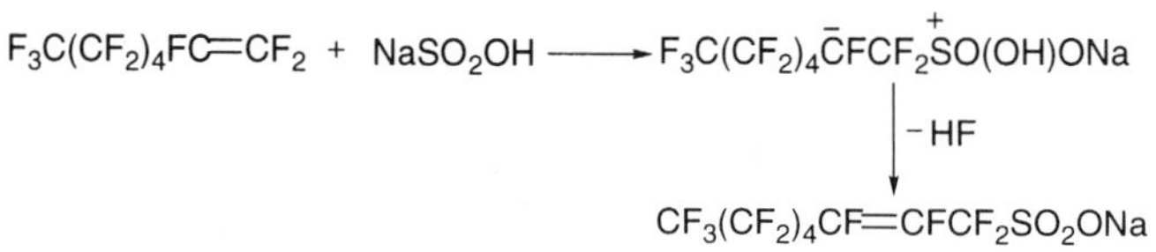

Scheme 50

7
Carbon Nucleophiles

Oligomerisations of fluorinated alkenes, via carbanion formation by addition of fluoride ion to the double-bond, was discussed in Sect. 5.3 (Scheme 51).

$$C{=}C + F^- \rightleftharpoons FC{-}C^- + C{=}C \xrightarrow{etc} \text{oligomers}$$

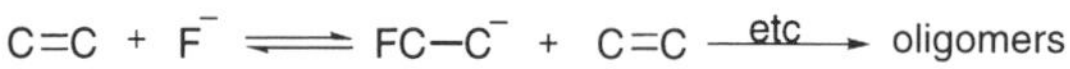

Scheme 51

However, ylides may be generated by reactions of amines with various perfluorinated -alkenes or -cycloalkenes [110] and attack of these ylides also promotes oligomerisation [62, 111, 112] (Scheme 52).

Note the different structure of the trimer (**93**), obtained by this route, in comparison with the trimer obtained by reaction of fluoride ion with (**68**) (Sect 5.3, Scheme 30). Other processes involving reactions with ylides have been described [113, 114] (Scheme 53).

Lithium or magnesium derivatives react readily and in predictable ways but the allyl system (**94**) surprisingly leads to attack at the 3-position, (Scheme 54) [115– 117].

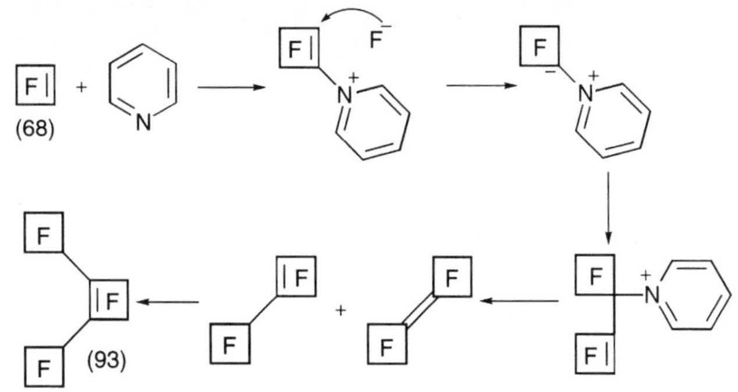

Scheme 52

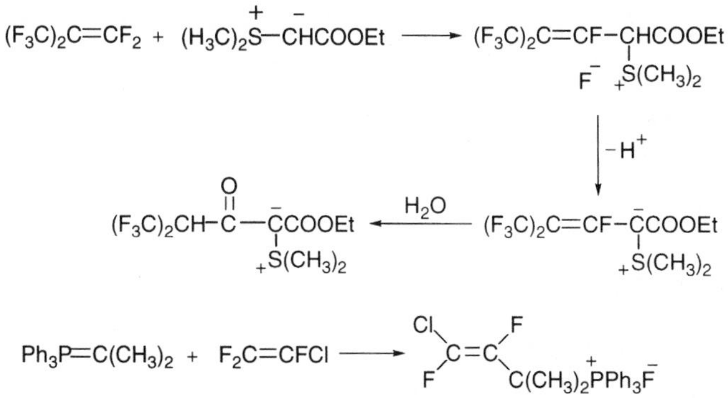

$(F_3C)_2C{=}CF_2$ + $(H_3C)_2\overset{+}{S}{-}\overset{-}{C}HCOOEt$ $\longrightarrow$ $(F_3C)_2C{=}CF{-}CHCOOEt$
$\phantom{(F_3C)_2C{=}CF{-}CHCOOEt}$ $\underset{+\text{S(CH}_3)_2}{\overset{-}{F}}$

$\Big\downarrow -H^+$

$(F_3C)_2CH{-}\overset{\overset{O}{\|}}{C}{-}\overset{-}{C}COOEt$ $\overset{H_2O}{\longleftarrow}$ $(F_3C)_2C{=}CF{-}\overset{-}{C}COOEt$
$\underset{+\text{S(CH}_3)_2}{}$ $\underset{+\text{S(CH}_3)_2}{}$

$Ph_3P{=}C(CH_3)_2$ + $F_2C{=}CFCl$ $\longrightarrow$ $\underset{F}{\overset{Cl}{}}C{=}C\underset{C(CH_3)_2\overset{+}{P}Ph_3\overset{-}{F}}{\overset{F}{}}$

Scheme 53

CH_3MgBr + $F_2C{=}CF{-}CF_2Br$ $\longrightarrow$ $CH_3CF_2CF{=}CF_2$

$\boxed{F\,|}$ + CH_3CH_2MgBr $\longrightarrow$ $\boxed{F\,|}\overset{CH_2CH_3}{}$ (75%)

$CH_3CH{=}CH{-}\overset{+}{C}H_2MgBr$ $\longrightarrow$ $CH_3H\overset{}{C}{-}CF{=}CFCl$
(94) $F_2C{=}CFCl$ $\underset{H}{\overset{C{=}CH_2}{}}$

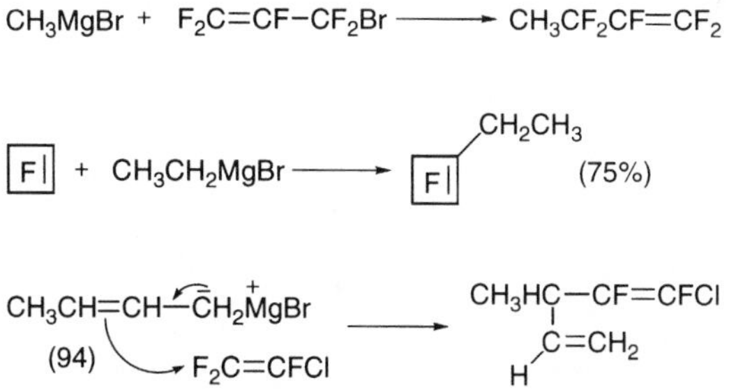

Scheme 54

Carbanions generated by proton-abstraction, using fluoride ion, have led to cyclic systems with fluorinated dienes [118] (Scheme 55).

Reactions of cyanide with unsaturated fluorocarbons have been reported and an example is shown below [119, 120] (Scheme 56).

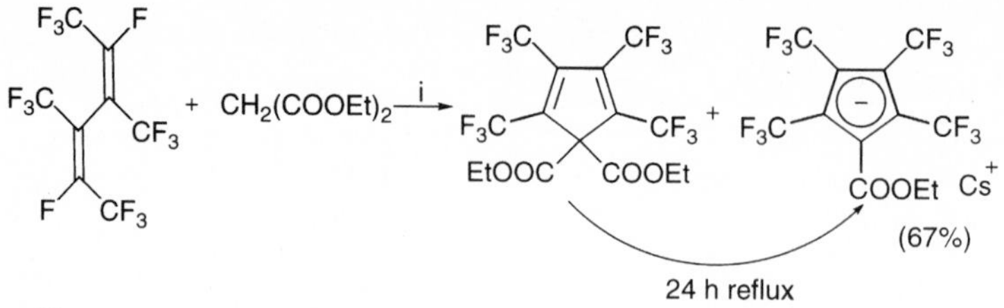

Scheme 55

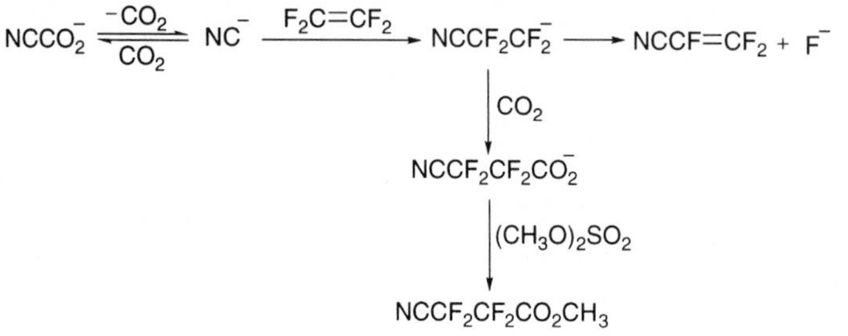

Scheme 56

8
Hydride Reductions

It has been established that complex hydrides react via nucleophilic displacement of either vinylic or allylic halogen by hydrogen [121–124] (Scheme 57).

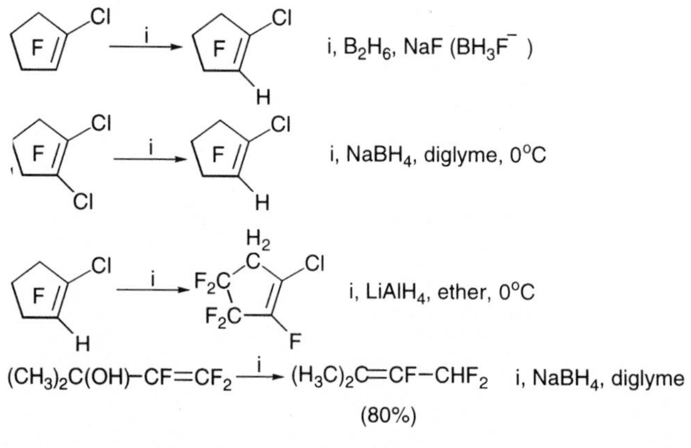

Scheme 57

However, systems with vinylic iodine do not undergo nucleophilic substitution when reacted with $LiAlH_4$ but form aluminium complexes in solution. Hydrolysis of these mixtures with D_2O is a convenient route to deuterated fluoro-alkenes [125] (Scheme 58).

Scheme 58

9
Reactions of Oligomers

Various oligomers of fluorinated alkenes and cycloalkenes have been prepared by fluoride ion induced oligomerisation of various monomers (Sect. 5.3), and the chemistry of these systems provides some unique reactions. The oligomers of special interest here may be described as of types (**95**) or (**96**) (Scheme 59), i.e. systems with either four (**95**) or three (**96**) perfluoro-alkyl or -cycloalkyl groups attached to the double bond, whereas systems with two perfluoroalkyl groups attached, i.e. (**97**) and (**98**), have a chemistry more similar to fluorinated alkenes that may be derived from other sources.

$(R_F)_2C{=}C(R_F)_2$	$(R_F)_2C{=}CFR_F$	$R_FFC{=}CFR_F$	$(R_F)_2C{=}CF_2$
(95)	(96)	(97)	(98)

Scheme 59

Reactions of compounds of type (**96**) with fluoride ion have already been described in Sect 5. Compounds of type (**96**) are, generally, more reactive than type (**95**).

The unique feature of type (**95**) is that the double bond is very much activated to nucleophilic attack but reactions proceed with allylic displacement of fluoride ion, often producing an intermediate that is highly reactive towards further attack by nucleophiles, including intramolecular processes. For example, the furan derivative (**99**) may be obtained directly from (37) (Scheme 60, and also see Scheme 66, later). Other nucleophiles give products of di- and polysubstitution [126] (Scheme 61). The bicyclobutylidine system (**41 a**) is electronically very similar to the tetramer (**37**), but (**41 a**) is particularly reactive, as a consequence of strain (Scheme 62, [111]).

In fact, reactions involving the tetramer (**37**) can be quite complex [127] because the situation is as outlined in Scheme 63, where the observed rate constant for reaction depends on an equilibrium $K_2 - K_3$ between isomers (**37 a, b**), in the presence of fluoride ion, and a rate constant $k_1 - k_3$, where $k_3 \gg k_2 \gg k_1$.
The products observed depend on the ability of the nucleophile to generate active fluoride ion to promote the equilibration, as well as the reactivity k_1, with the most abundant isomer. Some products are shown in Scheme 64, including reactions with carbanions [128].

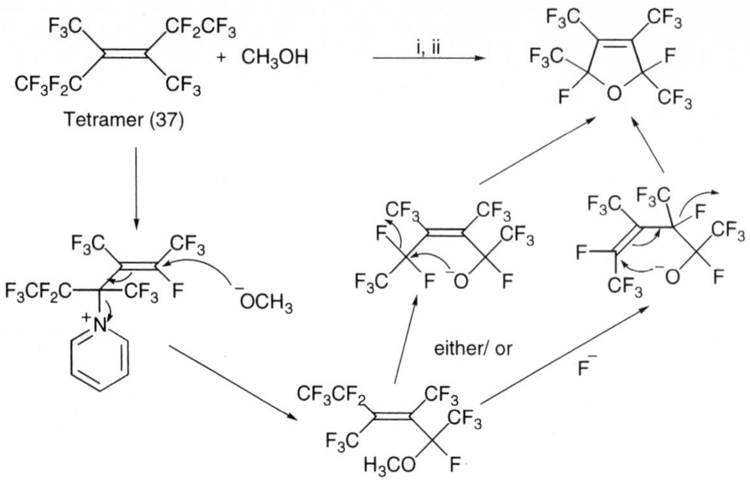

i, Pyridine; ii, CH_3OH, tetraglyme, reflux

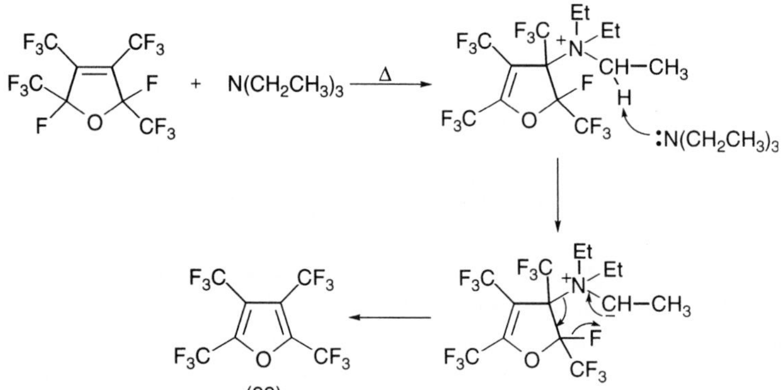

(99)

Scheme 60

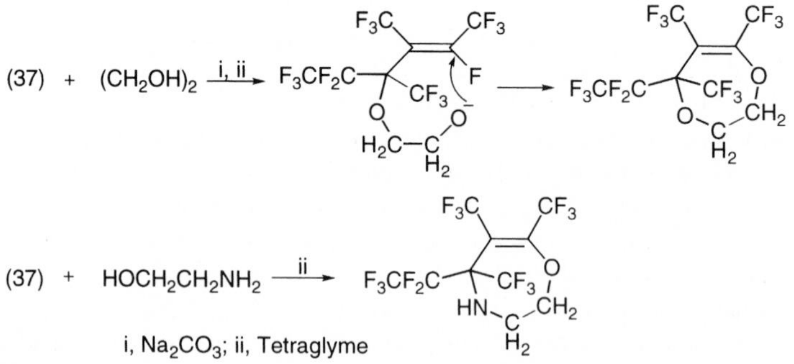

Scheme 61

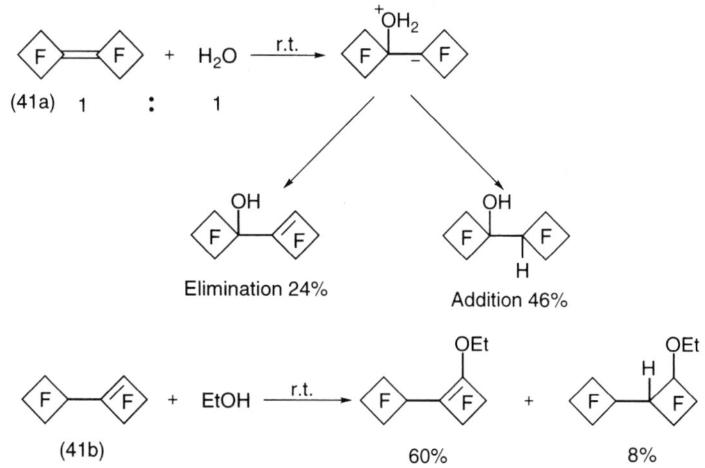

Scheme 62

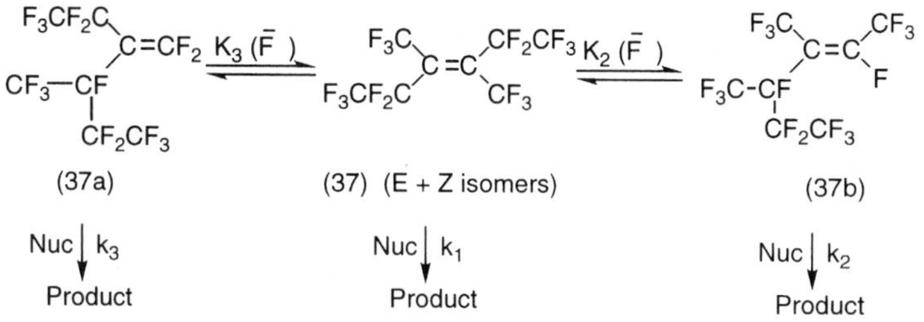

Scheme 63

The pentamer of tetrafluoroethene (**66**) (Scheme 29) is an unusual example of type (**96**) and reacts readily with nucleophiles [129] (Scheme 65). In contrast, (**66**) undergoes a remarkable reaction with aqueous triethylamine, producing the dihydrofuran derivative (**101**) and the process formally involves a direct intramolecular displacement of fluorine from a saturated site and a mechanism has been advanced (Scheme 66) which accounts for the product formed [126]. Understandably, this process is not easily accepted [3, 130] because it has essentially no precedent. Indeed, it is well established that nucleophilic displacement from saturated sites in fluorocarbons occurs only in exceptional circumstances. Consequently, other mechanisms have been advanced, which seem no more convincing [3, 130]. It should be remembered that the major point in favour of the step (**100**) to (**101**) (Scheme 66) is that the nucleophile is generated in *close proximity* to the reaction centre because of the special geometry of this situation. Consequently, much of the otherwise high energy/entropy barrier has already been overcome in this case.

Further reactions of (**101**) have been described [130] (Scheme 67) and the products depend on a pre-equilibrium with fluoride ion [126]. Cyclisation occurred with malonic ester.

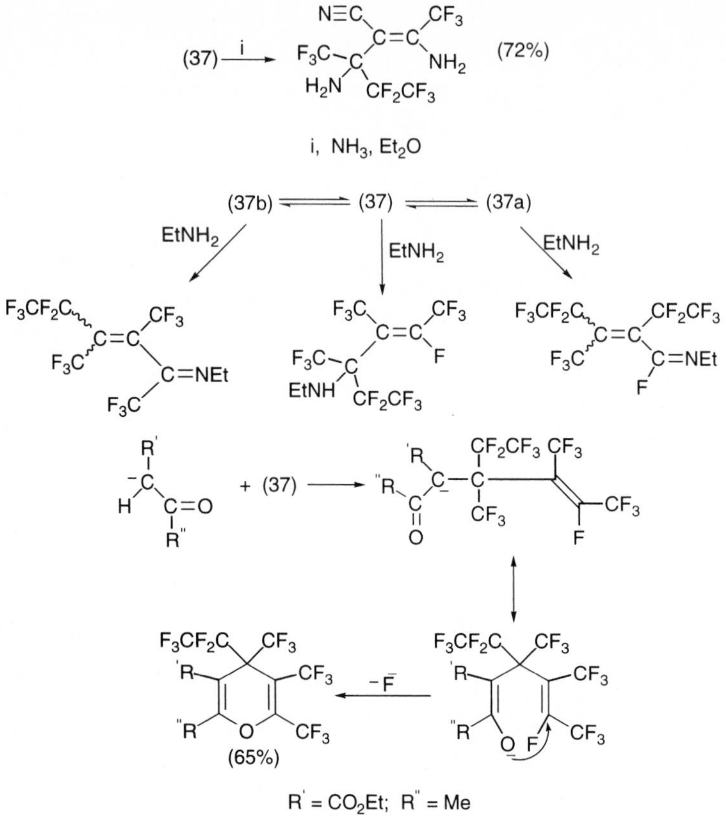

Scheme 64

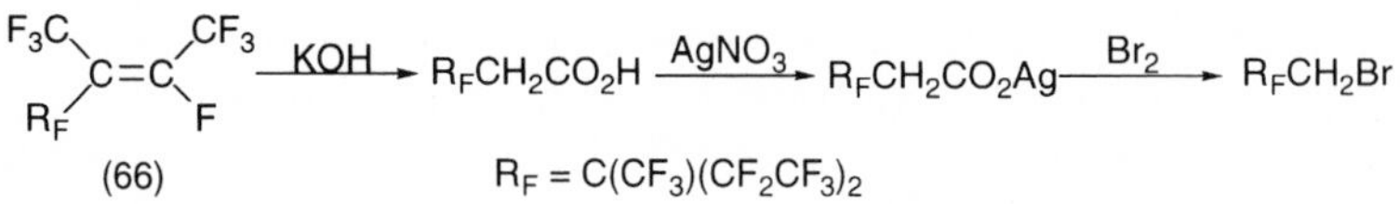

Scheme 65

When the pentamer (**66**) reacts with alkoxide anions at low temperatures (–30 to –40 °C), then the products of kinetic control (**102**) are isolated, whereas at higher temperatures, thermodynamic control prevails and the products (**103**) are obtained [131, 132] (Scheme 68). Similar observations have been made with sulphur nucleophiles [132], and complex products are obtained with amines, including the formation of heterocycles [132]. Reaction of (**66**) with ethyl acetoacetate gave a pyran derivative (**104**) in a reaction that may be rationalised as shown in Scheme 69 [133]. In an analogous way, furan derivatives are formed from perfluoro-2-butene and -cyclohexene in base-induced reaction with 1,3-dicarbonyl derivatives [133].

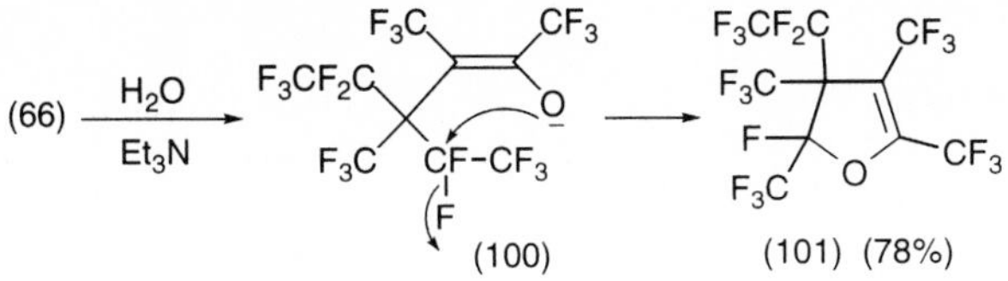

Scheme 66

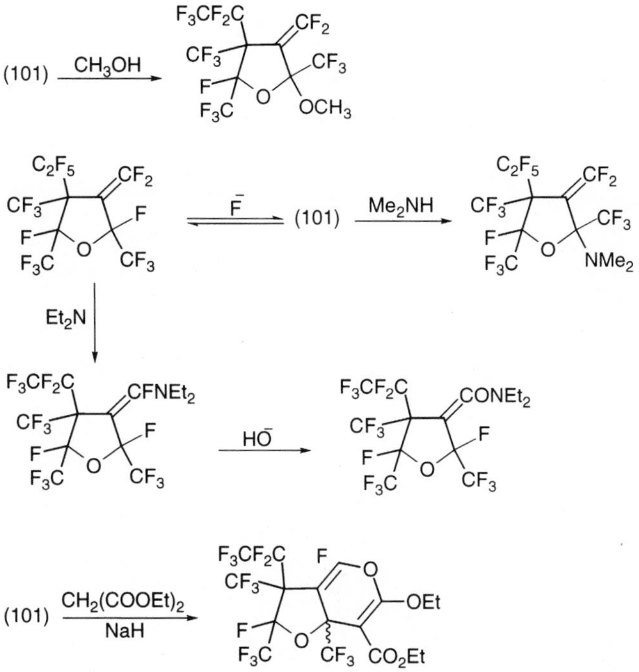

Scheme 67

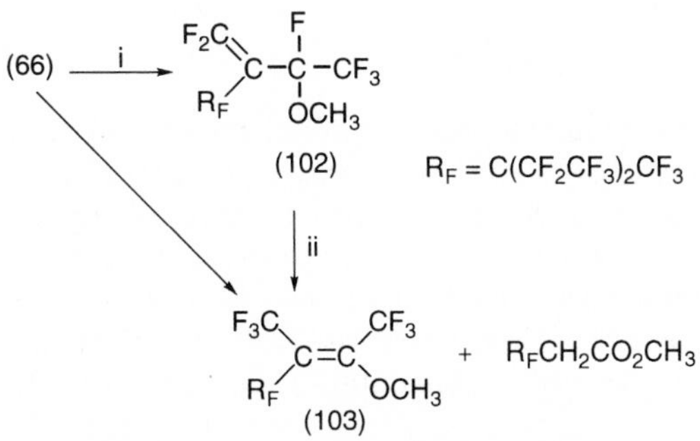

i, $NaOCH_3$, -30° to -40°C, $CF_2ClCFCl_2$; ii, Et_3N, r.t.

Scheme 68

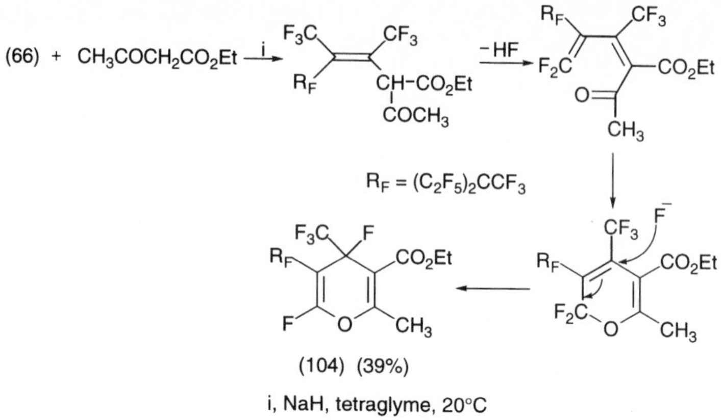

Scheme 69

Tetrafluoroethene hexamer (**67**) also gave a range of products with amines, dependent on the reagent [134, 135], including the formation of ketenimines (Scheme 70).

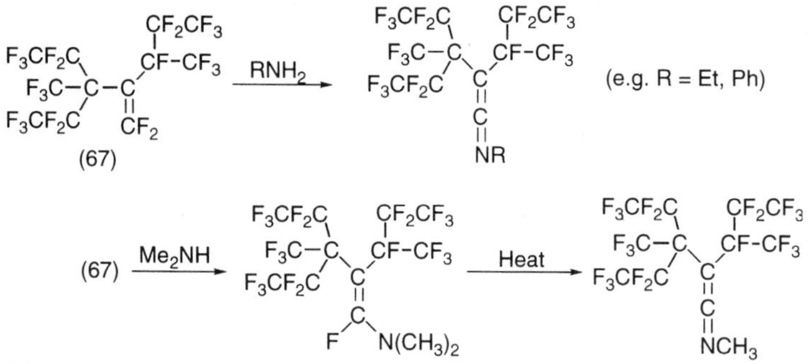

Scheme 70

Hexafluoropropene trimer is a mixture containing (**105a**) and (**105b**) and the ratio of product formation depends very much on the nucleophile [136] (Schemes 71 and 72).

Remarkably, with thiophenol isomer (**105a**) gives the expected product (**106**) (Scheme 71), while the isomer (**105b**) gives a product (**107**), which can be rationalised as involving a stage with nucleophilic attack on sulphur [137]. However, reactions of other sulphur nucleophiles with the dimer (**108**) have revealed an interesting and unusual elimination of RSF, leading to diene intermediates which react further [138]. For example, reaction of (**108**) with $PhCH_2SH$, gave a variety of products, i.e. (**109**), (**110**) and (**111**), with the latter arising via an unusual elimination.

Hexafluoropropene dimers and trimers can give a variety of products if a pre-equilibrium is established with fluoride ion before reaction with the nucleo-

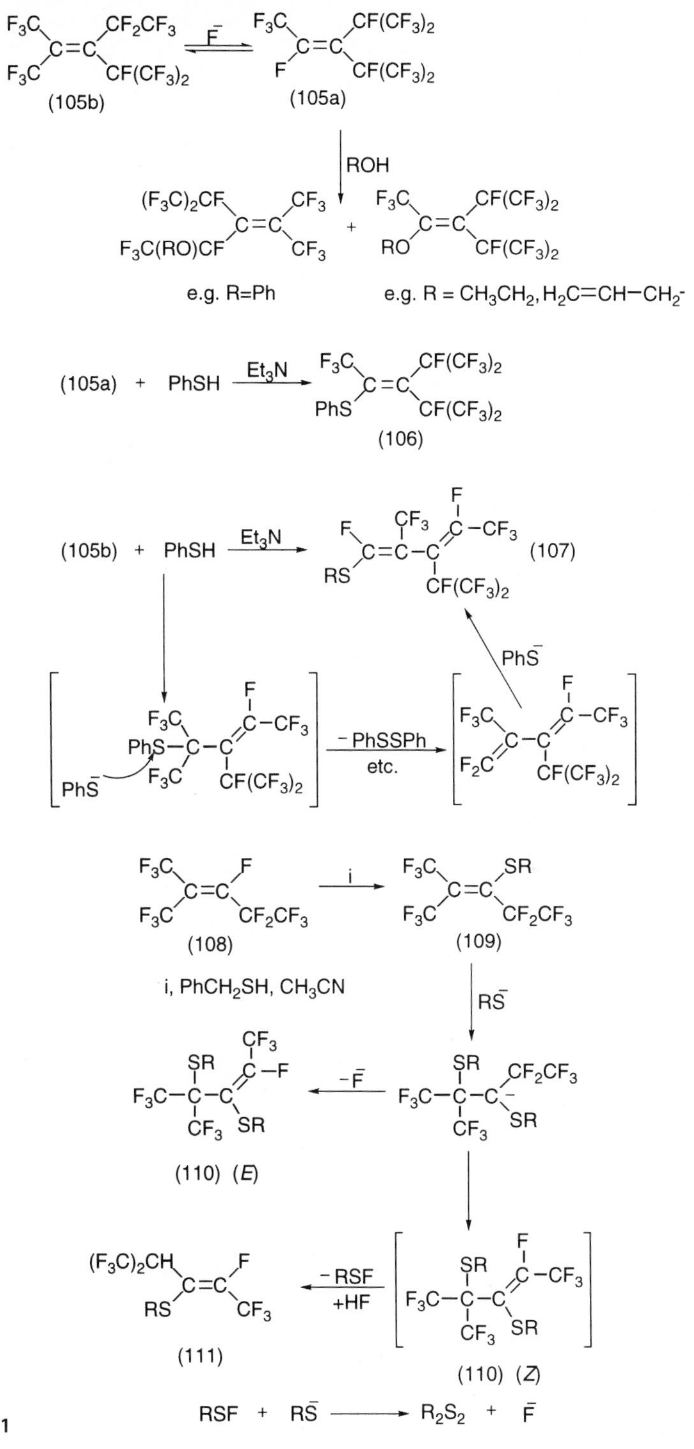
(105b)
(105a)
ROH
e.g. R=Ph
e.g. R = CH3CH2, H2C=CH−CH2-
(105a) + PhSH
Et3N
(106)
(105b) + PhSH
(107)
PhS
− PhSSPh
etc.
(108)
i
(109)
i, PhCH2SH, CH3CN
RS
−F
(110) (E)
− RSF
+HF
(111)
(110) (Z)
RSF + RS → R2S2 + F

Scheme 71

phile [61]. For example, trimer (**105a**) is in equilibrium with the isomer (**105b**), in the presence of triethylamine and products are derived from the most reactive isomer (**105a**) with phenol, although the ratio of (**112**) and (**113**) depends significantly on the solvent [139] (Scheme 72).

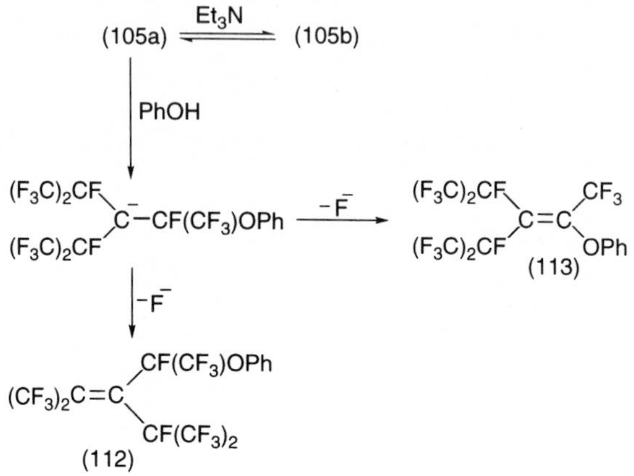

Scheme 72

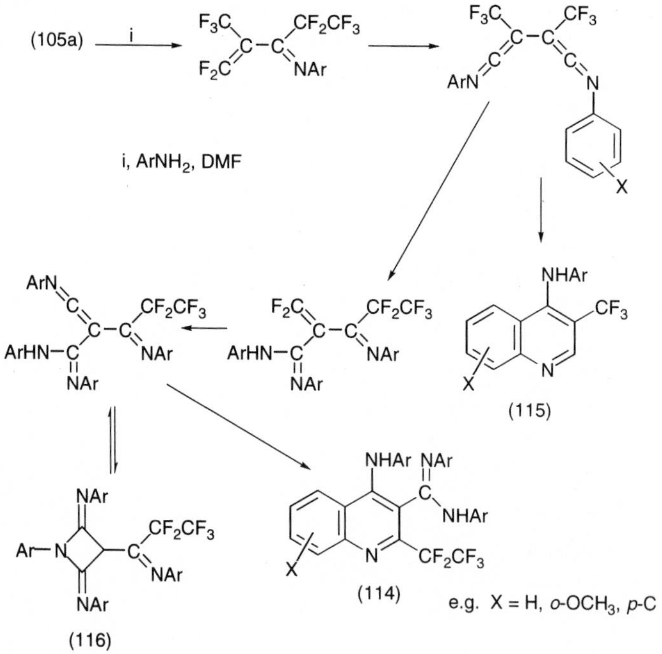

Scheme 73

Reactions with nitrogen nucleophiles are also characterised by formation of intermediates with further sites of unsaturation and these can lead to a variety of heterocyclic systems [61, 140, 141] (Scheme 73). Products corresponding to (**114**)–(**116**) have been identified, depending on the solvent and substituent X.

Aromatic amines react uniquely with various oligomers derived from cyclic systems as *carbon* nucleophiles and the result is a remarkable array of annelation processes [142] as illustrated by the reaction with "Proton-Sponge" (**117**), shown in Scheme 74.

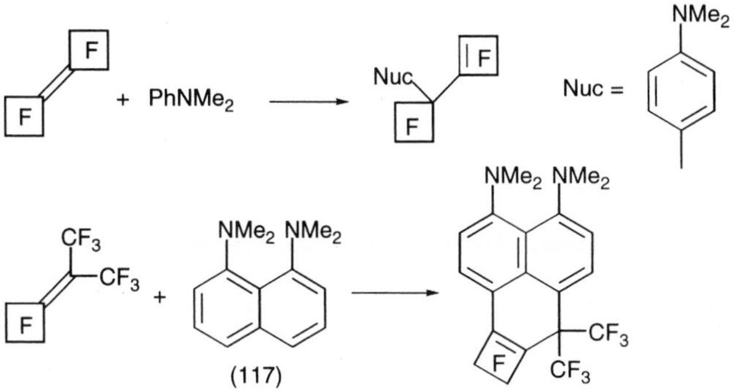

Scheme 74

References

1. Chambers RD, Mobbs RH (1965) Adv Fluorine Chem 4:50
2. Chambers RD (1973) Polyfluoroalkanes, polyfluoroalkenes, polyfluoroalkynes and derivatives. Fluorine in organic chemistry. Wiley-Interscience, New York, p 138
3. Chambers RD, Bryce MR (1987) Fluoro-carbanions. In: Buncel E, Durst T (eds) Comprehensive carbanion chemistry, part c, vol 6. Elsevier, Amsterdam, p 271
4. Park JD, McMurty RJ, Adams JH (1968) Fluorine Chem Rev 2:55
5. Hudlicky M, Pavlath AE (1995) Chemistry of organic fluorine compounds II. American Chemical Society, Washington
6. Smart BE (1983) Fluorocarbons. In: Patai S, Rapport Z (eds) Chemistry of functional groups, supplement d. Wiley, New York, p 603
7. Chambers RD (1992) Unusual fluorinated alkenes and dienes via fluoride ion induced processes. In: Olah GA, Chambers RD, Prakash GKS (eds) Synthetic fluorine chemistry. Wiley, New York, p 359
8. Chambers RD, Vaughan JFS, Mullins SJ, Nakamura T, Roche AJ (1995) J Fluorine Chem 72: 231
9. Farnham WB (1996) Chem Rev 96:1633
10. Chambers RD, Gribble MY (1973) J Chem Soc, Perkin Trans 1:1411
11. Xikui J, Guozhen J, Yiqun S (1995) Youji Huaxue: 400
12. Xikui J, Guozhen J, Yiqun S Chem Abstr 104:185933p
13. Heilbronner E (1977) Proc. inst. petroleum conf. In: West RS (ed) Molecular spectroscopy. Heyden & Sons, London and references contained
14. Domelsmith LM, Houk KN, Piedrahita C, Dolbier WJ (1978) J Am Chem Soc 100:6908 and references contained
15. Rozhkov IN, Borisov TA (1990) Izv Akad Nauk SSSR, Ser Khim (Engl Transl):1649
16. Rozhkov IN, Borisov YA (1993) Izv Akad Nauk SSSR, Ser Khim:1041
17. Dixon DA, Smart BE (1989) J Phys Chem 93:7780

18. Chambers RD (1973) Fluorine in organic chemistry. Wiley-Interscience, New York
19. Smart BE (1994) Characteristics of C-F systems. In: Banks RE, Smart BE, Tatlow JC (eds) Organofluorine chemistry. Plenum, New York, p 57
20. Reed AE, Schleyer PVR (1990) J Am Chem Soc 112:1434 and references contained
21. Dixon DA, Fukunaga T, Smart BE (1986) J Am Chem Soc 108:4027
22. Farnham WB, Smart BE, Middleton WJ, Calbrese JC, Dixon DA (1985) J Am Chem Soc 107:4565
23. Chambers RD, Waterhouse JS, Williams DLH (1974) Tetrahedron Lett 9:743
24. Rozhkov IN, Stepanoff AA, Borisov YA (1985) The fourth regular meeting of soviet-japanese fluorine chemists. Electronic structure and polarographic reduction potentials of fluoroolefins, Kiev, p 125
25. Corvaga C, Farnia G, Formenton G, Navarrini W, Sandonia G, Tortelli V (1994) J Phys Chem 94:2307
26. Doyle AM, Patrick CR, Pedler AE (1968) J Chem Soc (C):2740
27. Doyle AM, Pedler AE (1971) J Chem Soc (C):282
28. Briscoe MW, Chambers RD, Silvester MJ, Drakesmith FG (1988) Tetrahedron Lett 29:1295
29. Stepanov AA, Rozhkov IN (1983) Izv Akad Nauk SSSR, Ser Khim (Engl Transl):809
30. Rozhkov IN, Stepanoff AA (1981) The second regular meeting of soviet-japanese fluorine chemists. Reductive coupling of fluoroolefins Synthesis of higher perfluorinated 1,3-dienes and allenes, Moscow, p 133
31. Stepanov AA, Bekker GY, Kurbakova AP, Leites LA, Rozhkov IN (1981) Izv Akad Nauk SSSR, Ser Khim (Engl Transl):2285
32. MacNicol DD, Robertson CD (1988) Nature (London) 332:59
33. Ter-Gabrielyan EG, Gambaryan NP, Knunyants IL (1981) Izv Akad Nauk SSSR, Ser Khim (Engl Transl):301
34. Howells MA, Howells RD, Baenziger NC, Burton DJ (1973) J Amer Chem Soc 95:5366
35. Burton DJ, Valk PDV (1981) J Fluorine Chem 18:413
36. Briscoe MW, Chambers RD, Mullins SJ, Nakamura T, Vaughan JFS, Drakesmith FG (1994) J Chem Soc, Perkin Trans 1:3115
37. Wiberg N (1968) Angew Chem, Int Ed Engl 7:766
38. Bock H, Borrmann H, Haulas Z, Oberhammer H, Ruppert K, Simon A (1991) Angew Chem, Int Ed Engl 30:1678
39. Koch HF (1987) In: Buncel E, Durst T (eds) Comprehensive carbanion chemistry, part c. Elsevier, Amsterdam, p 321
40. Koch HF, Koch JG, Donovan DB, Toczko AG, Kielbania AJ (1981) J Am Chem Soc 103:5417
41. Chambers RD, Lindley AA, Fielding HC (1978) J Fluorine Chem 12:85
42. Rozhkov IN, Borisov YA (1988) Relative reactivity of F-alkenes in regard to the hard-soft nucleophiles. The fifth regular meeting of soviet-japanese fluorine chemists, Tokyo, p 75
43. Bryce MR, Chambers RD, Taylor G (1984) J Chem Soc, Perkin Trans 1:509
44. Briscoe MW, Chambers RD, Mullins SJ, Nakamura T, Vaughan JFS (1994) J Chem Soc, Perkin Trans 1:3119
45. Olah GA, Prakash GKS, Sommer J (1985) Superacids. John Wiley, New York
46. Bayliff AE, Chambers RD (1988) J Chem Soc, Perkin Trans 1:201
47. Farnham WB, Dixon DA, Calbrese JC (1988) J Am Chem Soc 110:9453
48. Ikeda H, Watanabe H, Hirao T, Kurosawa H (1994) J Fluorine Chem 69:97
49. Smart BE, Middleton WJ, Farnham WB (1986) J Amer Chem Soc 108:4905
50. Dmowski W, Wozniacki F (1987) J Fluorine Chem 36:385
51. Vilenchik YM, Lekontseva GI, Semerikova LS (1981) Zh Vses Khim Ob-va im Mendeleeva (Engl Transl) 2:40
52. Millauer H, Schwertfeger W, Siegemund G (1985) Angew Chem, Int Ed Engl 24:161
53. Ishikawa N, Awamoto K, Ishiwata T, Kitazume T (1982) Bull Chem Soc Jpn 55:2956
54. Chambers RD, Gray WK, Korn SR (1995) Tetrahedron 51:13167
55. Fielding HC, Rudge AJ (1967) Brit Pat 1,082,127
56. Deem WR (1973) Brit Pat 1,302,350
57. Graham DP (1966) J Org Chem 31:955

58. Chambers RD, Jackson JA, Partington S, Philpot PD, Young AC (1975) J Fluorine Chem 6:5
59. Brunskill W, Flowers WT, Gregory R, Haszeldine RN (1970) J Chem Soc, Chem Commun: 1444
60. Chambers RD, Jones CGP (1981) J Fluorine Chem 17:581
61. Ishikawa N (1979) The first regular meeting of soviet-japanese fluorine chemists. Nucleophilic reactions of hexafluoropropene oligomers, Tokyo, p 125 and references contained
62. Chambers RD, Taylor G, Powell RL (1980) J Chem Soc, Perkin Trans 1:426
63. Chambers RD, Gribble MY, Marper E (1973) J Chem Soc, Perkin Trans 1:1710
64. Chambers RD, Lindley AA, Philpot PD (1978) Isr J Chem: 150
65. Scherer KV, Terranova TF (1979) J Fluorine Chem 13:89
66. Postovoi SA, Zeifman YV (1981) Izv Akad Nauk SSSR, Ser Khim (Engl Transl): 1434
67. Chambers RD, Mullins SJ, Roche AJ, Vaughan JFS (1995) J Chem Soc, Chem Commun: 841
68. Chambers RD, Roche AJ, Vaughan JFS (1996) Can J Chem 74:1925
69. Chambers RD, Gray WK, Vaughan JFS, Korn SR, Médebielle M, Batsanov AS, Lehmann CW, Howard JAK (1997) J Chem Soc, Perkin Trans 1:135
70. Miller WT, Fried JH, Goldwhite H (1960) J Am Chem Soc 82:3091
71. Miller WT, Frass W, Resnick PR (1961) J Am Chem Soc 83:1767
72. Banks RE, Braithwaite A, Haszeldine RN, Taylor DR (1969) J Chem Soc (C): 996
73. Kurykin MA, German LS (1982) Izv Akad Nauk SSSR, Ser Khim (Engl Transl): 2202
74. Feiring AE, Hovey MC, Arthur SD (1984) J Fluorine Chem 24:125
75. Chambers RD, Kirk JR, Taylor G, Powell RL (1982) J Chem Soc, Perkin Trans 1:673
76. Chambers RD, Cheburkov YA, Tanabe T, Vaughan JFS (1995) J Fluorine Chem 74:227
77. Apsey GC, Chambers RD, Odello P (1996) J Fluorine Chem 77: 127
78. Dmowski W, Voellnagel-Neugebauer H (1984) Bull Pol Acad Sci, Chem 32:39
79. Gross U, Storek W (1984) J Fluorine Chem 26:457
80. Doroginskii VA, Kolomiets AF, Sokol'skii GA (1985) Zh Org Khim (Engl Transl) 21:742
81. Rokhlin EM, Abduganiev EG, Utebaev U (1976) Usp Khim (Engl Transl) 45:1177
82. Maksimov AM, Platonov VE, Yakobson GG (1986) Izv Akad Nauk SSSR, Ser Khim (Engl Transl): 138
83. Platonov VE, Maksimov AM, Yakobson GG (1977) Izv Akad Nauk SSSR, Ser Khim (Engl Transl): 2387
84. Krespan CG (1986) J Org Chem 51:332
85. Zapevalov AY, Filyakova TI, Peschanskii NV, Kodess MI, Kolenko IP (1990) Zh Org Khim (Engl Transl) 26:265
86. Coe PL, Sellars A, Tatlow JC (1983) J Fluorine Chem 23:102
87. Coe PL, Mott AW, Tatlow JC (1982) J Fluorine Chem 20:243
88. Coe PL, Mott AW, Tatlow JC (1990) J Fluorine Chem 49:21
89. Zapevalov AY, Filyakova TI, Kolenko IP (1979) Izv Akad Nauk SSSR, Ser Khim (Engl Transl): 2812
90. Kolenko IP, Filyakova TI, Zapevalov AY, Lur'e ÉP (1979) Izv Akad Nauk SSSR, Ser Khim (Engl Transl): 2509
91. Clark C, Hermans P, Meth-Cohn O, Moore C, Taljaard HC, Vuuren GV (1986) J Chem Soc, Chem Commun: 1378
92. Meth-Cohn O, Moore C, Taljaard HC (1988) J Chem Soc, Perkin Trans 1:2663
93. Chambers RD, Vaughan JFS, Mullins SJ (1995) J Chem Soc, Chem Commun: 629
94. Chambers RD, Vaughan JFS, Mullins SJ (1996) Res Chem Intermed 22:703
95. Krespan CG (1982) US Pat. 4,576,752
96. Feiring AE, Wonchoba ER (1992) J Org Chem 57:7014
97. Hu C-M, Long F, Xu Z-Q (1990) J Fluorine Chem 48:29
98. Dedek V, Linhart I, Kovac M (1985) Collect Czech Chem Commun 50:1714
99. Linhart I, Trska P, Dedek V (1985) Collect Czech Chem Commun 50:1727
100. Dmowski W (1980) J Fluorine Chem 15:299
101. Dmowski W (1982) J Fluorine Chem 21:201
102. Dmowski W (1986) Pol J Chem 60:129

103. Hu CM, Tu MH (1992) Chinese Chem Lett 3:87
104. Doroginskii VD, Kolmiets AF, Sokol'skii GA (1984) Zh Org Khim (Engl Transl) 19:1564
105. Farnham WB (1992) Silicon mediated reactions in organofluorine chemistry. In: Olah GA, Chambers RD, Prakash GKS (eds) Synthetic fluorine chemistry. Wiley, New York, p 247
106. Chambers RD, Gray WK, Mullins SJ, Korn SR (1997) accepted for publication in the J Chem Soc, Perkin Trans 1
107. Chambers RD, Roche AJ, Batsanov AS, Howard JAK (1994) J Chem Soc, Chem Commun: 2055
108. Hayashi S-I, Nakai T, Ishikawa N (1980) Chem Lett: 651
109. Gross V, Engler G (1985) J Fluorine Chem 29:425
110. Pruett RL, Bahner CT, Smith HA (1952) J Am Chem Soc 74:1638
111. Chambers RD, Taylor G (1980) J Fluorine Chem 16:161
112. Chambers RD, Taylor G, Powell RL (1980) J Chem Soc, Perkin Trans 1:429
113. Zeifman YV, Lantseva LT (1983) Izv Akad Nauk SSSR, Ser Khim (Engl Transl): 2149
114. Burton DJ, Lee TM (1976) J Fluorine Chem 8:189
115. Tarrant P, Warner DA (1954) J Am Chem Soc 76:1624
116. Park JD, Fontanelli R (1963) J Org Chem 28:258
117. Tarrant P, Hetes J (1965) J Org Chem 30:1485
118. Chambers RD, Greenhall MP (1990) J Chem Soc, Chem Commun: 1128
119. Krespan CG, Van-Catledge FA, Smart BE (1984) J Am Chem Soc 106:5544
120. Krespan CG (1986) J Org Chem 51:326
121. Johnson RL, Burton DJ (1965) Tetrahedron Lett 46:4079
122. Burton DJ, Johnson RL (1964) J Am Chem Soc 86:5361
123. Burton DJ, Johnson RL (1966) Tetrahedron Lett 24:2681
124. Tellier F, Sauvétre R, Normant J-F (1987) Tetrahedron Lett 28:3335
125. Burton DJ, Mettille FJ (1982) J Fluorine Chem 20:157
126. Chambers RD, Lindley AA, Philpot PD, Fielding HC, Hutchinson J, Whittaker G (1979) J Chem Soc, Perkin Trans 1:214
127. Bartlett S, Chambers RD, Lindley AA, Fielding HC (1980) J Chem Soc, Perkin Trans 1: 1551
128. Bartlett S, Chambers RD, Kirk JR, Lindley AA, Fielding HC, Powell RL (1983) J Chem Soc, Perkin Trans 1:1235
129. Coe PL, Sellers SF, Tatlow JC, Fielding HC, Whittaker G (1983) J Chem Soc, Perkin Trans 1:1957
130. Coe PL, Ray NC (1991) J Fluorine Chem 53:15
131. Chen LF (1983) Act Chim Sinica 41:375
132. Chen LF (1994) J Fluorine Chem 67:95 and references contained
133. Chambers RD, Kirk JR, Powell RL (1983) J Chem Soc, Perkin Trans 1:1239
134. Coe PL, Sellars A, Tatlow JC (1985) J Chem Soc, Perkin Trans 1:2185
135. Coe PL, Sellars A, Tatlow JC (1986) J Fluorine Chem 32:135
136. Makarov KN, Gervits LL, Cheburkov YA, Knunyants IL (1977) J Fluorine Chem 10:323
137. Makarov KN, Nikolaeva EE, Snegirev VF (1990) J Fluorine Chem 48:133
138. Maruta M, Ishikawa N (1979) J Fluorine Chem 13:111
139. Maruta M, Ishikawa N (1979) J Fluorine Chem 13:421
140. Flowers WT, Haszeldine RN, Owen CR, Thomas A (1974) J Chem Soc, Chem Commun: 134
141. Ishikawa N, Nagashima A (1976) Bull Chem Soc Jpn 49:1085
142. Chambers RD, Korn SR, Sandford G (1994) J Chem Soc, Perkin Trans 1:71

Reactions of Electrophiles with Polyfluorinated Olefins

Viacheslav A. Petrov[1] and Vadim V. Bardin[2]

[1] DuPont Central Research and Development, Experimental Station, PO Box 80328, Wilmington, DE, 19880-0328, USA. *E-mail: Petrovva@a1.esvax.umc.dupont.com*
[2] Institute of Organic Chemistry Russian Academy of Sciences, Novosibirsk, 630090, Russia

This chapter seeks to cover recent developments in the area of electrophilic reactions of polyfluorinated olefins. It contains sections on methods of generation, types and relative reactivity of electrophilic reagents, and reactivity of fluoroolefins. It also summarises recent achievements in generation and characterisation of long-lived polyfluorinated carbocations. The section dealing with electrophilic reactions of fluoroolefins is divided in two parts. The first part covers electrophilic addition reactions of fluoroolefins, e.g., halogenation, oxidative fluorination, addition of hypohalides, nitration, oxidation, addition of sulfur halides, along with alkylation, alkenylation, and acylation reactions and the second part covers processes involving activated C-F bonds in fluoroolefins, such as double bond migration, cleavage of vinyl ethers, and reactions with sulfur trioxide and boron triflate.

Keywords: Electrophilic; polyfluoroolefin; Lewis acid; polyfluorinated carbocations.

Dedicated to Dr. Carl G. Krespan on the occasion of his 70th birthday.

1
Introduction

Any reaction of electrophilic reagents with an unsaturated system consists of several steps including an attack of the electrophile on the double bond as a first step of the process. In other words, in electrophilic reactions of olefins, reactivities of both counterparts are equally important. Due to specific electronic and steric properties of the F-alkyl groups, polyfluorinated olefins, particularly those containing one or two R_f groups, are extremely resistant towards an electrophilic attack. However, although lower reactivity, compared to hydrocarbon analogs, of polyfluoroolefins can be compensated by higher reactivity of electrophilic reagents. Actually, such factors as higher stability of the C=C bond in polyfluoroolefins towards strong oxidizers and electrophilic reagents and the resistance of carbocationic intermediates to secondary processes, e. g., migration of alkyl group or elimination reactions, are responsible for much cleaner reactions of fluoroolefins, even with extremely reactive electrophiles.

Electrophilic reagents in general can be divided into two groups: charged and neutral. Charged electrophiles are ionized compounds in which the cation plays an active role. Some of them, for instance nitronium tetrafluoroborate $NO_2^+ BF_4^-$, form stable and fully ionized salts [1]. Such reagents as the antimonate salt of the F-allyl cation (1) [2] are not stable enough to be isolated, but they still can be generated in situ and used for further reactions:

$$CF_2=CFCF_3 + n\,SbF_5 \rightleftharpoons \text{(1)}$$

HFP

1

Neutral electrophiles contain a highly polarized E-X bond; therefore, they are able to participate in electrophilic reactions. Interhalogen compounds

(ClF, BrF$_3$, IF$_5$), perfluoroalkyl hypohalites, halogen fluorosulfates, and trifluoromethanesulfonates XOSO$_2$Y (X=Cl, Br, I, Y=F; X=Cl, Br, Y=CF$_3$) are examples.

Based on mechanism, electrophilic reactions involving a C=C bond could be divided into two groups. The first group includes reactions having a stepwise mechanism. This is typical of charged electrophiles. According to this mechanism, initial attack on the double bond results in generating linear or bridged carbocations:

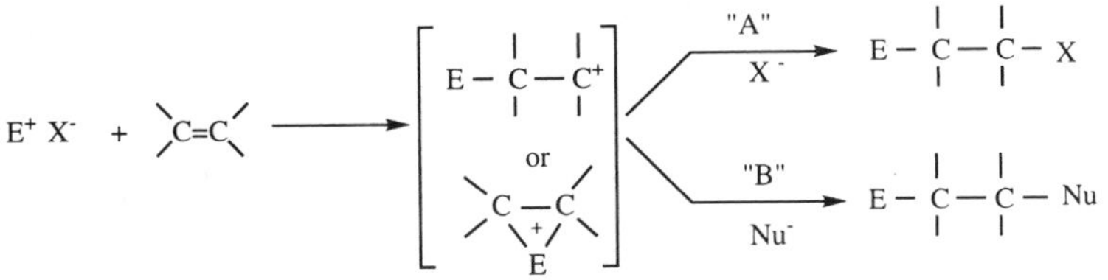

$$E^+ X^- \ + \ \underset{}{\overset{}{C{=}C}} \ \longrightarrow \ \left[\begin{array}{c} E-\overset{|}{\underset{|}{C}}-\overset{|}{\underset{|}{C}}^+ \\[2mm] \text{or} \\[2mm] \underset{\overset{|}{+}}{\overset{}{C}}{-}\overset{}{\underset{}{C}} \\ E \end{array} \right] \begin{array}{c} \overset{\text{"A"}}{\underset{X^-}{\longrightarrow}} \ E-\overset{|}{\underset{|}{C}}-\overset{|}{\underset{|}{C}}-X \\[4mm] \overset{\text{"B"}}{\underset{Nu^-}{\longrightarrow}} \ E-\overset{|}{\underset{|}{C}}-\overset{|}{\underset{|}{C}}-Nu \end{array} \qquad (2)$$

The intermediate further reacts with the counter anion (or the anion derived from it, usually F$^-$), giving a product of addition of E-X to the double bound (pathway A). However, carbocations may also react with another nucleophile present in the reaction media to give a product in which two fragments not connected to each other in the starting material are added across the double bond (pathway B, so called conjugate addition [3]). Carbocations are thus the key intermediates in both processes.

In general, the addition of neutral electrophiles across a C=C bond, such as SO$_3$ [4], proceeds in accordance with polarization of reagents as a concerted process going through a highly organized cyclic transition state, exemplified by Eq. (3):

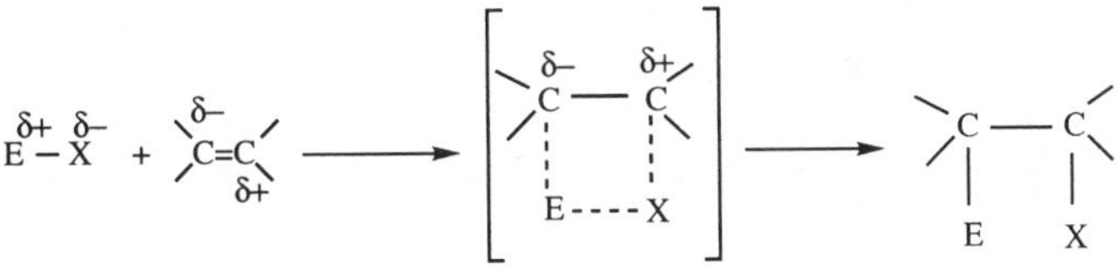

$$\overset{\delta+ \ \delta-}{E-X} \ + \ \underset{\delta+}{\overset{\delta-}{C{=}C}} \ \longrightarrow \ \left[\begin{array}{c} \overset{\delta-}{C}{-}\overset{\delta+}{C} \\ \vdots \quad \vdots \\ E{-}{-}{-}{-}X \end{array} \right] \ \longrightarrow \ \underset{E \quad X}{\overset{}{C{-}C}} \qquad (3)$$

Finally, along with electrophilic reactions involving a C=C bond, typical of both hydrocarbon and polyfluorinated olefins, there is another group of transformations which is specific for fluoroolefins only – reactions involving an allylic C-F bond. This group of reactions includes (but is not limited to) such processes as migration of the double bond in fluoroolefins or insertion reactions, as shown in Eq. (4), leading to F-allyl fluorosulfate (2) [5]:

$$CF_2{=}CFCF_3 \ + \ SO_3 \ \xrightarrow[\text{50°C, 6h}]{\text{BF}_3} \ CF_2{=}CFCF_2OSO_2F \qquad (4)$$

$$\textbf{2}, 60\%$$

The reactions of electrophilic reagents with olefins containing more than two fluorines in the molecule – *polyfluoroolefins* – are the subject of this review, which is organized as follows. Methods of generation, types, and relative activities of electrophiles are discussed in Sects. 2 and 3; reactivity of fluoroolefins and orientation of electrophilic addition across a C=C bond are described in Sect. 4. Since carbocations are often formed in these reactions, Sect. 5 deals with the question of existence and stability of carbocationic intermediates. Finally, Sect. 6 presents different types of electrophilic reactions, which are divided into two groups: addition across the C=C bond and insertion into a C-F bond.

The last comprehensive review on electrophilic reactions of fluoroolefins was published in 1969 [6]. Since then, several reviews and papers dealing with different aspects of this chemistry, such as alkylation and alkenylation reactions [7], addition of halogen fluorosulfates [8], trifluoromethanesulfonates [9] and halogen fluorides [10] to fluoroolefins have been published. Additional information on the reactions involving carbocations could be found in two recent review articles [11, 12]; some data on the subject are scattered in several books and journals [13–19].

In the last 27 years, tremendous progress has been made in this field of organic chemistry. Since the area is broad and the subject quite complex, no attempt has been made to present comprehensive coverage of the related literature. Instead, the authors of this review attempt to describe electrophilic reactions of fluoroolefins with particular stress on the mechanistic and synthetic aspects, to make it as comprehensive as possible in the main types of electrophilic reactions and to give the most important examples. Inevitably, this approach has left out some material and we offer in advance our regrets to all whose work was omitted.

2
Methods of Generation of Electrophiles

Electrophilic reagents can be generated in situ using several procedures. Ionization of a C-halogen bond by the action of strong protic or Lewis acids is generally used for generation of short-lived electrophilic species, for example carbocations. Oxidative processes are mostly used for the preparation of neutral but highly electrophilic materials, such as halogen fluorosulfate.

2.1
Protonation

Despite the fact that H^+ (in terms of electrophilic reactions of polyfluorinated compounds only) is a relatively mild electrophile, strong protic acids ($HOSO_2F$, $HOSO_2CF_3$, anhydrous HF) are widely used for generating electrophilic species. Thus, protonation of fluorine in HgF_2 by anhydrous HF results in formation of the corresponding metal centered cation **3** [17, 20]:

$$\text{F-Hg-F} + H^+ \longrightarrow \underset{\mathbf{3}}{\text{F-Hg}^+} + \text{HF} \qquad (5)$$

It should be noted that in the presence of some Lewis acids a drastic increase of acidity (which correlates with an increase of electrophilicity of H^+) of protic acids may be observed. "Superacids" were successfully used in preparation of a large number of onium (oxonium, sulfonium, azonium, and halonium) cations and homo- and hetero- species, such as polyatomic cations of halogens [I_n^+ (n = 2, 3, 5), Br_n^+ (n = 2, 3), Cl_3^+] and interhalogens (for instance, BrF_2^+ and ClF_2^+). More information on the preparation, handling, and use of superacids for generation of cationic species can be found in the book [1] describing the groundbreaking work of George Olah and coworkers.

A combination of electrophilic properties of H^+ and relatively low nucleophilicity of the counter anion in superacids makes these materials convenient media for the generation of different types of carbocations as a result of protonation of the carbon-element bond:

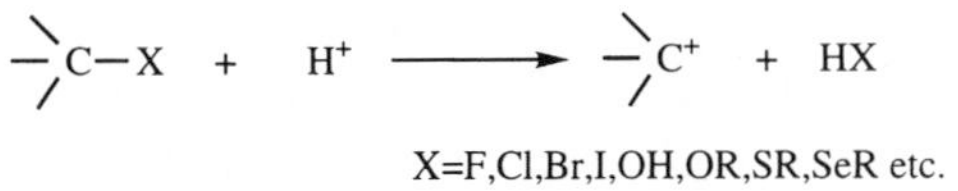

$$\text{\textbackslash}C-X \quad + \quad H^+ \longrightarrow \quad C^+ \quad + \quad HX \tag{6}$$

$$X = F, Cl, Br, I, OH, OR, SR, SeR \text{ etc.}$$

$$E = C \quad + H^+ \longrightarrow H-E-C+ \tag{7}$$

However, the low basicity of polyfluorinated materials usually limits this method to the carbocations originating from nonfluorinated compounds (some alcohols, aromatics, etc.) or such relatively "basic" fluorinated substrates as mono- and dihaloalkanes, and fluoroethylenes, e.g., $CH_2=CF_2$ [21].

2.2
Direct Interactions with Lewis Acids

Generation of electrophilic species from organic *polyfluorinated* materials usually requires direct interaction of a strong Lewis acid with a substrate, which can result in formation of a cation either by abstraction of a halogen anion (usually F^-) from a substrate (Eq. 8) or by formation of a zwitterionic intermediate as a result of coordination of the Lewis acid with an unshared electron pair on a multiply bonded heteroatom (Eq. 9) [12]:

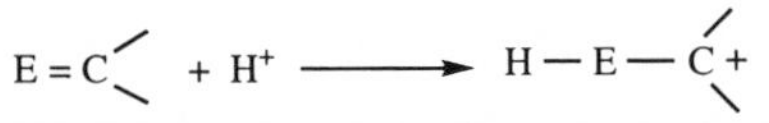

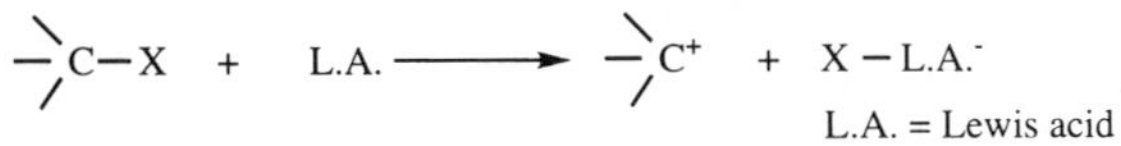

$$\text{\textbackslash}C-X \quad + \quad L.A. \longrightarrow \quad C^+ \quad + \quad X-L.A.^- \tag{8}$$

$$L.A. = \text{Lewis acid}$$

$$X = C \quad + L.A. \longrightarrow \quad {}^-L.A.-X-C+ \tag{9}$$

$$L.A. = \text{Lewis acid}$$

Although SbF_5 is probably the most studied and widely used reagent and catalyst for electrophilic reactions, it is not the only Lewis acid employed in fluoroorganic synthesis. Boron trifluoride and triflate, tantalum and niobium pentafluoride, chlorofluorides of antimony(V), aluminum chloride and bromide along with recently discovered, highly effective aluminum chlorofluoride should be added to this list. Unfortunately, quantitative data on the relative activity of Lewis acids are scattered in the literature and are often contradictory and it is difficult to make a "scale" of strength for different Lewis acids. However, a qualitative picture is more or less clear and we can arrange most Lewis acids used in

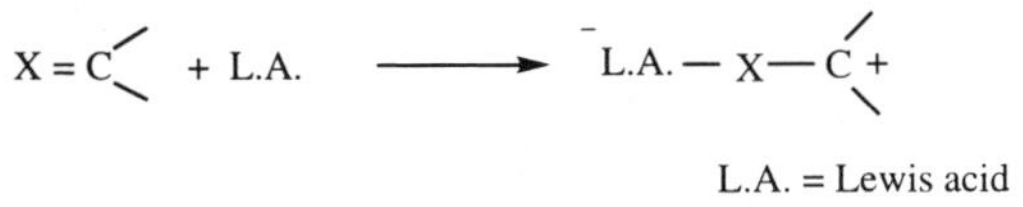

electrophilic reactions of polyfluoroolefins in the following sequence, where their activity, decreasing from left to right, is based on the ability of the Lewis acid to abstract the fluoride anion from a substrate:

$$AlClxFy > SbF_5, > B(OSO_2CF_3) > AlBr_3 > AlCl_3 > SbClxF_{5-x} > AsF_5 > TaF_5 \geq BF_3 > NbF_5 > BiF_5 > VF_5. \quad (10)$$

Aluminum chlorofluoride $AlCl_xF_y$ (ACF) is probably the most effective Lewis acid known; several reactions have been reported so far, such as condensation of *F*-pentene-2 with tetrafluoroethylene (TFE), which are catalyzed by ACF only. However, ACF is easy to deactivate by traces of water or other proton sources (for instance, HF or HCl) [22, 23]; this limits the use of the catalyst to perfluoro-, mono-, or dihydroperfluorocarbons. Antimony pentafluoride, being one of the strongest Lewis acids, was used as a catalyst for a large number of reactions [7, 11, 12, 24]. Boron triflate seems to be not quite as powerful as SbF_5, although it also allows "functionalization" of polyfluorocarbons [25] as a result of further transformations of the -OSO_2CF_3 group introduced. Unfortunately, the reagent has limited shelf life and is not stable at elevated temperatures. Aluminum halides were probably the first among Lewis acids to be employed for isomerization of halofluorocarbons, such as $ClCF_2CCl_2F$ (CFC-113) [13, 26], and later for condensation of halomethanes with fluoroethylenes [27]. Antimony chloro- and bromofluorides, usually generated in situ, for example by the reaction of SbF_3 with Cl_2 or Br_2 or $SbCl_5$ with HF, are widely used for fluorination of chlorocarbons by chlorine exchange (Swarts Reaction) [13]. Arsenic pentafluoride is a fairly strong Lewis acid, but its use in fluoroorganic synthesis is limited. TaF_5 and BF_3 are weaker Lewis acids than SbF_5, but stronger than NbF_5 [28]. All three are often used in combination with such protic acids as HF or $HOSO_2F$.

BiF_5 and especially VF_5 are weak Lewis acids, but both of them very potent oxidizing agents. VF_5 was employed for fluorination of the C=C bond in a variety of fluorinated materials [19, 29]. Both of them react violently with most organic materials and, therefore, most of the above-mentioned compounds should be handled with care and by trained personnel.

It should be kept in mind that the sequence at Eq. (6) represents only relative activities of Lewis acids, and the order of activity could change significantly depending on the substrate and reaction conditions. ACF or $AlCl_3$, for example, completely lose their activity in HF, in contrast to SbF_5 which usually works well in the presence of small amounts of HF or even when HF is used as reaction media.

A recent review, along with numerous examples of electrophilic reactions of polyfluorocarbons, has a section summarizing data on physical properties of some Lewis acids and recommendations on their use in synthesis [12].

2.3 Oxidative Processes

The relationship between Lewis acidity and oxidative properties of different compounds is neither straightforward nor well-defined; many materials possess both properties. For example, SbF_5, one of the strongest Lewis acids, is also a very strong fluorooxidizer [19]. Oxidative properties of this fluoride were used for generating polynuclear sulfur, selenium, and telurium cations [30]. Sulfur trioxide is another example of a Lewis acid being at the same time a potent oxidizer. Oxidation of iodine by an excess of oleum or SO_3 results in formation of the I_2^+ cation [31]:

$$2I_2 + 5SO_3 + H_2S_4O_{13} \longrightarrow 2I_2^+ + 2HS_4O_{13}^- + SO_2 \tag{11}$$

A mechanism postulating formation of a carbocation as an intermediate during oxidative fluorination of the C=C bond was proposed for the reaction of CoF_3 with F-benzene and F-pyridine [32], and VF_5 with fluoroolefins and polyfluoroaromatic compounds [33]:

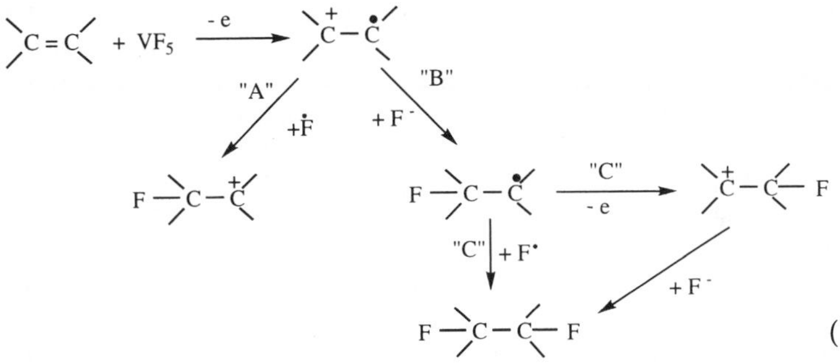

$$\tag{12}$$

Neutral electrophilic reagents are usually prepared by reactions involving strong oxidants. For example, the most practical route to relatively stable hypohalites $XOSO_2F$ (X=Cl, Br, I) [8] is based on the reaction of halogens with peroxide FSO_2OOSO_2F. This material can be made on a large scale by catalytic fluorination of SO_3 [34]:

$$SO_3 + F_2 \xrightarrow{AgF_2} FSO_2OOSO_2F \xrightarrow{X_2} XOSO_2F \tag{13}$$
$$X = Cl, Br, I$$

Chlorine trifluoromethanesulfonate, which is much less stable than the corresponding fluorosulfate was synthesized by the reaction of triflic acid with ClF [35]. In the synthesis of bromine trifluoromethansulfonate, the reaction of $ClOSO_2CF_3$ with Br_2 was used [36].

Perfluoroalkylhypochlorites are usually prepared by the reaction of corresponding carbonyl compounds with ClF in the presence of metal fluorides [37]:

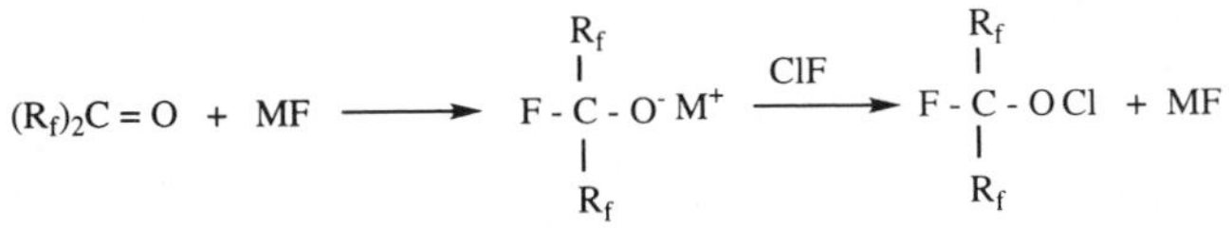

$$(R_f)_2C=O + MF \longrightarrow F\text{-}\underset{\underset{R_f}{|}}{\overset{\overset{R_f}{|}}{C}}\text{-}O^-M^+ \xrightarrow{ClF} F\text{-}\underset{\underset{R_f}{|}}{\overset{\overset{R_f}{|}}{C}}\text{-}OCl + MF \tag{14}$$

However, the addition of ClF across the C=O bond is also catalyzed by HF and Lewis acids such as AsF_5 or BF_3 [38]. Only a few reported examples of perfluoroalkyhypobromites [39] were made by the reaction of corresponding alkoxides with relatively stable $BrOSO_2F$:

$$R_fC(CF_3)_2O^-Na^+ + BrOSO_2F \xrightarrow[-25^\circ C]{} R_fC(CF_3)_2OBr + NaOSO_2F \quad R_f = CF_3, C_2F_5 \qquad (15)$$

3 Types of Electrophiles and Relative Activity

As mentioned above, electrophilic reagents could be divided into two large groups: charged (salts of nitronium, nitrosonium, carbonium, or allyl cations) and neutral. Most charged electrophiles have a limited lifetime and are usually generated in situ in the course of reaction. *F*-Allyl cation (**1**) is generated by abstraction of F^- from *F*-propylene (HFP, Eq. 1). Despite the fact that this cation was characterized by NMR, even at −40 °C it is in equilibrium with the starting material. The reaction of *F*-cyclobutene with fluoroolefins, catalyzed by SbF_5, is believed to proceed via the formation of *F*-cyclobutenyl cation [7]; however, several attempts to characterize this intermediate by NMR spectroscopy have failed [21, 40], which probably means that the lifetime of this carbocation is too short for the NMR scale:

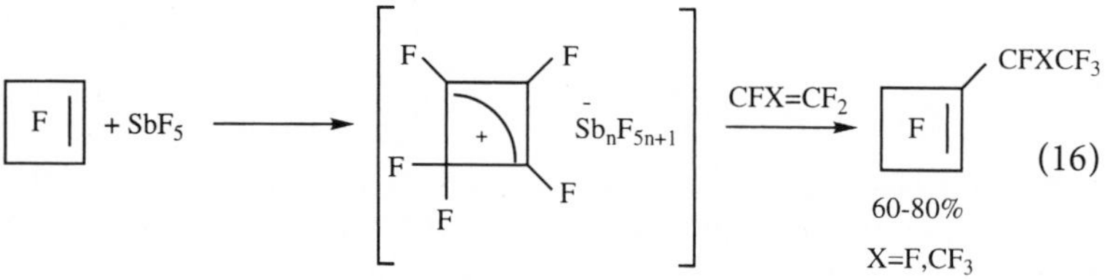

(16)

On the other hand, some of these species are stable enough to be isolated as salts. Nitronium cation NO_2^+ exists in equilibrium with nitric acid at ambient temperature; however, more than 15 crystalline nitronium salts with a variety of counter ions have been isolated and characterized [1]. The most widely used salt $NO_2^+BF_4^-$ is made by the reaction of HNO_3 and BF_3 in anhydrous hydrogen fluoride [41]:

$$HNO_3 + HF + 2BF_3 \longrightarrow NO_2^+BF_4^- + BF_3 \cdot H_2O \qquad (17)$$

Compounds containing a polarized bond form another group of electrophilic reagents. Polyfluorohypohalites (hypochlorites, hypobromites, but not *hypofluorites*), interhalogen compounds (ClF, BrI, IF, BrF_3, IF_5 etc.), hydrogen halides, sulfur trioxide, and hexafluoroacetone are representatives of this large group of electrophilic reagents.

Reactivity of these materials may vary from quite low for HX (X=F, Cl, Br) and hexafluoroacetone to extremely high in the case of BrF_3, and chlorine and bromine fluorosulfates. It should also be noted that the reactivity of many of these compounds could be significantly increased by adding Lewis acids and/or increasing polarity of the media.

Unfortunately, quantitative data on the relative reactivity of electrophilic reagents are rare and scattered in the literature, which makes it difficult to build a precise "scale" of reactivity. However, on the basis of data available, different electrophilic reagents could be arranged in the following sequence in the order of declining activity:

$$H^+ < F_n\text{-}E^+, \;\overset{\cdot}{\to}C^+, < E^+, \leq NO_2^+, Hal^+, SO_3, IF < IOSO_2F, BrF \qquad (18)$$

$$ClOR_f, ClF, XOSO_2CF_3, (X = Cl, Br, I), XOSO_2F (X = Cl, Br)$$

A proton is the weakest electrophile. For example, anhydrous HF does not add to tetrafluoroethylene at ambient temperature or to hexafluoropropene, even at 200 °C [6]. Arsenic trifluoride in the presence of SbF_5 (this mixture is a source of the F_2As^+ cation [42, 43]) reacts with tetrafluoroethylene at 20 °C to form a mixture of F-(diethyl)-(4) and F-(triethyl)(5) arsines:

$$AsF_3 + SbF_5 \longrightarrow \left[F_2As^+ \cdot \overset{-}{Sb_nF_{5n+1}}\right] \xrightarrow{n\,CF_2=CF_2} (C_2F_5)_2AsF + (C_2F_5)_3As$$

$$\qquad\qquad\qquad\qquad\qquad\qquad\qquad\qquad\qquad\qquad\quad \underset{30\text{-}40\%}{4} \qquad \underset{30\text{-}40\%}{5}$$

$$(19)$$

However, under similar conditions, AsF_3 does not reacts with HFP [7].

In general, polyfluorinated carbocations are more active electrophiles, for instance cation 1 attacks F-propene (HFP) at a temperature as low as 70 °C to yield a dimer 6 [44]

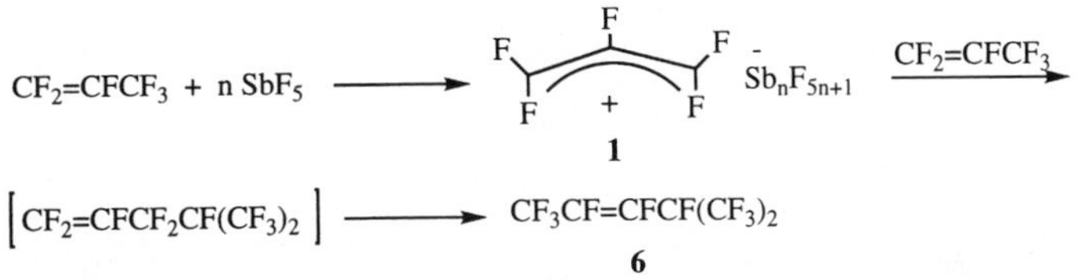

$$\left[CF_2=CFCF_2CF(CF_3)_2\right] \longrightarrow \underset{6}{CF_3CF=CFCF(CF_3)_2} \qquad\qquad (20)$$

Protonation of mercuric fluoride results in formation of highly electrophilic cation 3 (Eq. 5), which is able to attack the C=C bond of HFP at 85 °C to give the corresponding mercurial [6]:

$$HgF_2 + 2\,CF_2=CFCF_3 \xrightarrow{HF, 85°C} Hg[CF(CF_3)_2]_2 \qquad\qquad (21)$$

Polynuclear cations of certain elements, nitronium cation, halogen mono- and polynuclear cations and dications, such as S_n^{2+} or Se_n^{2+}, sulfur trioxide, and some interhalogen compounds (IF) form a second group of electrophilic reagents. All members of this group interact with fluoroethylenes and hexafluoropropene under mild condtions (24 – 70 °C), although these reagents, as a rule, are not able to attack a highly electron deficient C=C of internal fluoroolefins $R_f CF=CFR_f'$.

The third group contains the most powerful electrophilic reagents. Despite the fact that all of these materials are neutral, they have very high activity; therefore, handling of these materials requires special training. Reagents such as chlorine monofluoride, fluoroalkyl hypochlorites, bromine and chlorine fluorosufates react with fluoroethylenes and even HFP at temperatures well below 0 °C. Some halogen fluorosulfates, for example, are able to add not only to internal fluoroolefins $R_f CF{=}CFR_f'$, but even to *F*-cyclobutene and *F*-cyclopentene having low reactivity towards electrophiles [8]. Reactivity of halogen fluorosulfates decreases along the series $ClOSO_2F > BrOSO_2F > IOSO_2F$. The first member of the family readily adds to both *F*-isobutene (PFIB) and even *F*-cyclopentene, but $IOSO_2F$ only slowly adds to PFIB at ambient temperature, and does not form an additional product with *F*-cyclobutene and *F*-cyclopentene [8].

4 Reactivity of Fluoroolefins and Orientation of Addition in Electrophilic Reactions

Data on the reactivity of fluoroolefins in reaction with electrophiles are scattered. By combining data from several different sources [3, 6–8, 11, 12, 15–17], the following conclusions could be drawn. In general, the reactivity of fluoroethylenes decreases with an increasing fluorine content in the molecule. Highly fluorinated alkenes are more resistant to electrophilic attack, particularly when one, two, or more perfluoroalkyl groups are present [15]. Thus, reactivity of fluoroolefins in reactions with electrophilic reagents decreases going from $CH_2{=}CF_2$ to tetrakis(trifluoromethyl)ethylene:

$$\begin{gathered} CH_2{=}CF_2 > CFH{=}CF_2 > CFCl{=}CF_2 \geq CF_2{=}CF_2 > CF_2{=}CHCF_3 > CF_2{=}CFCF_3 > \\ R_fCH{=}CHR_f' \geq CF_2{=}C(CF_3)_2 > CF_3CF{=}CFCF_3,\ CF_3CF{=}CFC_2F_5 \geq R_fCF{=}CFR_f' \geq \\ c\text{-}C_4F_6 > c\text{-}C_5F_8 > (CF_3)_2C{=}CFCF_3,\ (CF_3)_2C{=}CFC_2F_5 > (CF_3)_2C{=}C(CF_3)_2 \end{gathered} \quad (22)$$

The first member of this family – vinylidene fluoride – readily adds fluorosulfonic acid, even in the absence of the catalyst, and it is nitrofluorinated by a HNO_3/HF mixture six times faster than $CFH{=}CF_2$ [15]; however, HF/HSO_3F superacid does not react at ambient temperature with $ClCF{=}CF_2$ or $CF_2{=}CF_2$. These two olefins show a similar reactivity in electrophilic reactions. Interaction of polyfluorinated propylenes $CF_2{=}CXCF_3$ (X=H,F) with trifluoro- and tetrafluoroethylenes proceeds at room temperature in the presence of a Lewis acid catalyst to give the corresponding polyfluoropentenes-2:

$$CF_2{=}CXCF_3 + CFY{=}CF_2 \xrightarrow[25^\circ C]{SbF_5} \underset{70\text{-}90\%}{CF_3CX{=}CFCFYCF_3} \qquad X,Y=F,H \quad (23)$$

In the absence of ethylene at 20 °C, $CF_2{=}CHCF_3$ readily forms a dimer in the presence of SbF_5, although electrophilic dimerization of hexafluoropropylene proceeds only at 70–80 °C and is followed by a side reaction – fluorination of the C=C bond of the olefin by SbF_5 [44]:

$$CF_2=CXCF_3 \quad \Big\langle \begin{array}{c} \xrightarrow[X=H]{SO_2,\ 20°C} \quad \begin{array}{c} CF_3CH=CFCH(CF_3)_2 \\ 56\% \end{array} \\[2em] \xrightarrow[X=F]{70°C} \quad \begin{array}{c} \textbf{6} \ + \ CF_3CF_2CF_3 \\ 30\% \end{array} \end{array} \qquad (24)$$

The double bond in *F*-1,2-dialkylethylenes is very unreactive towards electrophiles. Reaction with a mixture of X_2/SO_3 (X=Cl, Br, I) is one of very few examples of such transformations [45]:

$$C_4F_9CH=CHC_4F_9 \xrightarrow{1)Br_2\,/\,SO_3\ 2)\,H_2O} \underset{\underset{Br\quad OH}{|\quad\ |}}{C_4F_9CH - CHC_4F_9} \quad \textbf{8},\,78\% \qquad (25)$$

Whereas the difference in the reactivity of $R_fCH=CHR_f'$ and $CF_2=C(CF_3)_2$ is marginal, the double bond in PFIB is significantly less reactive towards E$^+$ compared to *F*-propylene. Most electrophilic reactions of this olefin proceed in a temperature range of 100–200 °C. Among known reactions are those with HF, HgF_2/HF, sulfur/SbF_5, and iodine monofluoride [46]:

$$(CF_3)_3CH$$
$$\uparrow HF$$
$$[(CF_3)_3C]_2Hg \xleftarrow{HgF_2\,/HF} CF_2=C(CF_3)_2 \xrightarrow[200°C]{S\,/\,SbF_5} [(CF_3)_3CS-]_2 \qquad (26)$$
$$130°C \Big\downarrow IF_5\,/\,I_2$$
$$(CF_3)_3CI$$

A combination of two effects of *F*-alkyl groups – a large sterical volume and the strong electron-withdrawing effect – are probably the factors responsible for the extremely low reactivity of the C=C bond in cyclic and internal *F*-olefins, although these olefins still have the ability to interact with very strong electrophiles such as chlorine and bromine fluorosulfates [8]:

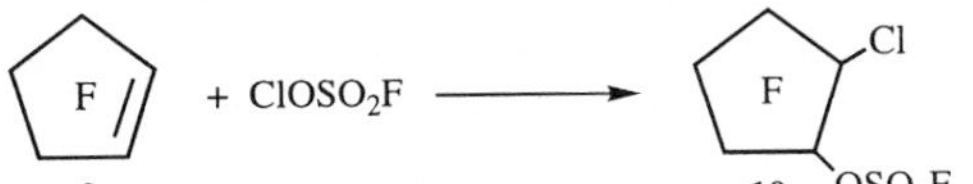

$$\underset{\textbf{9}}{\boxed{F}} \ + \ ClOSO_2F \ \longrightarrow \ \underset{\textbf{10}}{\boxed{F}}\overset{Cl}{\underset{OSO_2F}{}} \qquad (27)$$

$$CF_3CF=CFC_2F_5 \ + \ ClOSO_2F \xrightarrow{-65°C} \underset{\underset{OSO_2F}{|}}{CF_3CF\,CFClC_2F_5} + \underset{\underset{OSO_2F}{|}}{CF_3CFCl\,CFC_2F_5} \qquad (28)$$
$$\textbf{11} \qquad\qquad\qquad\qquad\qquad\qquad 2 \quad : \quad 1$$
$$\text{yield } 76.5\ \%$$

Reactions of non-symmetrical internal *F*-olefins are not usually regiospecific. Fluoroolefins carrying three or four fluoroalkyl groups have a highly electron-deficient C=C bond, so it is not surprising that very few, if any, reactions of these materials with electrophilic reagents have been reported.

An important subject that needs to be addressed in this section is the orientation of electrophilic addition across the C=C bond of fluoroolefins and the factors affecting it. In general, addition of electrophilic reagents proceeds in accordance with polarization of the double bond of the fluoroolefin. This polarization is the result of interplay of two major factors: the electron withdrawing effect of R_f groups (σ-inductive effect, I_σ, transmitted through a σ-bonded system) and the electronic effects of fluorine substituents attached directly to a π-system, which are rather complex. In addition to the I_σ and classic field effects (through spacial electrostatic interaction [16]), fluorine substituents have the ability to interact with π-density of C=C by a resonance effect. A quantitative measure of the electron-withdrawing effect (I_σ) is the substituent constant σ_I (0.52 for F and 0.44 for CF_3) [15]. On the other hand, the constant σ_R that reflects the ability of the substituent to donate electron density (−0.34 for F and 0.10 for CF_3) [15] is a measure of the resonance effect. A negative value of σ_R for the fluorine substituent indicates that it has the pronounced ability to donate an unshared p-electron pair. Interaction between p-electrons of fluorine substituent and π-electrons of the C=C bond leads to the shift of π-electrons (p-π repulsion) and polarization of the double bond. In general, the resonance effect dominates the inductive effect when fluorine is directly connected to the C=C bond, a carboanionic or a carbocationic center.

As a result of the overlaping inductive and resonance effects of substituents (acting in the same direction as in the molecules of HFP and PFIB), the central carbon in these olefins has a much higher negative charge than that of the terminal atom. It means that the attack of electrophile E^+ on these substrates is aimed at the *central* carbon in contrast to the attack of nucleophile Nu^- which is directed toward *terminal* carbon carying a positive charge:

(29)

In the molecule of internal olefin *F*-butene-2, the inductive and resonance effects of CF_3 and F substituents compensate for each other due to symmetry of the molecule. As a result of the much lower polarization of the C=C bond (along with steric shielding of it by a bulky CF_3 group), it is responsible for the significantly lower reactivity of this olefin towards electrophilic reagents:

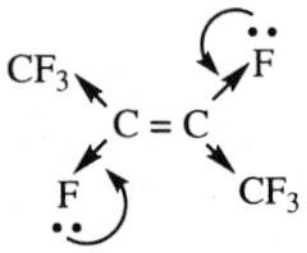

(30)

This is also true for all internal fluoroolefins carrying two or more *F*-alkyl substituents at the C=C bond.

Carbocations are important intermediates in most electrophilic reactions, since the attack of an electrophile on the C=C bond often results in the formation of these species (see Eq. 2). Factors affecting the stability of carbocations are important for the prediction of orientation of an electrophilic attack (more detailed discussion on carbocations is given in Sect. 5). A fluorine substituent located at the α-position to a cation center is able to stabilize it by the "back donation" mechanism represented by Eq. (31):

$$R - \overset{/F}{\underset{\backslash F}{C}} + \quad \longleftrightarrow \quad R - \overset{\overset{+}{F}}{\underset{\backslash F}{C}} \tag{31}$$

However, due to a pronounced I_σ effect, fluorines in β-positions strongly destabilize carbocations. $FCH_2CH_2^+$ is calculated to be 29.6 kcal/mol less stable than isomeric CH_3CHF^+. Unlike other halogens, fluorine does not form bridged cyclic alkylfluoronium cations [15, 22].

According to the rule formulated in [15], "the combined α- and β-effects (*of fluorine substituents*) imply that fluoroolefins will react with electrophiles so as to minimize the number of fluorines β to electron-deficient carbon in the transition state." In accordance with this rule, reaction of $CH_2=CF_2$ with HF starts as an attack of electrophile (H^+) on the CH_2 group of ethylene (Eq. 32, pathway A), since this process leads to carbocation **12** stabilized by two α-fluorines in contrast to the much less stable intermediate **13** containing two β-fluorines and derived from the initial attack of H^+ on the CF_2 group of the olefin:

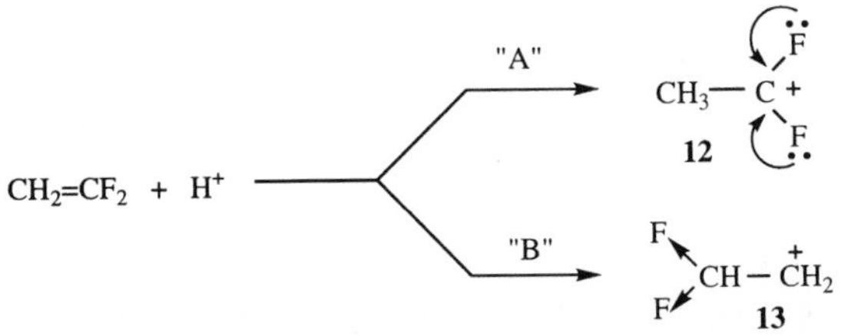

$$CH_2=CF_2 + H^+ \quad \begin{cases} \xrightarrow{\text{"A"}} & CH_3-\overset{..}{\underset{..}{C}}\overset{F}{\underset{F}{+}} \quad \mathbf{12} \\ \\ \xrightarrow{\text{"B"}} & \overset{F}{\underset{F}{\diagdown}}CH-\overset{+}{C}H_2 \quad \mathbf{13} \end{cases} \tag{32}$$

The orientation of addition of HF to $CH_2=CF_2$ is in agreement with data from [47], where it was demonstrated that in the gas phase the $CH_3CF_2^+$ cation is at least 20 kcal/mol more stable than FCH_2CHF^+.

The best known reaction of $CFH=CF_2$ with such electrophiles as H^+, FHg^+, F_2As^+, hydrocarbon, and polyfluoroalkyl cations [6, 7] starts with the attack of electrophile on the CFH group of the olefin and proceeds as Markovnikov addition:

$$E^+ + CFH=CF_2 \longrightarrow \left[E\text{-}CFH\text{-}CF_2^+ \right] \xrightarrow{+F^-} E\text{-}CFH\text{-}CF_2\text{-}F \tag{33}$$

On the other hand, most reactions of chlorotrifluoro- and bromotrifluoroethylenes are not regioselective and usually proceed with the formation of approximately equal amounts of both isomers. This is exemplified by the the mixture obtained from the reaction of CF_3CH_3 with $CFCl=CF_2$ [48] and by cycloaddition of hexafluoroacetone (HFA) to $CFBr=CF_2$ catalyzed by $AlCl_xF_y$ [49]:

$$CH_3CF_3 + CFCl=CF_2 \xrightarrow[50°C, 8h]{SbF_5} CH_3CF_2CFClCF_3 + CH_3CF_2CF_2CF_2Cl \tag{34}$$
$$ 60 \quad : \quad 40 \quad \text{yield 50\%}$$

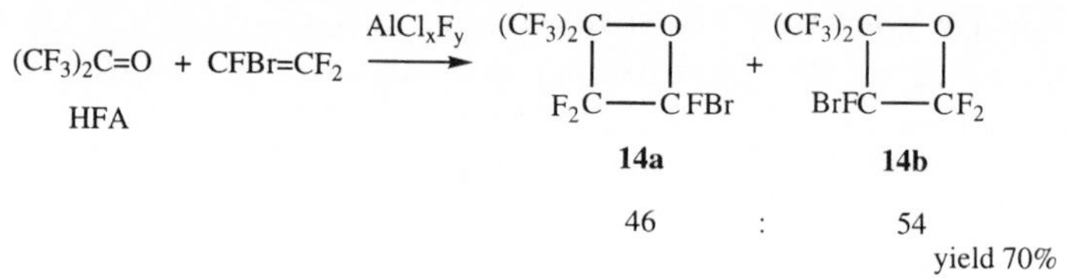

$$(CF_3)_2C=O \ + \ CFBr=CF_2 \ \xrightarrow{AlCl_xF_y} \ \underset{\substack{F_2C\text{---}CFBr}}{(CF_3)_2C\text{---}O} \ + \ \underset{\substack{BrFC\text{---}CF_2}}{(CF_3)_2C\text{---}O} \quad (35)$$

$$\text{HFA}$$

$$\textbf{14a} \qquad\qquad \textbf{14b}$$

$$46 \qquad : \qquad 54$$

$$\text{yield } 70\%$$

In accordance with polarization of the C=C bond, iodofluorination of $CF_2=CXCF_3$ [50] and $(CF_3)_2C=CF_2$ [49] starts with the attack of "I^+" electrophile on the central carbon of the olefin and results in each case in formation of a single regioisomer, $(CF_3)_2CFI$ and $(CF_3)_3CI$, respectively:

$$"I^+" \ + \ CF_3CX=CF_2 \ \longrightarrow \ \left[\underset{}{\overset{I}{CF_3CX\text{---}CF_2^+}}\right] \ \xrightarrow{+F^-} \ \underset{}{\overset{I}{CF_3CX\text{---}CF_2\text{-}F}} \quad (36)$$

$$X = H, F$$

$$CF_3CX=CF_2 \ + \ ICl \ \xrightarrow[50°C, 18h]{HF/BF_3} \ (CF_3)_2CXI \quad (37)$$

$$\begin{array}{ll} X = H & \textbf{15}, 59\% \\ X = F & \textbf{16}, 79\% \end{array}$$

$$(CF_3)_2C=CF_2 \ + \ I_2 / IF_5 \ \xrightarrow{AlX_3} \ (CF_3)_3CI \quad (38)$$

$$70\text{-}80\%$$

The same orientation was observed in reaction of both olefins with HF, HgF_2/HF, S/SbF_5 mixtures and $XOSO_2F$ [6–8].

Despite the fact that orientation of addition across the double bond in polyfluoroolefins is usually unambiguous and in most cases regiospecific, it is not always straightforward; several exceptions have been reported, mostly in the reactions involving neutral electrophiles. For instance, formation of a significant amount of regioisomer **17b** was observed in reaction of $CFH=CF_2$ with $IOSO_2F$:

$$IOSO_2F \ + \ CFH=CF_2 \ \longrightarrow \ ICFHCF_2OSO_2F \ + \ ICF_2CFHOSO_2F \quad (39)$$

$$\textbf{17a} \qquad\qquad \textbf{17b}$$

$$85 \qquad : \qquad 15$$

In contrast to most reported electrophilic reactions of $CF_2=CFCF_3$, addition of CF_3OCl to this olefin is not regiospecific, and it results in formation of two isomers [9], which may be an indication that a competetive radical pathway also operates in this reaction:

$$CF_3CF=CF_2 \ + \ CF_3OCl \ \xrightarrow[-196 \text{ to } 22°C]{CFCl_3} \ \underset{\substack{Cl}}{CF_3CF\text{-}CF_2OCF_3} \ + \ \underset{\substack{OCF_3}}{CF_3CF\text{-}CF_2Cl} \quad (40)$$

$$71.5 \qquad : \qquad 28.5$$

In contrast, interaction between CF_3SO_2OCl and $CFCl{=}CF_2$ which is not expected to be selective, actually is and leads to formation of a single isomer [9] (see also Sect. 6.1.4):

$$CF_3SO_2OCl + CFCl = CF_2 \longrightarrow CF_3SO_2OCFClCF_2Cl \qquad (41)$$

Reactions of internal *F*-alkenes with electrophiles such as BrF [51] or $ClOSO_2F$ [52] are not regioselective and both regio isomers are formed, but the addition of $ClOSO_2F$ was reported to be sensitive to the size of the *F*-alkyl substituents attached to the double bond [52] (see also Eq. 28):

$$CF_3CF{=}CFR_f + ClOSO_2F \longrightarrow CF_3CF(OSO_2F)\text{-}CFClR_f + CF_3CFCl\text{-}CFR_f(OSO_2F) \qquad (42)$$

R_f	Ratio			Yield %
n-C_3F_7	3	:	1	71.4
i-C_3F_7	2	:	3	83.3

5 Polyfluorinated Carbocations as Intermediates in Electrophilic Reactions

Fluorinated carbocations play an important role as intermediates in electrophilic reactions of fluoroolefins and other unsaturated compounds. For example, *F*-allyl cation **1** was proposed as a reactive intermediate in reactions of HFP with fluoroolefins catalyzed by Lewis acids [7]. The difference in stability of the corresponding allylic cations was suggested as the explanation for regiospecific electrophilic conjugated addition to $CF_2{=}CClCF{=}CF_2$ [11]. Allylic polyfluorinated carbocations were proposed as intermediates in the reactions of terminal allenes with HF [53] and BF_3 [54], ring-opening reactions of cyclopropanes [55]. Carbocations are also an important part of the classic mechanism of electrophilic addition to olefins (see Eq. 2). This section deals with the questions of existence and stability of poly- and perfluorinated carbocations.

The ability of α-fluorine substituents to stabilize the carbocationic center was discovered by Olah's group. The first long-lived carbocation containing fluorine – dimethylfluorocarbenium cation – was reported by this group [56] in 1967 and was prepared either by ionization of 2,2-difluoropropane by SbF_5 in SO_2 or protonation of $CH_3CF{=}CH_2$ with "magic acid." Latter methylfluoro- (**18**) and methyldifluoro- (**12**) carbenium cations, generated at low temperature by the reaction of SbF_5 with corresponding fluoroethanes in SO_2ClF as a solvent, were characterized by ^1H and ^{19}F spectroscopy [57]:

$$CH_3CF_3 + SbF_5 \xrightarrow[-80^{\circ}C]{SO_2ClF} CH_3\overset{+}{C}F_2 \;\; \overset{-}{Sb_nF_{5n+1}} \qquad (43)$$

12

$$CH_3CF_2H + SbF_5 \xrightarrow[-78^{\circ}C]{SO_2ClF} CH_3\overset{+}{C}FH \;\; \overset{-}{Sb_nF_{5n+1}} \qquad (44)$$

18

Further accumulation of fluorine substituents at carbon carrying positive charge results in significant destabilization, the trifluoromethyl cation being the least stable among halomethyl cations [58]. Recently, a series of trihalomethyl cations CX_3^+ (X=Cl, Br, I) has been generated in solution and characterized using NMR techniques; however, all attempts to prepare trifluoromethyl cation have failed [58]:

$$CX_4 + SbF_5 \xrightarrow[-78^{\circ}C]{SO_2ClF} \overset{+}{C}X_3\ \overset{-}{Sb}_nF_{5n+1} \qquad X=Cl, Br, I \tag{45}$$

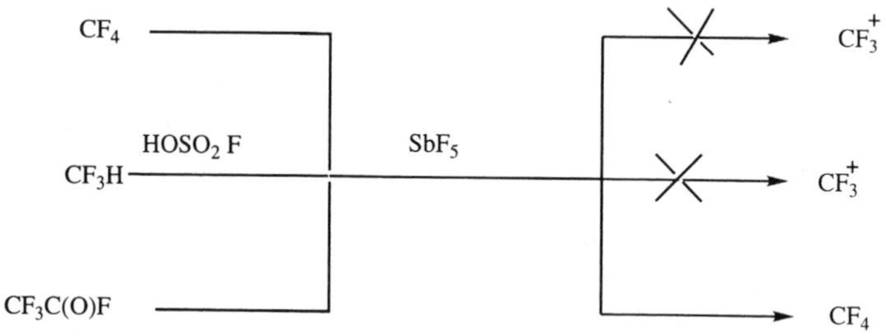

Scheme 1

On the other hand, for RCX_2^+ and R_2CF^+ the cation stability increases along with the increase of resonance (σ_R) effect of a halogen F > Cl > Br > I [59]. The significant stabilizing effect of fluorine substituent was explained as a result of back-donation of an unshared electron pair of F on the vacant orbital of carbon. Stability of substituted fluoromethyl cations in gas phase increases going from CF_3^+ to **12** [15]:

$$CH_3^+ < CF_3^+ < HCF_2^+ < CFH_2^+\ CH_3CH_2^+ \ll \mathbf{12} \le \mathbf{18};$$
$$^+CH(CH_3)_2 < {}^+CF(CH_3)_2$$

The fact that the α, α-difluoroethyl cation **12** has much higher stability than $CH_3CH_2^+$ means that, although alkyl groups stabilize carbocations better than α-fluorine, the latter has a much more pronounced stabilizing effect than hydrogen. Interestingly, cations **12** and **18** in gas phase have reverse stability compared to condensed phase. In solution, cation **12**, having two α-fluorine substituents, seems to be more stable than monofluoro cation **18**, since the former has been observed as a long-lived species; however no monofluoroalkyl cations $RCHF^+$ have been reported as being in solution under non-exchanging conditions [21].

The first example of a *perfluorinated* cation – *F*-cylopropenium – was reported by Sargeant and Krespan in 1969 [60], and it was followed by the discovery of *F*-benzyl [61] and *F*-benzoyl cations [21]. Since several unsuccessful attempts to generate *F*-allyl carbocations were reported [21, 40, 62], another "stepwise" approach to polyfluorinated delocalized cations has been developed. Stability of a polyfluorinated allylic system could be significantly increased by introduction of electron-donating groups in position 1 or 3 of the allylic triad, since the positive charge in this type of cations is localized at terminal carbons [63]. Using this methodology, a number of long-lived substituted analogs of F-methallyl **19–21** [64–66] and F-allyl **22** [62] cations were prepared:

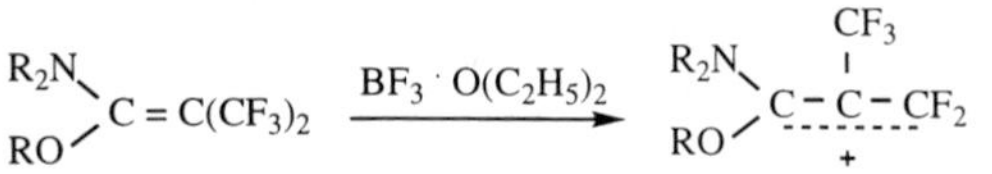

$$
\begin{array}{c}
R_2N \\
\diagdown \\
C = C(CF_3)_2 \\
\diagup \\
RO
\end{array}
\xrightarrow{BF_3 \cdot O(C_2H_5)_2}
\begin{array}{c}
CF_3 \\
| \\
R_2N \diagdown \\
C - C - CF_2 \\
RO \diagup \quad + \\
\textbf{19}
\end{array}
\tag{46}
$$

$$
CH_3OCF = CFCF_3
\xrightarrow[SbF_5]{SO_2ClF}
\begin{array}{c}
F \\
| \\
F \diagdown \\
C - C - CF_2 \\
CH_3O \diagup \quad + \\
\textbf{20}
\end{array}
\tag{47}
$$

$$
\begin{array}{c}
C_2F_5 \\
\diagdown \\
C = C(CF_3)_2 \\
\diagup \\
CH_3O
\end{array}
\xrightarrow{SbF_5}
\begin{array}{c}
CF_3 \\
| \\
C_2F_5 \diagdown \\
C - C - CF_2 \\
CH_3O \diagup \quad + \\
\textbf{21}
\end{array}
\tag{48}
$$

$$
p\text{-}CH_3OC_6H_4CF = CFCF_3
\xrightarrow{SbF_5}
\begin{array}{c}
F \\
| \\
p\text{-}CH_3OC_6H_4CF - C - CF_2 \\
+ \\
\textbf{22}
\end{array}
\tag{49}
$$

The alkoxy group (usually CH_3O- or C_2H_5O-) has a strong ability to stabilize carbocations. For example, $(CH_3O)_3C^+$ is stable in solution at ambient temperature [67], in contrast to the elusive CF_3^+. Existence of **20** and **21** as long-lived cations is another illustration of the significant stabilizing effect of the CH_3O group. Introduction of CH_3O-groups into fluorinated molecules leads to a significant increase in the stability of corresponding carbocations. A number of stabilized cyclic allylic cations, including derivatives of F-cyclobutenyl **23** [40], F-cyclopentenyl **24** and F-cyclohexenyl **25** [68] cations, were observed as long-lived cations and characterized by NMR spectroscopy:

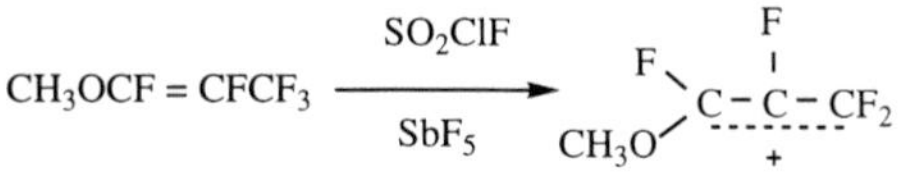

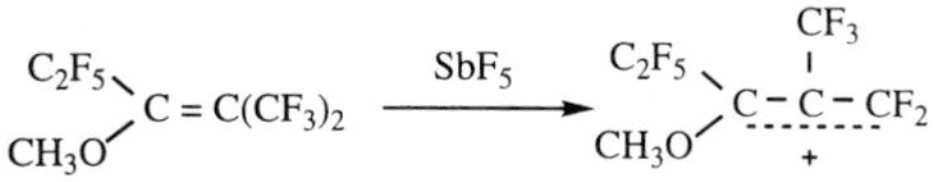

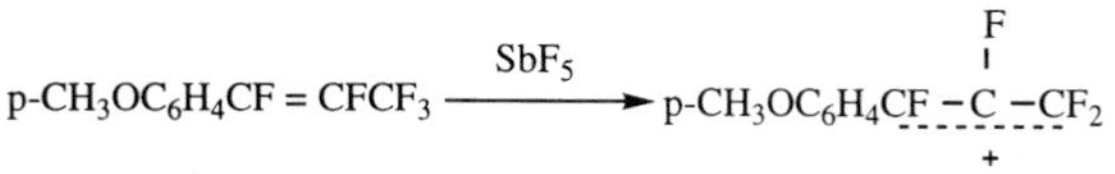

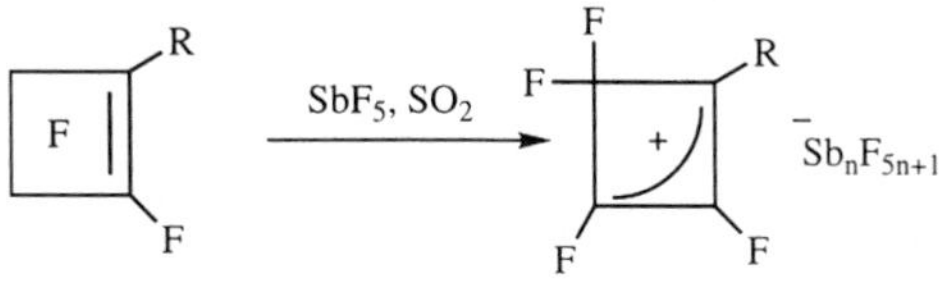

$$
\tag{50}
$$

R = CH_3O-, CH_3S-   **23**

$$
\tag{51}
$$

24, n=0, R_f=CF_3
25, n=1, R_f=C_2F_5

Pentafluorophenyl has an excellent ability to enchance the stability of allylic linear and cyclic carbocations. Species **26** and **27** were observed as long-lived species in the reaction of SbF_5 with corresponding olefins [69]:

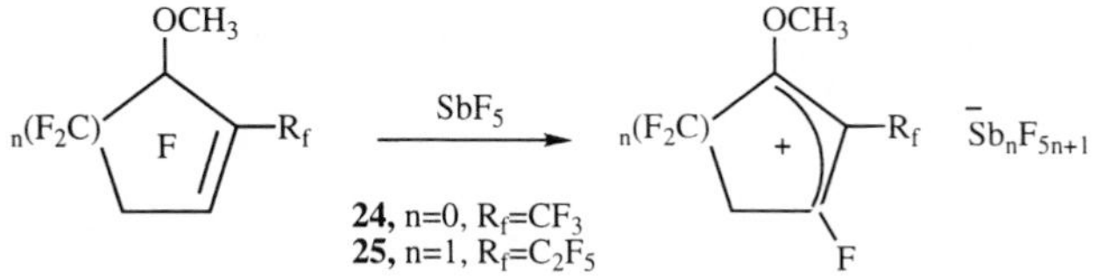

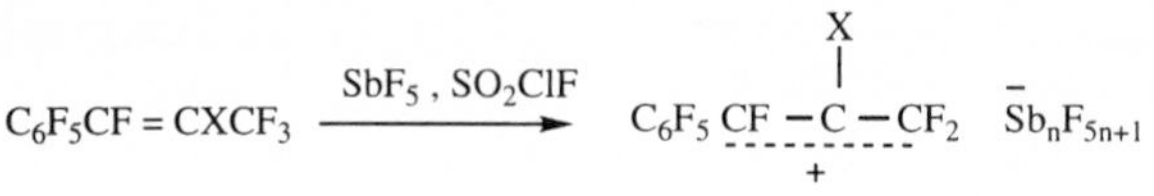

$$C_6F_5CF = CXCF_3 \xrightarrow{SbF_5,\ SO_2ClF} C_6F_5 \underset{+}{CF} \overset{X}{\underset{|}{-C}} -CF_2 \quad \bar{Sb}_nF_{5n+1} \tag{52}$$

26, X=F
27, X=CF$_3$

Based on ^{19}F NMR data and MNDO calculations, the conclusion was drawn that the C_6F_5- group in these cations is partially removed from the plane of conjugation with the allylic system [69]. The stabilizing effect of the C_6F_5-group is not as pronounced as that of a CH_3O substituent, and although F-1-benzocyclobutenyl cation was observed in equilibrium with the precursor at low temperature, an attempt to generate corresponding cations by action of SbF$_5$ on F-indan or F-tetralin failed [70] unless carbocation benefitted from stabilization by chlorine (Eq. 54) [71]:

$$\tag{53}$$

$$\tag{54}$$

A chlorine substituent has a stabilizing effect on the allylic system. A stable complex of hexachloropropene and AlCl$_3$ known for a long time was found to be a salt of the pentachloroallyl cation [72]. Recently, the formation of long-lived allylic cations **28** and **29** in the reaction of SbF$_5$ with corresponding chlorofluoroolefins has been reported [73]:

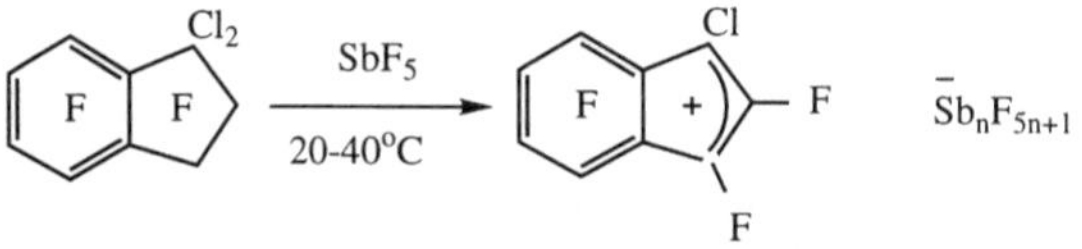

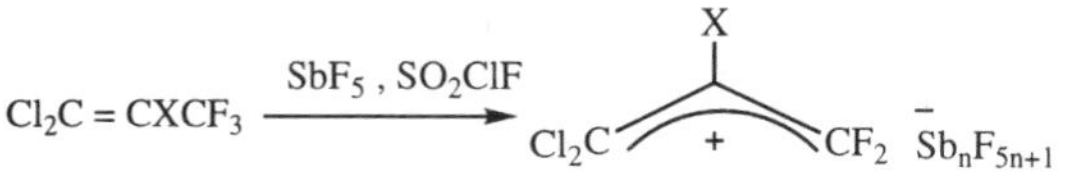

$$Cl_2C = CXCF_3 \xrightarrow{SbF_5,\ SO_2ClF} \text{} \quad \bar{Sb}_nF_{5n+1} \tag{55}$$

28, X=Cl
30, X=CF$_3$

For a long time it was believed that the trifluorovinyl group is not "basic" enough to stabilize an α-carbocationic center, and perfluorinated allylic cations were thought not to be stable [15, 21, 59]. However, in 1984 the formation of the F-methallyl cation **30** in the reaction of PFIB with SbF$_5$ in SO$_2$ClF solvent was reported [74], followed by the F-allyl cation (**1**) [75]:

$$CF_2=CXCF_3 + n\ SbF_5 \rightleftharpoons \text{} \quad \bar{Sb}_nF_{5n+1} \tag{56}$$

1, X=F
30, X=CF$_3$

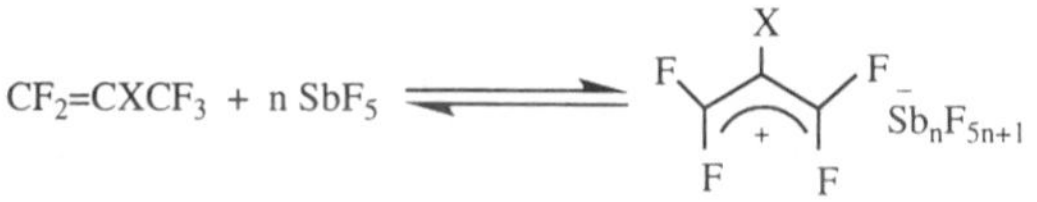

The choice and the purity of the solvent along with purity of antimony pentafluoride was reported to be important in generation of cations [63], since the presence of even a small amount of HF (resulting for example, from partial hydrolysis of SbF_5 during handling) significantly reduces the Lewis acidity of antimony pentafluoride and it may be responsible for several unsuccessful attempts to generate cation **1** using this method [21, 62]. Despite the fact that cations **1** and **30** have a lifetime long enough to be characterized by NMR, their stability is low and at temperatures as low as $-50\,°C$ both are in equilibrium with the starting olefin.

Introduction of a second $CF_2=CF$-group to the cationic center results in a striking increase of stability, and F-pentadienyl cation **31** was observed as a long-lived species even at ambient temperature:

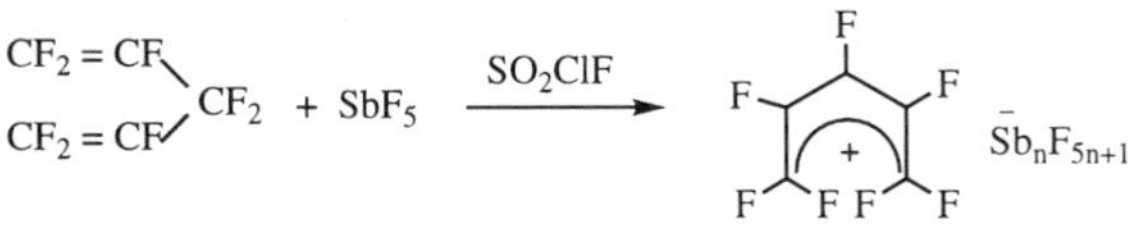

$$
\begin{array}{c}
CF_2=CF \\ \\ CF_2=CF
\end{array}\!\!\diagup\!CF_2 \; + \; SbF_5 \xrightarrow{\;SO_2ClF\;} \quad \text{(ring structure)} \quad \bar{S}b_nF_{5n+1}
\tag{57}
$$

Generation of several other allylic and polyenylic cations as a result of a rather unusual dehydrofluorination reaction of the corresponding polyfluoro-alkane by SbF_5 has been recently reported [76]:

$$
(CF_3)_2CFCH_2CF_2(CH_2CF_2)_nCF_2CF_3 \xrightarrow[HF]{SbF_5} (CF_3)_2CFCH_2\overset{+}{C}F(CHCF)_nCF_2CF_3\bar{S}b_nF_{5n+1}
\tag{58}
$$

For reasons that are not well understood, the cation **32**, prepared by the same procedure, exists not as pentadienyl, but as a stable bis(allyl) dication [76]:

$$
[(CF_3)_2CFCH_2CF_2CH_2CF_2]_2 \xrightarrow[HF]{SbF_5} [(CF_3)_2CFCH_2\overset{+}{C}F\,CH\,CF]_2\,\bar{S}b_nF_{5n+1}
\tag{59}
$$

A substantial number of substituted allylic cations were prepared using the reaction of SbF_5 and fluoroolefin and characterized by NMR spectroscopy. Spectroscopic data along with data on relative stabilities of polyfluorinated allylic cations were summarized and thoroughly discussed in a comprehensive review [63]. Stability of polyfluorinated allylic cations bearing a substituent at central carbon decreases in the following order [63]:

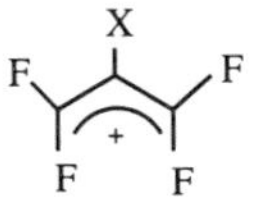

$$
\text{(structure)} \qquad CF_3 < CF_3O < F < Br < Cl < H
\tag{60}
$$

$$
 \mathbf{30} \quad \mathbf{33} \quad \mathbf{1} \quad \mathbf{34} \quad \mathbf{35} \quad \mathbf{36}
$$

Lower stability of cations **30** ($X=CF_3$) and **33** ($X=OCF_3$) compared to F-allyl cation **1** (in contrast to CF_3 and CF_3O (both have $\sigma_I=0.39$) fluorine has a higher electron-withdrawing effect ($\sigma_I=0.52$)) and reverse order of stability of cations **34** and **35** were explained in terms of steric repulsion of a bulky group attached to C-2 carbon with fluorines at neighboring carbons. This conclusion was supported by semi-empirical quantum-mechanical calculations indicating a shorter distance between two *cis*-fluorine substituents for cations containing bulky substituents at central carbon [63]:

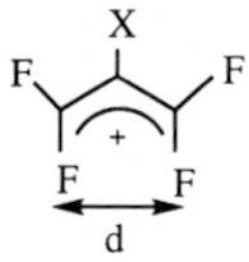

X	d (Å)
H	2.975
F	2.907
Br	2.821
CF_3	2.775

(61)

The absence of electronwithdrawing properties and small steric volume of hydrogen substituents are the reasons for the higher stability of cation **36**. For 1-substituted analogs of *F*-allyl and *F*-methallyl cations (existing as a mixture of *cis*- and *trans*-isomers, except **43**, X=H for which only *cis*-isomer was found in solution), stability correlates well with electronegativity of the substituent [63]:

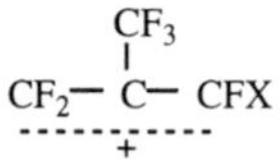

(62)

CH_3O > I > Br > Cl > F
37 **38** **39** **40** **1**

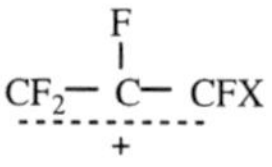

(63)

Br > Cl > F > H
41 **42** **1** **43**

Cation **43** has the lowest stability and is at least 1.5 kcal/mol less stable than *F*-allyl cation. This is in good agreement with the above-mentioned experimental data on the stability of CH_3^+, CH_2F^+, CHF_2^+, $(CH_3)_2CH^+$, and $(CH_3)_2CF^+$ cations in the gas phase.

6 Reactions

6.1 Electrophilic Addition Across the C=C bond

6.1.1 *Halogenation and Addition of Interhalogen Compounds*

Addition of halogens to fluoroolefins is a common reaction which usually proceeds by a radical mechanism. However, some halogenations are catalyzed by Lewis acids. This reaction has been reviewed [6, 14, 77]. Among recently reported reactions are ionic addition of Br_2 to $RC_6H_4CF{=}CFX$ that was found not to be stereoselective [78], and the ratio of *syn*- and *anti*-addition products was thought to be determined by a combination of steric and electronic factors.

Very interesting data were obtained for the reaction of polyfluorinated conjugated dienes with halogens (Cl_2, Br_2 and I_2) at low temperature [79]:

$$
\begin{array}{c}
RCH=CF - CF=CF_2 \\[2em]
CF_2=CH - CF=CF_2
\end{array}
\quad
\xrightarrow[\substack{-5°C}]{X_2,\ solvent}
\quad
\begin{array}{c}
XCH(R) - CF=CF -CF_2X \\[2em]
XCF_2 - CH=CF -CF_2X
\end{array}
\tag{64}
$$

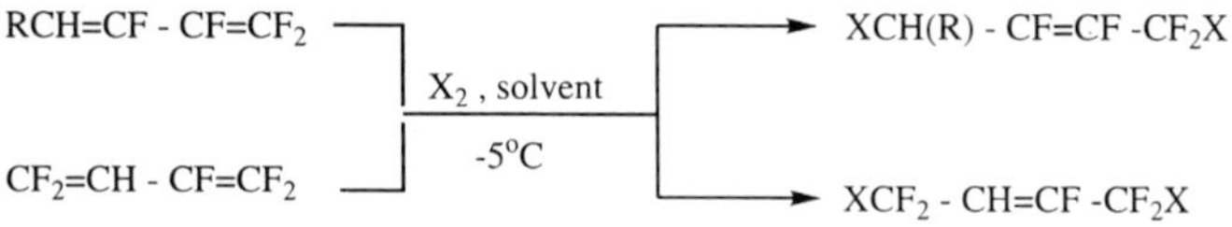

$$X_2 = Cl_2,\ Br_2,\ I_2\ ;\ R = H,\ CH_3,\ CF_3$$

In an inert solvent (CH_2Cl_2 or $CHCl_3$) the reaction proceeds as 1,4-addition and in nearly all cases the *trans*-isomer is formed exclusively. Addition with iodine monochloride under similar conditions also proceeds as 1,4-addition, but it is complicated by formation of two regioisomers in the case of nonsymmetrical dienes, for example:

$$
CF_2=CH - CF=CF_2 + ICl \xrightarrow[-5°C]{CH_2Cl_2} ICF_2-CH=CF-CF_2Cl
\tag{65}
$$
$$66\%$$

In general, addition of more polar interhalogen compounds (ICl, IF, BrF, ClF) to fluoroolefins proceeds as an ionic process, starting with attack of halogen bearing positive charge on the C=C bond and formation of a cyclic halonium cation (Eq. 2) [10, 18].

Instead of unstable fluorides, such as IF or BrF, "stoichiometric equivalents" – mixtures of IF_5/I_2 or BrF_3/Br_2 – are often used [10]. Chlorine monofluoride and "BrF" (same is true for BrF_3, BrF_5, ClF_3, and ClF_5) are much more powerful reagents than "IF" or IF_5. Since they are quite strong oxidizers and most of them react explosively with most organic compounds, they require special precautions in handling [10, 18] Reactions of these materials with fluoroolefins should be carried out with diluted reagents and at low temperature. For instance, ClF could be added even to ethylene and butadiene if diluted ClF is introduced into a solution of an unsaturated compound in an inert solvent at low temperature [18]. Mixture of BrF_3/Br_2 was shown to add to TFE vigorously [80]. Hexafluoropropene adds BrF at ambient temperature. Bromo-*F*-cyclohexane was isolated in the reaction of *F*-cyclohexene and BrF_3/Br_2 at elevated temperature:

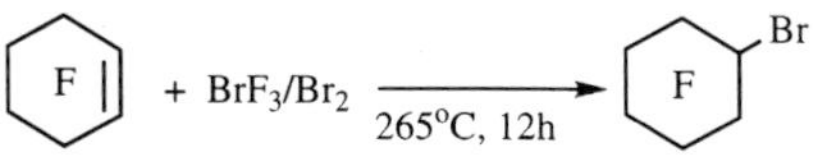

$$
\begin{array}{c}
\text{F} \\
\bigcirc\!\!\!\parallel
\end{array}
+ BrF_3/Br_2 \xrightarrow[265°C,\ 12h]{}
\begin{array}{c}
\text{Br} \\
\bigcirc\!\!\!\!\text{F}
\end{array}
\tag{66}
$$
$$78\%$$

Bromine fluoride adds exothermally across the C=C bond of *F*-heptene-1, even *F*-heptene-2 and *F*-4-methylpentene-2 (6), but not isomeric *F*-2-methylpentene-2 [51]:

$$
\begin{array}{c}
\mathbf{6} \\[2em]
(CF_3)_2C=CFC_2F_5
\end{array}
\quad
\xrightarrow[0°C]{BrF_3/Br_2}
\quad
\begin{array}{c}
(CF_3)_2CFCF_2CFBrCF_3\ +\ (CF_3)_2CFCFBrC_2F_5 \\
1\ \ :\ \ 1 \\[1em]
\text{Conversion 75\%,} \\
\text{selectivity for } trans\ 100\%
\end{array}
$$

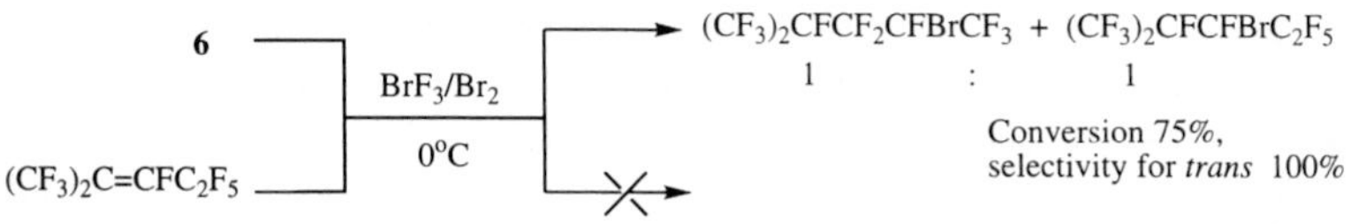

Iodine monofluoride has much lower reactivity. It adds to TFE at ambient temperature, but reaction with *F*-propene proceeds at elevated temperature [80]:

$$\begin{array}{l} CF_2{=}CF_2 \text{ (TFE)} \xrightarrow[]{IF_5/I_2,\ 25^\circ C} C_2F_5I \quad 89\% \\ CF_2{=}CFCF_3 \text{ (HFP)} \xrightarrow[150^\circ C]{IF_5/I_2} (CF_3)_2CFI \quad 99\% \end{array} \tag{68}$$

PFIB reacts with IF_5 in the presence of KF at 200 °C to give *t*-*F*-butyl iodide [81]:

$$(CF_3)_2C{=}CF_2 + IF_5 \xrightarrow[200^\circ C,\ 1h]{KF} \underset{73\ \%}{(CF_3)_3CI} \tag{69}$$

Recently it has been found that a combination of ICl/HF in the presence of a Lewis acid is effective for iodofluorination of fluoroolefins [50]. For example, iodopentafluoroethane was isolated in 91 % yield in a reaction catalyzed by BF_3:

$$TFE + ICl + HF \xrightarrow[25^\circ C,\ 16h]{BF_3} \underset{78\text{-}91\%}{CF_3CF_2I} + \underset{1\%}{ClCF_2CF_2I} + HCl$$

$$CFX{=}CF_2 + ICl + HF \xrightarrow{BF_3} CF_3CFXI + XCF_2CF_2I \tag{70}$$

X=H	25°C, 12h	73%	
X=Cl	25°C, 12h	31%	15%
X=CF_3	50°C, 18h	79%	

Niobium and tantalum pentafluorides are also effective catalysts for iodofluorination of TFE, while $AlCl_3$ and $ZnCl_2$ show no activity.

Some terminal polyfluoroalkenes and trifluorovinyl ethers can be converted into corresponding iodides using this procedure:

$$CF_2{=}CFOR_f + ICl + HF \longrightarrow ICF_2CF_2OR_f \tag{71}$$

$R_f = CF_3$	50°C, 20h	84%
$R_f = C_3F_7$	70°C, 24h	82%
$R_f = FSO_2(CF_2)_2CF(CF_3)CF_2$-	90°C, 24h	82%

The double bond in *F*-pentene-2 is very stable to iodofluorination, and the starting material was recovered in the reaction of this olefin with $ICl/HF/BF_3$ mixture after 24 h at 200 °C. Iodofluorination of polyfluoroolefins by the ICl/HF system [50] proceeds as conjugate addition (see Eq. 2) where the carbocation generated as a result of the electrophilic attack of "I^+" on the C=C bond is stabilized by addition of F^- (pathway A). Formation of small amounts of $ClCF_2CF_2I$ in this reaction could be a result of the reaction of the intermediate cation with Cl^-, since hydrogen chloride forms in this process as a by-product.

The combination of X_2/protic acid ($X_2=Cl_2, Br_2, I_2$) is usually ineffective for reactions of conjugate addition, due to formation of significant amounts of a halogenation product. However, addition of a Lewis acid may alter the course of the reaction. For example, fluorosulfate **44** was isolated in 84% yield in the reaction of HFP with Cl_2/$HOSO_2F$ catalyzed by SbF_5, but in the absence of the catalyst no formation of **44** was observed [82]:

$$CF_3CF=CF_2 + Cl_2 \xrightarrow[20°C, 48h]{HOSO_2F/SbF_5} CF_3CFClCF_2OSO_2F + (CF_3)_2CFCl + CF_3CFClCF_2Cl$$
$$\textbf{44, 84\%}$$
$$(72)$$

Conjugated bromo- and chloro-fluorination of polyfluoroalka-1,3-dienes is performed at room temperature using N-bromosuccinimide (NBS) and hexa-chloromelamine in anhydrous HF, respectively [83]. While 2-chloropentafluoro-1,3-butadiene is stable to the action of strong protic acids (HF, HSO_3F, $HOSO_2CF_3$ at 20–150 °C), it readily adds antimony pentafluoride to give a mixture of fluorinated alkenylantimony fluorides (Eq. 75):

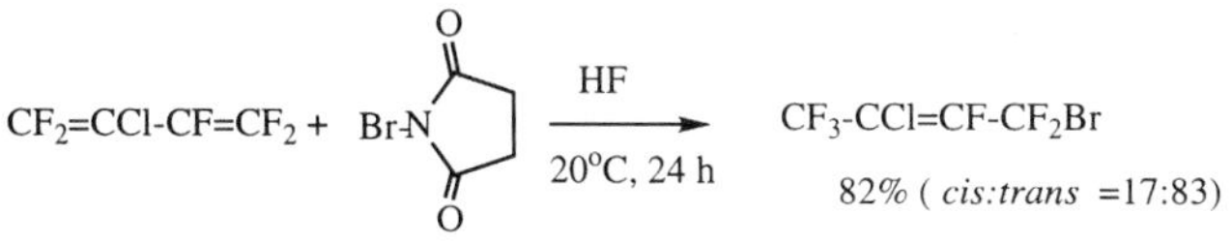

$$CF_2=CCl-CF=CF_2 + Br-N\left(\begin{array}{c}O\\O\end{array}\right) \xrightarrow[20°C, 24\,h]{HF} CF_3-CCl=CF-CF_2Br \qquad (73)$$
$$82\% \,(\,cis\!:\!trans\,=17\!:\!83)$$

NBS

$$(74)$$

$$CF_2=CCl-CF=CF_2 + C_3Cl_6N_6 \xrightarrow[20°C, 24\,h]{HF} CF_3-CCl=CF-CF_2Cl$$
$$74\% \,(\,cis\!:\!trans\,=40\!:\!60)$$

$$CF_2=CCl-CF=CF_2 + SbF_5 \xrightarrow[]{SO_2ClF} \underset{Cl}{\overset{CF_3}{\diagdown}} C = C \underset{SbF_4}{\overset{CF_3}{\diagup}} + \underset{Cl}{\overset{CF_3}{\diagdown}} C = C \underset{SbF_3}{\overset{CF_3}{\diagup}}$$

$$\underset{CF_3}{\overset{Cl}{\diagdown}} C = C \underset{CF_3}{\overset{SbF_3}{\diagup}} \qquad (75)$$

$$72\% \qquad\qquad 12\%$$

6.1.2
Oxidative Fluorination

As found recently, a stoichiometric equivalent of BrF – a BrF_3/Br_2 mixture – behaves as a powerful fluorinating agent for polyfluorinated aromatic compounds C_6F_5X ($X=F, CF_3, H, OCH_3$) [84]. Its reaction with F-benzene proceeds at a temperature as low as 0 °C to give a mixture of F-cyclohexadiene-1,4 and 4-Br-F-cyclohexane:

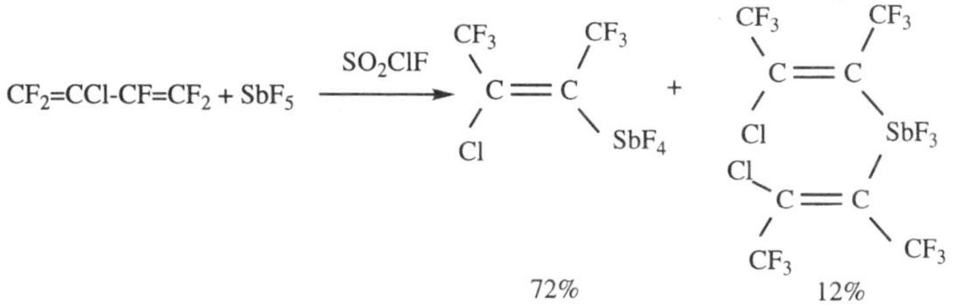

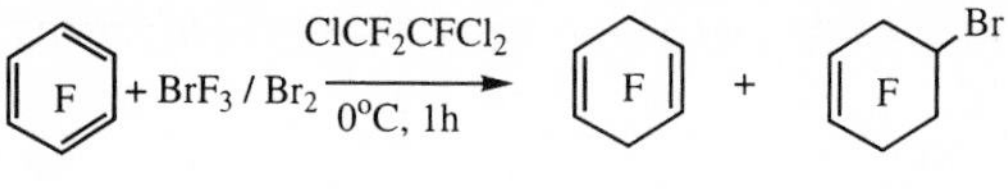

13% 87% (76)

As in halogens (see above), the reactivity of halogen fluorides (BrF_3, BrF_5, ClF_3) in fluorination reactions could be significantly increased in the presence of protic or Lewis acids [85]. For example, salt $BrF_2^+SbF_6^-$ reacts with *F*-benzene even at –80 °C with the formation of *F*-cyclohexadiene-1,4:

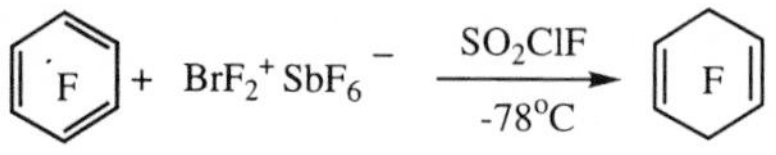

77% (77)

More data on the reaction of oxidative fluorination of polyfluoroaromatic compounds with halogen fluorides can be found in [86].

Finding the above-mentioned reaction was followed by the discovery of another, new and extremely effective fluorinating agent: vanadium pentafluoride [29]. The terminal double bond of polyfluoroalkenes is easily fluorinated by vanadium pentafluoride at or below room temperature to give the corresponding saturated compounds [85, 86]. Reactions are performed at atmospheric pressure in $CFCl_3$ or SO_2FCl or in the absence of solvent. It requires two moles of VF_5 per C=C unit; stable and insoluble VF_4 is formed as a by product:

$$CF_2{=}CXR_f + 2\ VF_5 \xrightarrow[-20^\circ C,\ 30\ min]{CFCl_3} CF_3CXFR_f + 2\ VF_4 \qquad (78)$$

Rf = C_2F_5, X = Cl (81 %),
Rf = $-(CF_2)_4CFClCF_2Cl$, X = F (83 %)

$$CF_2{=}CFOCF_3 + 2\ VF_5 \xrightarrow[40^\circ C,\ 3\ h]{} CF_3CF_2OCF_3 + 2\ VF_4 \qquad (79)$$

40 % at 50 % conversion

The fluorination of internal *F*-alkenes and cycloalkenes proceeds at higher temperature; the replacement of vinylic fluorine by chlorine in olefin leads to a remarkable acceleration of the reaction (Eq. 82):

$$(CF_3)_2CFCF{=}CFCF_3 + 2\ VF_5 \xrightarrow[60^\circ C,\ 4\ h]{} (CF_3)_2CFCF_2CF_2CF_3 \qquad (80)$$

94 %

59 % (81)

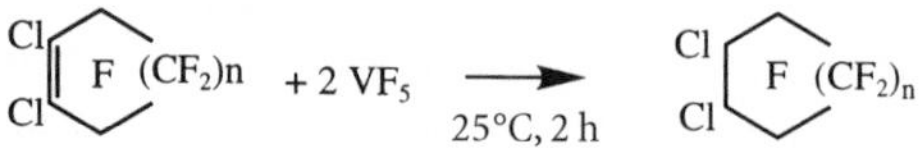

(82)

n = 1 72 %
n = 2 89 %

Fluorination of butadiene **45** is not selective and gives a mixture of butene **46** and butane **47**:

$$\underset{\mathbf{45}}{CF_2{=}CCl{-}CF{=}CF_2} + 2\,VF_5 \xrightarrow[-20^{\circ}C,\ 20\ min]{CFCl_3} \underset{\substack{\mathbf{46}\\56\%}}{CF_3CCl{=}CFCF_3} + \underset{\substack{\mathbf{47}\\8\%}}{CF_3CFClCF_2CF_3} \qquad (83)$$

Although polyfluoro-1,4-cyclohexadienes contain two isolated double bonds, fluorination of these compounds proceeds under milder conditions than those in fluorination of the corresponding cyclohexenes and usually leads to formation of regioisomers:

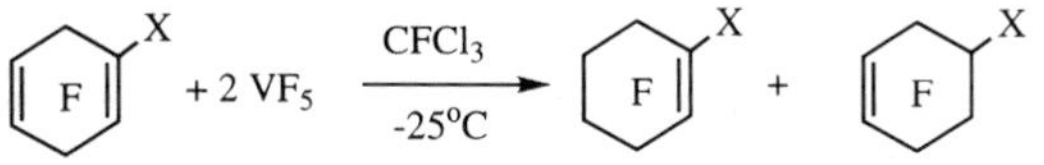

(84)

X	yield (%)	
F	87%	
H	75	4
Cl	59	27
CF_3	-	92

The proposed mechanism of oxidative fluorination of unsaturated compounds by halogen fluorides [84–86] and VF_5 [33] includes electron transfer from substrate to halogen fluoride or VF_5 as a first step, followed by addition of F^- to a radical-cation, leading to formation of a radical and its further oxidation to carbocation (see Eq. 12, pathways A,B). It should be pointed out that this is not the only direction, and the actual mechanism may depend strongly on the substrates and reaction conditions. Other mechanisms, such as a radical process (pathway C), cannot be ruled out.

In contrast to VF_5, antimony pentafluoride is a less powerful fluorinating agent [19].

Fluorination of the C=C bond of fluoroethylenes by SbF_5 at 50 °C has been reported [44]:

$$\underset{X=H,F}{CF_2{=}CFX + SbF_5} \xrightarrow[50^{\circ}C]{} CF_3CF_2X \qquad (85)$$

Polyfluorinated benzenes RC_6F_5 (R=F, Cl, Br, I, but not CF_3 or SO_2F) react with SbF_5 at high temperature producing a mixture of cyclohexenes and cyclohexanes [19]; often the fluorination of the aromatic ring is complicated by replacement of halogen by fluorine. In the presence of small amounts of bromine or iodine, SbF_5 was reported to fluorinate the aromatic ring of *F*-alkylindanes and *F*-tetralin [87]. The reaction proceeds as initial halofluorination of the ring, followed by replacement of halogen. A recently reported example [88] of reaction of alkenylbenzene **67** with the Br_2/SbF_5 system provides further proof for much higher reactivity of a pentafluorophenyl ring relative to the electron deficient C=C bond shielded by three *F*-alkyl groups:

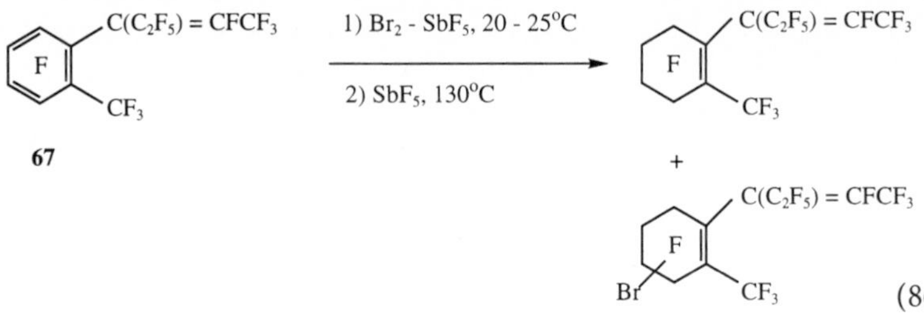

67

(86)

6.1.3
Addition of Hydrogen Halides and $HOSO_2F$

Electrophilic addition of hydrogen halides to polyhalogenated and polyfluorinated olefins is probably the most studied reaction of this kind. Reaction usually proceeds at elevated temperature and often in the presence of a Lewis acid catalyst. The literature on this reaction has been reviewed [6, 14]. Addition of hydrogen fluoride to halogen-containing olefins is an important process, since the products – chlorofluorocarbons or CFCs – are precursors for a number of commercial monomers and hydrofluorocarbon replacement of CFCs [89]. The work in this direction in the last years was mostly concentrated on finding new catalysts for both liquid and gas phase processes. Fluorosulfonic acid, being stronger than HF, reacts with vinylidene fluoride and trifluoroethylene at 20–30 °C without a catalyst. Addition proceeds in accordance with polarization of reagents and the combined α-, β-fluorine effect and leads to formation of a single regioisomer in each case [6]:

$$CHX{=}CF_2 + HOSO_2F \xrightarrow[X=H,F]{20\text{-}30^\circ C} XH_2CCF_2OSO_2F \tag{87}$$

HFP and higher terminal fluoroolefins add $HOSO_2F$ only at elevated temperature [6] to give the corresponding α,α-difluoroethers of fluorosulfonic acid:

$$R_fCF{=}CF_2 + HOSO_2F \xrightarrow{150\text{-}200^\circ C} R_fCFHCF_2OSO_2F \tag{88}$$

The electrophilic addition of CF_3COOH across the CF=CF moiety of difluoronorbornadiene has been reported [90]. Mercury acetate and trifluoroacetate also add to the fluorinated double bond rather than the non-fluorinated one [91].

6.1.4
Addition of Hypohalites, $XOSO_2CF_3$, $XOSO_2F$ and Related Compounds

Progress in the preparation and isolation of different hypohalites [92] results in a substantial number of publications exploring the addition of these materials to fluoroolefins. In contrast to addition of hypofluorites, which at this time is viewed as a radical process, a polar mechanism was proposed for addition of polyfluorinated hypochlorites and hypobromites to olefins. Reactions of fluoroolefins with CF_3OOCl [93], SF_5OOCl [94], CF_3OCl [95–98], $R_fC(O)OCl$ [99, 100], R_fOBr [39], $H(CF_2CF_2)_2CH_2OCl$ [101] have been reported. Since the reactivity of most of these compounds is high, reactions are carried out at low temperature, optionally in the presence of inert solvent. Yields of additional products are usually high. Based on the comparison of addition reactions of CF_3OOCl and CF_3OCl to C_2 and C_3 fluoroolefins and analysis of isomer ratios in products, the conclusion was drawn that these reactions proceed as electrophilic *syn*-addition [93, 98]. Additions of CF_3OCl [98] and R_fOBr [39] to $CH_2{=}CF_2$ and $R_fC(O)OCl$ to HFP [99, 100], which are regiospecific, occur according to the polarization of the double bond:

$$CF_2{=}CH_2 + R_fOX \xrightarrow[-88\text{ to }24^\circ C]{} R_fOCF_2CH_2X \qquad (89)$$

R_f	X	yield %
CF_3	Cl	50
$(CF_3)_3C$	Br	46
$C_2F_5C(CF_3)_2$	Br	25

$$XCF_2C(O)OCl + CF_2{=}CFCF_3 \xrightarrow[-78^\circ C]{} XCF_2C(O)OCF_2CFClCF_3 \qquad (90)$$

X	yield %
F	84
Cl	74

As expected, addition of CF_3OCl to chloro- and bromotrifluoroethylene gives a mixture of two regio isomers in both reactions [99, 100]:

$$CF_3OCl + CFX{=}CF_2 \xrightarrow[-140\text{ to }22^\circ C]{CFCl_3} CF_3OCF_2CFClX + CF_3OCFXCF_2Cl \qquad (91)$$

X	yield %	
Cl	45	55
Br	71	29

Several exceptions should be mentioned. For example, the reaction of HFP with CF_3OCl leads to formation of two regioisomers [100], but addition of $CF_3C(O)OCl$ to $CFCl{=}CF_2$ and $CCl_2{=}CF_2$ [93], in sharp contrast to addition of CF_3OCl [95, 98], is regiospecific:

$$CF_3OCl + CF_2{=}CFCF_3 \xrightarrow[-140\text{ to }22^\circ C]{CFCl_3} \underset{71.5}{CF_3OCF_2CFClCF_3} + \underset{28.5}{CF_3OCF(CF_3)CF_2Cl} \qquad (92)$$

yield 66%

$$CF_3C(O)OCl + CClX{=}CF_2 \xrightarrow[-140\text{ to }22^\circ C]{CFCl_3} CF_3C(O)OCClXCF_2Cl \qquad (93)$$

X	yield %
F	65
Cl	81

On the other hand, addition of CF_3OOCl to $CFCl{=}CF_2$ is regioselective, but the orientation is exactly opposite to that in Eq. (93) [93]:

$$CF_3OOCl + CFCl{=}CF_2 \xrightarrow[-78^\circ C,\ 21\text{ h}]{} \underset{>50\%}{CF_3OOCF_2CFCl_2} \qquad (94)$$

No good explanation exists for deviations in orientation of addition of hypochlorites to fluoroolefins; however, it could be an indication of either a different mechanism of addition, such as radical, or a compromise between two mechanisms: electrophilic and radical.

Unlike most known fluoroalkylhypohalites, which have low thermal stability, α,α-dihydropolyfluoroalkyl hypochlorites prepared by the reaction of $ClOSO_2F$ with the corresponding alcohols are stable at ambient temperature and can even be distilled at reduced pressure [101]:

$$R_fCH_2OH + ClOSO_2F \xrightarrow[-20\text{ to }47^{\circ}C]{CFC\text{-}113} \underset{\mathbf{48a\text{-}d},\ 30\text{-}40\%}{R_fCH_2OCl} \qquad (95)$$

$R_f = H(CF)_2$ - (**48a**); $H(CF_2)_4$ - (**48b**); $H(CF_2)_6$ -(**48c**); O_2NCF_2 - (**48d**)

$$\mathbf{48b\text{-}d} + TFE \xrightarrow[exothermic]{} \underset{\mathbf{49b\text{-}d}}{R_fCH_2OCF_2CF_2Cl} \qquad (96)$$

49b 91%; **49c** 89%; **49d** 72.5%

Compounds **48b–d** react vigorously with TFE, producing the corresponding additional products in high yield, but **48a** does not react with HFP; the addition proceeds only in the presence of $HOSO_2F$ which acts as a catalyst:

$$\mathbf{48a} + HFP \xrightarrow[20^{\circ}C,\ 16h]{HOSO_2F} \underset{14.5\%}{H(CF_2)_2CH_2OCF_2CFClCF_3} \qquad (97)$$

Under similar conditions hypochlorite **48b** was added across the C=C bond of *F*-ketene which is extremely resistant to the action of electrophilic agents:

$$\mathbf{48b} + (CF_3)_2C{=}C{=}O \xrightarrow[20^{\circ}C,\ 4h]{HOSO_2F} H(CF_2)_4CH_2OC(O)C(Cl)(CF_3)_2 \quad 49\% \qquad (98)$$

The synthesis of chlorine (I) and bromine (I) trifluoromethanesulfonates (triflates) was reported by DesMarteau. Stability and reactivity of these materials are similar to those of perfluoroalkyl hypohalites. Both compounds readily react at low temperature with a variety of fluoroolefins. Based on NMR analysis of the products of adding CF_3SO_2OX to pure *cis*- or *trans*-isomers of 1, 2-difluoroethylene, it was concluded that the reaction proceeds as *syn*-addition [35]. This statement was later criticized [18], since the assignment of stereoisomers was found to be incorrect. According to [18], addition of CF_3SO_2OX to haloolefins, as well as reactions of ClF, BrF and IF proceed as *anti*-addition via cyclic halonium cationic intermediates.

Both chlorine and bromine triflates react with HFP to give only one, Markovnikov type, additional product, which is yet further evidence of the electrophilic mechanism of the process, since radical addition of CF_3OF to *F*-propene produces a mixture of two regio isomers [35]:

$$CF_3SO_2OX + HFP \xrightarrow[-111 \text{ to } 22^\circ C]{} CF_3SO_2OCF_2CFXCF_3 \quad (99)$$

X	yield %
Cl	78
Br	80

On the other hand, the addition of both compounds to chlorotrifluoroethylene produces one regioisomer (see also Eq. 41):

$$CF_3SO_2OBr + CFCl{=}CF_2 \xrightarrow[-111 \text{ to } -5^\circ C]{} CF_3SO_2OCFClCF_2Br \quad 88\% \quad (100)$$

Under similar conditions, chlorine and bromine triflates add to olefin **9**:

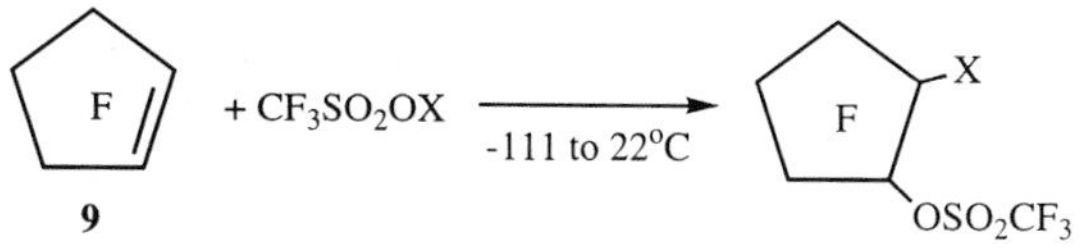

(101)

X	yield %
Cl	65
Br	96

Reactivity of halogen fluorosulfates $XOSO_2F$ (X=Cl, Br, I) is very similar to that of halogen triflates; however, all of them have higher thermal stability, and they are distillable liquids [8, 102]. The chemistry of halogen fluorosulfates, including addition to fluoroolefins, has been reviewed [8]; here only a short summary of these reactions is given.

The first publication on addition of $BrOSO_2F$ to several fluoroolefins, which appeared in 1966 [103], was followed by a communication on the reaction of chlorine and bromine fluorosulfate with HFP [97]. This initiated the appearance of a number of papers on the use of halogen fluorosulfates in fluoroorganic synthesis. Most of these publications came from Professor Fokin's group in the former Soviet Union.

Chlorine and bromine fluorosulfates have extremely high reactivity towards fluoroolefins. Addition to TFE proceeds vigorously even at –70 °C in CFC-113 as a solvent [104]. Reactions with $CCl_2{=}CCl_2$, $CFH{=}CF_2$, $CFCl{=}CF_2$, $CF_2{=}CFCF_3$ are carried out in a temperature range of –70 to –20 °C, producing the corresponding fluorosulfates in high yield [104, 105]. When gaseous PFIB is passed through liquid $ClOSO_2F$, exothermic reaction is observed, resulting in the high yield formation of α,α-difluoro fluorosulfate **50** [104]:

$$(CF_3)_2C{=}CF_2 + ClOSO_2F \xrightarrow[\text{exotherm.}]{} (CF_3)_2CClCF_2OSO_2F \quad \mathbf{50}, 96\% \quad (102)$$

Addition of $ClOSO_2F$ to internal *F*-olefins proceeds at –70 °C, producing a mixture of regioisomers (see Eq. 42). Reaction of $ClOSO_2F$ with pure *trans*-*F*-butene-2 is stereospecific and results in formation of a single stereoisomer [52].

In contrast to reaction of CF_3SO_2OX (see Eq. 100), addition of $ClOSO_2F$ or $BrOSO_2F$ to chlorotrifluoroethylene leads to formation of two regioisomers. Reaction of bromine tris (fluorosulfate) with fluoroolefins was reported to proceed rapidly and with the formation of equal amounts of the corresponding bromofluorosulfate and bis(fluorosulfonoxy)alkane [106], for example:

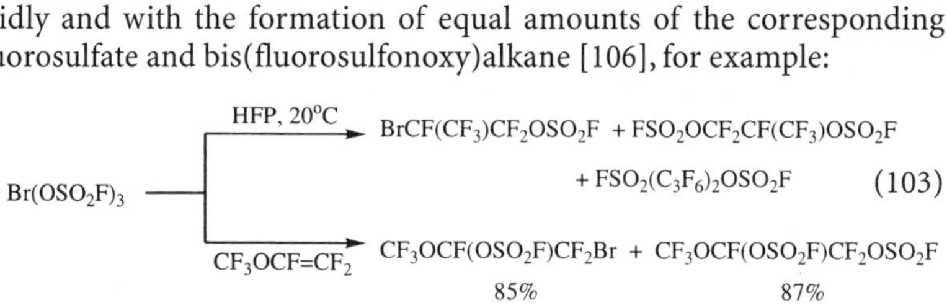

$$
Br(OSO_2F)_3
\begin{cases}
\xrightarrow{\text{HFP, 20°C}} BrCF(CF_3)CF_2OSO_2F + FSO_2OCF_2CF(CF_3)OSO_2F \\
\qquad\qquad\qquad + FSO_2(C_3F_6)_2OSO_2F \\
\xrightarrow{CF_3OCF=CF_2} CF_3OCF(OSO_2F)CF_2Br + CF_3OCF(OSO_2F)CF_2OSO_2F \\
\qquad\qquad\qquad\quad 85\% \qquad\qquad\qquad\qquad 87\%
\end{cases}
\tag{103}
$$

The formation of significant amounts of polymeric material, with even a slight excess of TFE, along with formation of dimeric products in reaction with HFP bromine tris (fluorosulfate) is strong evidence in favor of the radical mechanism, which in this reaction may compete with the electrophilic process.

Iodine fluorosulfate is significantly less reactive. Addition to fluorinated ethylenes and $CF_3CH=CH_2$ proceeds only at ambient temperature [108] and exclusively as Markovnikov addition:

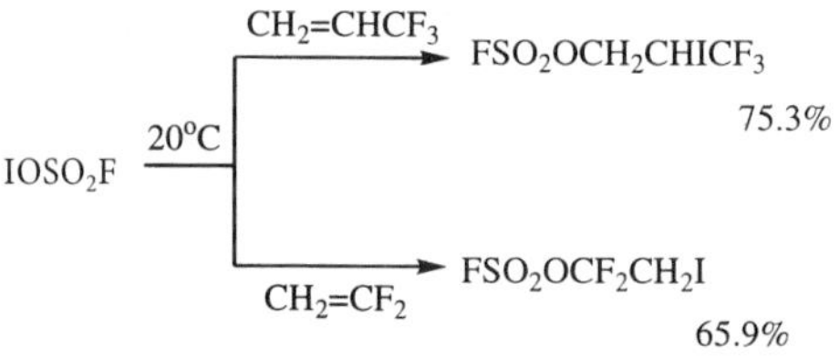

$$
IOSO_2F \xrightarrow{20°C}
\begin{cases}
\xrightarrow{CH_2=CHCF_3} FSO_2OCH_2CHICF_3 \\
\qquad\qquad\qquad\quad 75.3\% \\
\xrightarrow{CH_2=CF_2} FSO_2OCF_2CH_2I \\
\qquad\qquad\qquad 65.9\%
\end{cases}
\tag{104}
$$

Reaction with PFIB is slow, and even after three weeks at 20 °C the yield of additional product does not exceed 27.4 %. Iodine and chlorine fluorosulfates readily add across the C=C bond of F-ketene to give the corresponding acyl-fluorosulfates [8, 107], for example:

$$
IOSO_2F + (CF_3)_2C=C=O \qquad \underset{I}{(CF_3)_2CC(O)OSO_2F} \qquad 78.6\%
\tag{105}
$$

Reactivity of fluoroolefins in reaction with $XOSO_2F$ decreases going from TFE to PFIB [8]

$$
\underset{\text{TFE}}{CF_2=CF_2} > \underset{\text{HFP}}{CF_2=CFCF_3} > \underset{\text{PFIB}}{CF_2=C(CF_3)_2}
$$

Addition of halogen fluorosulfates $XOSO_2F$ (X=Cl, Br, I) to fluoroolefins is considered an electrophilic reaction [8]. However, the question of whether this process is concerted or the reaction proceeds via an independent carbocationic intermediate (Eqs. 2 and 3) is still open. Formation of carboxylic acid esters as byproducts in the reaction of HFP with $ClOSO_2F$, which was carried out in trifluoroacetic or heptafluorobutyric acids as solvents, could not be a solid pro-of of conjugate addition, since formation of esters may be a result of addition of $ClOC(O)R_f$ to olefin. These materials are known to be formed in the reaction of $ClOSO_2F$ with fluorinated carboxylic acids, even at low temperature [99].

An alternative and more convenient procedure for the preparation of poly-haloalkyl fluorosulfates, which excludes handling of potentially hazardous halogen fluorosulfates, is based on the conjugate addition of X^+ and FSO_2O^- to fluoroolefins. The exothermic reaction of a solution of N-halogen compound (NBS or hexachloromelamine) in fluorosulfonic acid with TFE or HFP results in formation of the corresponding fluorosulfates[108], although internal *F*-olefins, such as *F*-penetene-2 (**11**), are not active in this reaction [109]:

$$YCF{=}CF_2 + X\text{-}N\langle \xrightarrow{HOS_2F} XYCFCF_2OSO_2F \qquad (106)$$

X	Y	yield %
Cl	F	78
Cl	CF_3	70
Br	F	72
Br	CF_3	47

According to [108], no formation of the corresponding halogen fluorosulfates as intermediates was observed in the reaction. The mechanism includes protonation of N-halogen compound, which is able subsequently to transfer "Hal^+" fragment to fluoroolefin resulting in generation of carbocation followed by its reaction with FSO_2O^-.

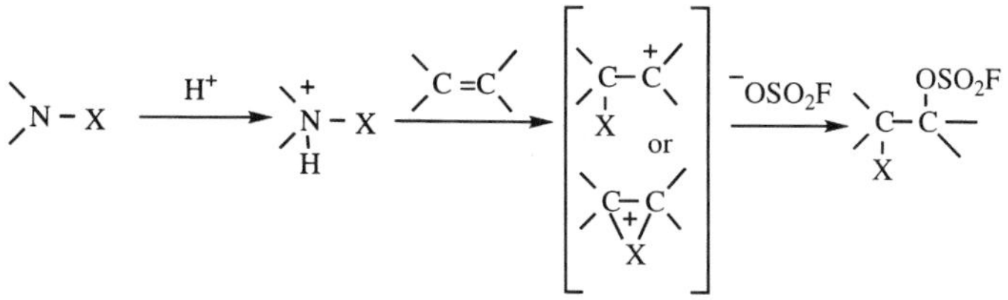

Scheme 2

A different mechanism, however, has recently been proposed [110] for a similar reaction. Solution of *N*-iodosuccinimide in CF_3SO_3H was found to be an effective reagent for iodination of deactivated aromatic materials. The mechanism includes generation of iodine (I) trifluoromethanesulfonate and a further reaction of the protonated form of this compound. At this point it is difficult to establish which of these mechanisms is correct. However, taking into account high energy of the O-Cl and O-Br bonds in the corresponding fluorosulfates, and the fact that internal fluoroolefins stay intact in solutions of *N*-chloro or *N*-bromo- compounds in $HOSO_2F$ (as mentioned above, *F*-pentene-2 reacts with $ClOSO_2F$ at −70 °C), the formation of $XOSO_2F$ as an intermediate in the reaction of fluoroolefins with a solution of hexachloromelamine or *N*-bromosuccinoimide is doubtful. A similar conclusion, i.e., that ClF, BrF or IF do not form as intermediates in reactions of haloolefins with mixtures of N-halogen compounds or iodine monochloride with hydrogen fluoride, was drawn in [10, 50].

Halooxygenation of the fluorinated C=C bond can also be achieved, for example, by conjugate addition of I_2 and SO_3. Reaction of TFE with I_2/SO_3 mixture followed by the treatment of an adduct with KF is a synthetic route for preparation of iododifluoroacetyl fluoride [111]. Reaction of iodine and

$Hg[OC(O)CH_3]_2$ with $CF_3CH{=}CH_2$ as a method of iodoacetylation of the C=C bond was discovered long ago [112] and later modified and extended to preparation of α-bromoacetates of perfluoroalkyethylenes [113]:

$$R_fCH{=}CH_2 \xrightarrow[Hg[OC(O)CH_3]_2]{Br_2,\ CH_3C(O)OH} R_f\,CHBrCH_2OC(O)CH_3 \quad 80\% \qquad R_f = C_4F_9,\ C_6F_{13},\ C_8F_{17} \tag{107}$$

A new development in this area has been reported recently [45]. It was found that reaction of X_2/SO_3 mixture with perfluoroalkyl or 1,2-bis(perfluoroalkyl)ethylenes followed by hydrolysis leads to formation of the corresponding halohydrins (Eq. 25)

6.1.5
Nitration and Related Reactions

Nitration of fluoroolefins can be achieved by several methods. Widely studied thermal reaction of N_2O_4 with fluoroolefins has a radical mechanism, although the low temperature reaction of nitrogen dioxide with polyfluorinated vinyl ethers proceeds as electrophilic addition of nitrosonium nitrate $NO^+\ NO_2^-$ across the C=C bond [6]:

$$XCF{=}CFOR + N_2O_4 \xrightarrow[-78^{\circ}C]{} ONXCFCF(OR)ONO_2 \xrightarrow{H_2O} ONXCFC(O)R \qquad X{=}Cl,\ F,\ CF_3;\ R{=}CH_3,\ C_2H_5 \tag{108}$$

Conjugate electrophilic nitrofluorination of fluoroolefins could be carried out by reaction with a mixture of concentrated nitric acid and anhydrous HF [3]:

$$CF_2{=}CFX + HNO_3 \xrightarrow{HF} CF_3CFXNO_2 \tag{109}$$

X	t (°C)	yield
H	-60	86%
CF_3	60	75%

This reaction has an electrophilic mechanism; it has been reviewed in [3, 6].

Nitrofluorination of *F*-indan by mixture of HNO_3/HF [114] or $NO_2^+BF_4^-$ in HF [114] is straightforward and leads to high yield formation of 2-nitro-*F*-indan; similar reactions of *F*-3-methyl- and *F*-2-methylene indans produce a mixture of products as a result of competitive electrophilic and radical reactions [115], for example:

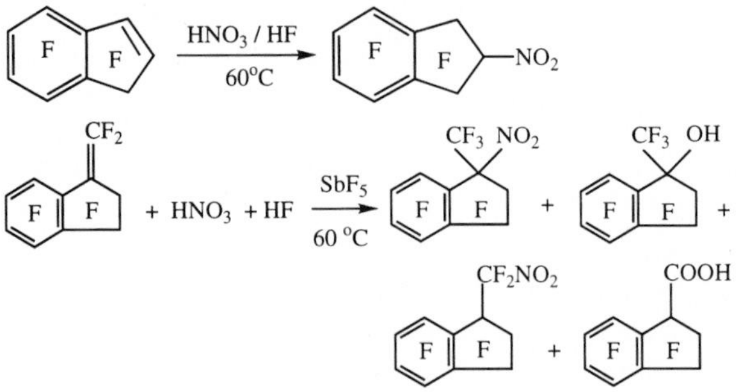

(110)

Introduction of nitro and fluorosulfonoxy groups can be carried out by reaction of ionic O_2NOSO_2F with fluoroolefins [8, 116 – 118]. Nitronium fluorosulfate is not as active as halogen fluorosulfates; addition to $CH_2=CF_2$ and $CFH=CF_2$ proceeds only at ambient temperature [117]. Reaction with TFE and chlorofluoroethylene proceeds at a noticeable rate at temperature above 50 °C, while addition to 2-H-F-propene only at temperatures higher than 120 °C [117]:

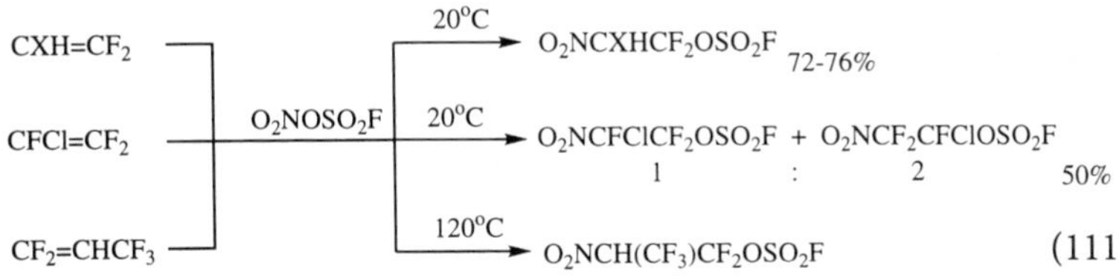

$$
\begin{array}{l}
CXH=CF_2 \\
CFCl=CF_2 \quad \xrightarrow{\quad O_2NOSO_2F \quad} \\
CF_2=CHCF_3
\end{array}
\qquad
\begin{array}{l}
\xrightarrow{20°C} O_2NCXHCF_2OSO_2F \quad {}_{72-76\%} \\[4pt]
\xrightarrow{20°C} O_2NCFClCF_2OSO_2F + O_2NCF_2CFClOSO_2F \\
\qquad\qquad\quad 1 \qquad\qquad : \qquad\qquad 2 \qquad\quad 50\% \\[4pt]
\xrightarrow{120°C} O_2NCH(CF_3)CF_2OSO_2F
\end{array}
\qquad (111)
$$

F-Butadiene reacts with O_2NOSO_2F at atmospheric pressure and ambient temperature to give the product of 1,4-addition:

$$
CF_2=CF\text{-}CF=CF_2 + O_2NOSO_2F \xrightarrow{20°C} O_2NCF_2\text{-}CF=CF\text{-}CF_2OSO_2F \qquad (112)
$$
$$
\text{> 90\% }trans \qquad\qquad \text{yield 79\%}
$$

On the other hand, F-1,4-pentadiene containing two isolated C=C bonds and F-1,1-dimethylallene do not react with O_2NOSO_2F even in the presence of $HOSO_2F$ at 140 °C but, suprisingly, O_2NOSO_2F adds across the C=C bond of F-ketene at 140 °C [118]. Rate of reaction of O_2NOSO_2F with fluoroolefins could be significantly increased by addition of a catalyst — fluorosulfonic acid. Reactions with $CH_2=CF_2$, $CFH=CF_2$, $CF_2=CF_2$, $CFCl=CF_2$, and $CF_3OCF=CF_2$ proceed even at 20 °C, producing the corresponding nitrocompounds in high yield [117], for example:

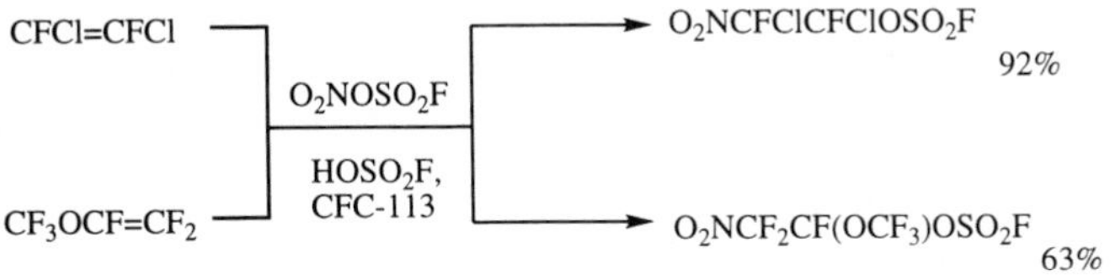

$$
\begin{array}{l}
CFCl=CFCl \\
\qquad\qquad \xrightarrow{\quad O_2NOSO_2F \quad} \\
\qquad\qquad \xrightarrow{\quad HOSO_2F, \quad CFC\text{-}113 \quad} \\
CF_3OCF=CF_2
\end{array}
\qquad
\begin{array}{l}
\longrightarrow O_2NCFClCFClOSO_2F \\
\qquad\qquad\qquad\qquad 92\% \\[10pt]
\longrightarrow O_2NCF_2CF(OCF_3)OSO_2F \\
\qquad\qquad\qquad\qquad\qquad 63\%
\end{array}
\qquad (113)
$$

Addition to HFP requires a more powerful catalyst and proceeds only when 10 – 20 mol% of SbF_5 is added to solution of O_2NOSO_2F in $HOSO_2F$ [117]:

$$
CF_2=CFCF_3 + O_2NOSO_2F \xrightarrow[50\text{-}60°C]{HOSO_2F/SbF_5} O_2NCF(CF_3)CF_2OSO_2F \qquad (114)
$$
$$
\qquad\qquad\qquad\qquad\qquad\qquad\qquad\qquad\qquad 78\%
$$

Olefins containing a more electron deficient C=C bond, such as PFIB and F-cyclobutene, do not react with solution of O_2NOSO_2F/SbF_5 in $HOSO_2F$ up to 150 °C [117]. However, reaction of O_2NOSO_2F with TFE and 2-H-pentafluoropropene at 60 °C in HF as a solvent proceeds as high yield conjugate nitrofluorination [118]:

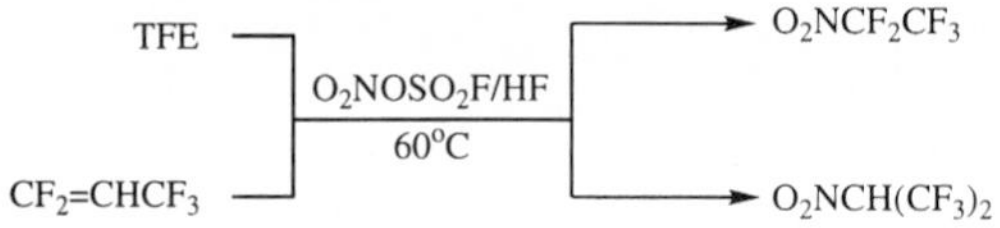

$$\text{TFE} \quad \underset{\begin{array}{c}O_2NOSO_2F/HF\\60°C\end{array}}{\underset{\displaystyle CF_2=CHCF_3}{\overline{}}} \quad \begin{array}{l} \longrightarrow O_2NCF_2CF_3 \\[2em] \longrightarrow O_2NCH(CF_3)_2 \end{array} \tag{115}$$

All experimental data are in agreement with the electrophilic mechanism of addition of O_2NOSO_2F to fluoroolefins.

Reactivity of nitrosonium fluorosulfate and trifluoromethansulfonate $ONOSO_2X$ ($X=F, CF_3$) is similar to that of nitronium fluorosulfate, although they are less active than the latter. Reaction, which occurs only with fluoroethylenes and perfluorinated vinyl ethers [119–121], is of an electrophilic nature; it results in low to moderate yields of nitrosoalkyl compounds when the olefin has a limited contact time with the reaction mixture:

$$ONOSO_2Y + CFX = CF_2 \xrightarrow{\text{solvent}} ONCFXCF_2OSO_2Y + \text{others}$$

X	Y	yield %
H	F	49
Cl	F	81
F	F	56
H	CF$_3$	27
Cl	CF$_3$	21

$$\tag{116}$$

As observed earlier for the reaction of ClNO and FNO with fluoroolefins [6], the nitroso compounds under reaction conditions can undergo further transformations by reaction with fluoroolefins to give oxazetidines and copolymers, which affects the yield of $R_f NO$. It should be emphasized that reaction of $CFCl=CF_2$ with both reagents produces only one regioisomer which has chlorine and nitroso group at the same carbon. HFP does not interact with $ONOSO_2F$ even in the presence of superacid catalyst $HOSO_2F/SbF_5$ at 100–110 °C [120].

6.1.6
Reactions with HgF$_2$, AsF$_3$, and SbF$_5$

Fluoroolefins add mercuric fluoride at 50–100 °C to give bis(polyfluoroalkyl)-mercury derivatives. Reaction is usually carried out in hydrogen fluoride as a solvent. This process was reviewed and the electrophilic nature of this process established [6]. Addition to olefin starts by the attack of cation 3 (Eq. 5) on the double bond of fluoroolefin.

As mentioned above, arsenic trifluoride does not react with fluoroolefins in the absence of a catalyst, and due to a convenient boiling point and high polarity it was often used as a solvent for an electrophilic reaction, such as mercuration. Later it was found that in the presence of a Lewis acid, such as SbF_5, this compound readily adds to different fluoroethylenes [7, 121, 122]:

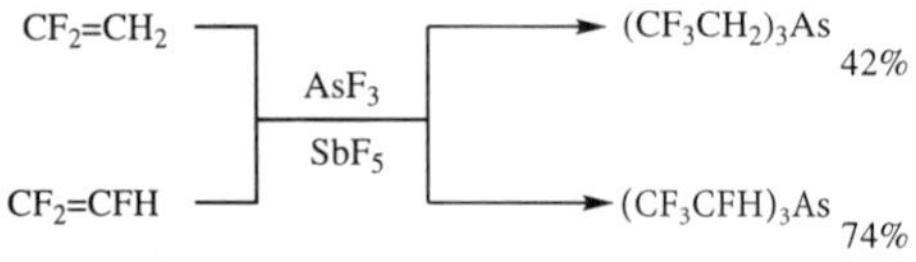

$$\text{CF}_2=\text{CH}_2 \quad \underset{\begin{array}{c}AsF_3\\SbF_5\end{array}}{\underset{\displaystyle CF_2=CFH}{\overline{}}} \quad \begin{array}{l} \longrightarrow (CF_3CH_2)_3As \quad 42\% \\[2em] \longrightarrow (CF_3CFH)_3As \quad 74\% \end{array} \tag{117}$$

Arsine 4 in the presence of a catalyst readily reacts with chlorotrifluoro-ethylene and acetylene[123]:

$$\left.\begin{array}{l} CF_2{=}CFCl \\ C_2H_2 \end{array}\right] \xrightarrow[SbF_5]{(C_2F_5)_2AsF} \begin{cases} (C_2F_5)_2AsCFClCF_3 + (C_2F_5)_2AsCF_2CF_2Cl \quad 3:7 \quad 63\% \\ (C_2F_5)_2AsCH{=}CFH \quad 43\% \end{cases} \tag{118}$$

Even antimony pentafluoride is able to add across the C=C bond. As shown [124], reaction with $CH_2{=}CF_2$ results in the formation of an additional product – tris(β,β,β-trifluoro-ethyl)antimony difluoride. Formation of the alkenyl derivatives of antimony in reaction with $CF_2{=}CCl\text{-}CF{=}CF_2$ (Eq. 75) is another example:

$$2SbF_5 \rightleftharpoons [\overset{+}{Sb}F_4]\,\overset{-}{Sb}F_6 \longrightarrow F_2Sb(CH_2CF_3)_3 \tag{119}$$

A similar reaction was reported [54] for the reaction of BF_3 with tetrafluoroallene resulting in insertion of allene into the B-F bond:

$$CF_2{=}C{=}CF_2 + BF_3 \xrightarrow[-78\text{ to }20^{\circ}C]{} \underset{35\%}{CF_2{=}C(BF_2)CF_3} \tag{120}$$

6.1.7
Electrophilc Oxidation and Oxidative Reactions of Sulfur and Selenium

Electrophilic oxidation of the C=C bond was intensively studied in the 1960s, and this reaction was reviewed [3]. Hydrogen peroxide or peroxyacids in HF as a solvent was shown to be moderately effective for oxidation of chloro- and fluoroethylenes. Vinylidene chloride and fluoride react with solution of $KMnO_4$/HF at –70 °C to give fluoroethanols. Interaction of TFE and HFP with $KMnO_4$/HF mixture yields the corresponding epoxides. The latter reaction is believed to proceed via the intermediate formation of $FMnO_4$, which further transfers an electrophilic oxygen atom to the C=C bond to form oxirane [3]. Chromium oxide in $HOSO_2F$ [125] is convenient for conversion of HFP into oxide. A modification of the latter procedure using CrO_3/Cr_2O_3 in $HOSO_2F$ makes it possible to convert HFP into fluorosulfonoxy-*F*-acetone directly:

$$CF_3CF{=}CF_2 \xrightarrow[45^{\circ}C,\ 3h]{CrO_3/Cr_2O_3,\ HOSO_2F} \underset{40\%}{CF_3C(O)CF_2OSO_2F} \tag{121}$$

Reaction of polyfluorinated indans with concentrated H_2O_2 in HF/SbF_5 has recently been reported [126]. For example, *F*-indene at –30 °C was readily oxidized to *F*-indanone-2, which at higher temperature reacts with H_2O_2 to give a mixture of isocumarine **51 a** and diacid **51 b**:

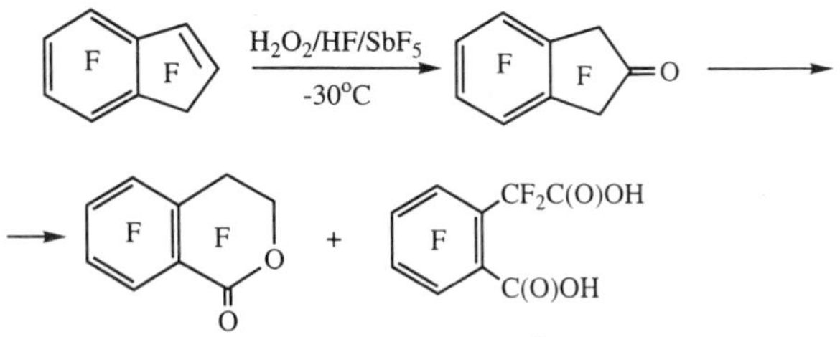

(122)

51 a **51 b**

A combination of xenon difluoride and water in HF was found to be a potent oxidizing agent. At ambient temperature this reagent is able to oxidize polyfluorobenzenes and derivatives of *F*-cyclohexadiene-1,4 . The oxidation is selective, and the C=C bond bearing Xe^+ fragment does not participate in the reaction [127, 128]:

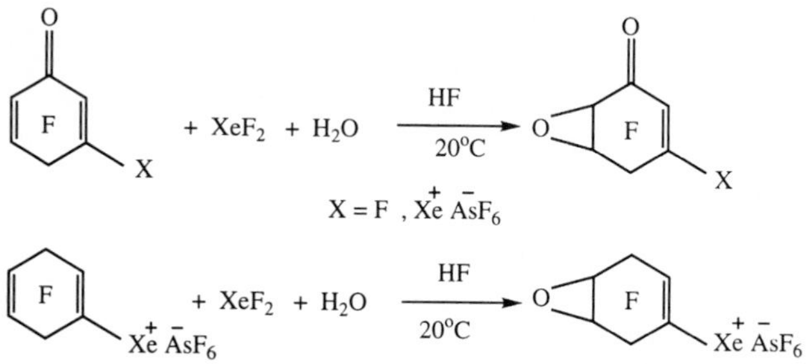

(123)

The well known ability of antimony pentafluoride to oxidize some elements, such as sulfur or selenium, to the corresponding polynuclear cations [1] found its application in the synthesis of fluorinated materials. Reaction of sulfur with TFE, HFP, and PFIB in the presence of SbF_5 was reported [129, 130]:

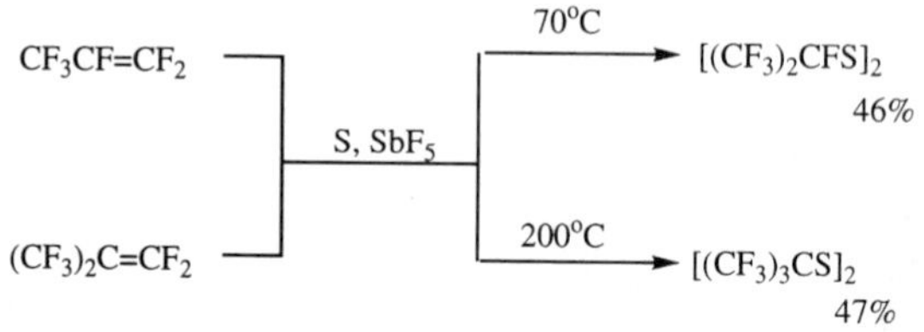

(124)

Selenium behaves in reaction with fluorinated olefins similarly, and bis (*F*-ethyl) diselenide was isolated in a 15% yield in reaction of TFE with selenium/SbF_5 mixture [130, 131].

In sharp contrast to *F*-alkyldisulfides, which are stable to the action of fluoroxidizing agents, such as SbF_5, even at elevated temperature (see Eq. 124), *F*-phenyldisulfide (**52**) reacts with SbF_5, producing electrophilic species [132]. The reaction generates *F*-phenylthiyl cation, which is probably stabilized by complex formation with starting material [7]. Disulfide **52** reacts with different fluoroolefins in the presence of SbF_5 catalyst [133]:

$$\underset{\mathbf{52}}{C_6F_5SSC_6F_5} \xrightarrow[SO_2]{TFE,\ SbF_5} \underset{5}{C_6F_5SC_2F_5} + \underset{1}{C_6F_5SCF_3} \quad 50\% \tag{125}$$

Formation of a by-product – trifluoromethylsulfide – was explained as a result of the reaction of carbocation **53** with solvent – sulfur dioxide – followed by two elimination reactions:

$$\mathbf{52} + TFE \longrightarrow \underset{\mathbf{53}}{[C_6F_5SCF_2CF_2^{+}]} \xrightarrow{SO_2} [C_6F_5SCF_2CF_2OS(O)F] \xrightarrow[SbF_5]{-SO_2F_2}$$

$$C_6F_5SCF_2C(O)F \xrightarrow[SbF_5]{-CO} C_6F_5SCF_3 \tag{126}$$

Similarly, formation of shorter chain disulfides was observed in the above-mentioned reaction of F-olefins with sulfur when SO_2 was used as a solvent [129]. In the absence of sulfur dioxide the side reaction does not occur, and reaction of HFP and PFIB with $(C_6F_5S)_2$ results in formation of expected products:

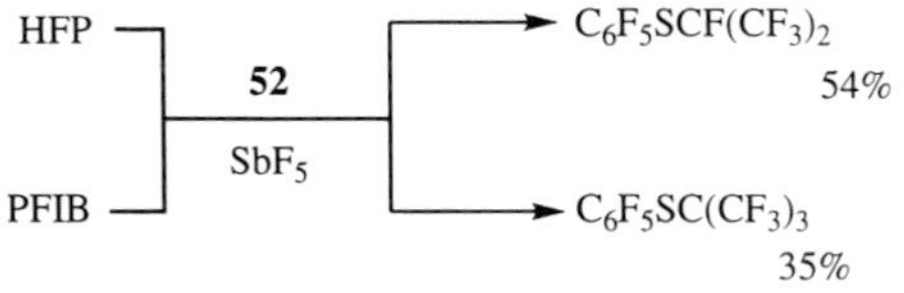

$$
\begin{array}{c}
\text{HFP} \quad\text{---} \\
\qquad \textbf{52} \\
\qquad \text{SbF}_5 \\
\text{PFIB} \quad\text{---}
\end{array}
\longrightarrow
\begin{array}{l}
C_6F_5SCF(CF_3)_2 \\
\qquad\qquad 54\% \\[1em]
C_6F_5SC(CF_3)_3 \\
\qquad\qquad 35\%
\end{array}
\tag{127}
$$

6.1.8
Addition of Sulfur Chlorides

Addition of sulfur chlorides and sulfenyl halides to hydrocarbon olefins is a classic example of electrophilic reaction which usually proceeds under mild conditions and results in stereospecific *trans*-addition via intermediate formation of cyclic episulfonium cation [134]. Ring-opening reactions of episulfonium cation with nucleophile is responsible for formation of regioisomers when nonsymmetrical olefins are used as substrates.

Polyfluorinated olefins are much more resistant to sulfur chlorides. For example, S_2Cl_2 adds to $CH_2{=}CF_2$ with an appreciable rate only at temperatures as high as $60-100\,°C$ to give disulfide **54** as a main product [135]:

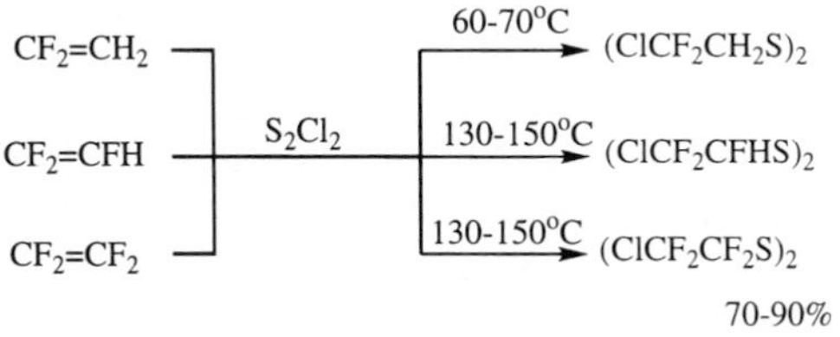

$$
\begin{array}{l}
CF_2{=}CH_2 \quad\text{---} \\[0.5em]
CF_2{=}CFH \qquad S_2Cl_2 \\[0.5em]
CF_2{=}CF_2 \quad\text{---}
\end{array}
\longrightarrow
\begin{array}{l}
\xrightarrow{60\text{-}70°C} (ClCF_2CH_2S)_2 \\[0.5em]
\xrightarrow{130\text{-}150°C} (ClCF_2CFHS)_2 \\[0.5em]
\xrightarrow{130\text{-}150°C} (ClCF_2CF_2S)_2 \\
\qquad\qquad\qquad 70\text{-}90\%
\end{array}
\tag{128}
$$

Addition of S_2Cl_2 to trifluoroethylene and TFE proceeds at $130-150\,°C$ giving the corresponding disulfides in high yield. The formation of small amounts $(5-10\%)$ of trisulfides is usually observed in all these reactions. Addition to $CFCl{=}CF_2$ was first found to be regiospecific, but later it was also shown to be sensitive to reaction temperature [136].

While only one isomer forms at $130\,°C$, reaction starts losing regioselectivity at $140\,°C$, and formation of both isomers was observed at $160\,°C$:

$$
CFCl{=}CF_2 + S_2Cl_2 \xrightarrow{\;t\;} (Cl_2CFCF_2S)_2 + (ClCF_2CFClS)_2
\tag{129}
$$

t (°C)	ratio (%)	
130	100	0
140	95	5
160	80	20

Addition to F-methylvinyl ether proceeds regioselectively, but only at high temperature producing a mixture of di- and trisulfides in high yield [136]:

$$CF_3OCF{=}CF_2 + S_2Cl_2 \xrightarrow{150\text{-}155^\circ C} \underset{n=2,3}{(CF_3OCFClCF_2)S_n} \; 82\% \qquad (130)$$

Although S_2Cl_2 adds to $CF_3CH{=}CF_2$ at 150–160 °C (see below), attempts to carry out addition to 1,1- and 1,2-dichloroethylenes and less reactive HFP and PFIB at temperatures up 200 °C have failed [135, 136].

Reactions of more electrophilic SCl_2 with fluoroolefins under thermolytic conditions are not selective; they always produce a mixture of sulfenyl and thiosulfenylchlorides, sulfides, and polysulfides, along with substantial amounts of products of alkane chlorination [135].

3,3,3-Trifluoropropene reacts with S_2Cl_2 under surpisingly mild conditions to give a product of anti-Markovnikov addition in high yield [138]:

$$CF_3CH{=}CH_2 \xrightarrow[20^\circ C]{S_2Cl_2} (CF_3CHClCH_2S)_2$$

$$CF_3CH{=}CH_2 \xrightarrow[20^\circ C]{SCl_2} \underset{52\%}{(CF_3CHClCH_2S)_2} + \underset{21\%}{(CF_3CHClCH_2)_2S} + \underset{18\%}{CF_3CHClCF_2Cl} \qquad (131)$$

Reaction with SCl_2, although more complex, also leads to exclusive formation of anti-Markovnikov adducts, along with an expected chlorination product. Addition of S_2Cl_2 to $CF_2{=}CHCF_3$ is another example[135, 136]:

$$CF_3CH{=}CF_2 + S_2Cl_2 \xrightarrow{150\text{-}160^\circ C} \underset{n=1\text{-}3}{CF_3CHClCF_2(S)_nCl} + \underset{m=1,2}{(CF_3CHClCF_2)_2S_m} \qquad (132)$$

Despite the fact that this reaction proceeds at much higher temperature and results in the formation of a mixture of products, all of them are derived from the attack of sulfur intermediately and exclusively on the *terminal* carbon of the olefin, which is not consistent with the orientation of electrophilic addition to polyfluropropenes where an electrophile attacks the *central* carbon of the double bond exclusively (see Eqs. 29, 37, 38).

Thermal addition of S_xCl_2 to fluoroolefins was often referred to as electrophilic addition [136, 137], although observed orientation of addition actually resembles that in radical reactions [139]. Regardless of the mechanism, this reaction is highly regioselective and undoubtedly has synthetic value as a high yield route to polyfluoroalkyldisulfides.

Activity of sulfur dichlorides could be significantly increased by modification of reaction media. As mentioned above, PFIB and *F*-ketene do not interact with sulfur chlorides even at high temperature. However, the reactions of S_2Cl_2 with PFIB and sulfur mono- and dichlorides with *F*-ketene proceed exothermally when dry acetonitrile is employed as a solvent, giving either thiosulfenylchlorides or disulfides in high yield [140–142]:

$$(CF_3)_2C{=}C{=}O + S_xCl_2 \xrightarrow[0\text{ to }20^\circ C]{CH_3CN} \underset{82\%}{(CF_3)_2C(C(O)Cl)SSCl} \xrightarrow[20^\circ C,\ 2d]{(CF_3)_2C=C=O} \underset{73\%}{[(CF_3)_2C(C(O)Cl)S]_2} \qquad (133)$$

Electrophilic activation of sulfur chlorides could be achieved by conducting the reaction with fluoroolefin in the presence of strong acids. For example, trifluoropropene readily reacts with the mixture $S_2Cl_2/HF/SF_4$ to produce the corresponding isopropylsulfide **55** in high yield [143]:

$$CF_3CH{=}CH_2 + SF_4 + S_2Cl_2 \xrightarrow{HF} (FCH_2\underset{\displaystyle CF_3}{\underset{|}{C}}H)_2S \qquad (134)$$

Formation of this compound was explained as an electrophilic attack of $ClSS^+$ (formed under the action of SF_4/HF mixture) on the central carbon of the olefin, resulting in formation of thiosulfenyl chloride **55**. Oxidation of this compound by a trace amount of chlorine produces sulfenyl chloride which reacts with a second mole of olefin to give a product. It is noteworthy that the orientation of addition is consistent with the electrophilic mechanism proposed:

$$SF_4 + HF + S_2Cl_2 \longrightarrow ClSS^+$$

$$CF_3CH{=}CH_2 + ClSS^+ \xrightarrow{HF} \underset{\mathbf{55}}{FCH_2CH(CF_3)SSCl} \xrightarrow[-SCl_2]{Cl_2} \qquad (135)$$

$$FCH_2CH(CF_3)SCl \xrightarrow{CF_3CH=CH_2} (FCH_2\underset{\displaystyle CF_3}{\underset{|}{C}}H)_2S$$

At elevated temperature, 1,2-dichlorodifluoroethylene gave the corresponding sulfenyl chloride in high yield [143, 144]:

$$CFCl{=}CFCl + SF_4 + S_2Cl_2 \xrightarrow[-SCl_2]{HF,\ 60^{\circ}C} \underset{74\%}{ClCF_2CFClSCl} \qquad (136)$$

Reactivity of sulfur chlorides significantly increases when reaction with fluoroolefin is carried out in fluoro- or chlorosulfonic acids, since strong protic acids are able to increase electrophilicity of the molecule by protonation [145]. For example, HFP reacts with solution of SCl_2 in $HOSO_2F$ at 40–60 °C and atmospheric pressure to give a mixture of products from which the corresponding sulfenyl chloride was isolated in 75% yield:

$$HFP + SCl_2 \xrightarrow{HOSO_2F} \underset{75\%}{CF_3CF(SCl)CF_2OSO_2F} + [FSO_2OCF_2CF(CF_3)]_2S_3 + CF_3CFClCF_2Cl \qquad (137)$$

Trisulfide **56** is formed as a major product when S_2Cl_2 reacts with HFP in chlorosulfonic acid, and the product is isolated in 85% yield:

$$HFP + S_2Cl_2 \xrightarrow{HOSO_2Cl} \underset{\mathbf{56},\ 85\%}{[ClSO_2OCF_2CF(CF_3)]_2S_3} + CF_3CF(SSCl)CF_2OSO_2Cl + CF_3CFClCF_2Cl \qquad (138)$$

6.1.9
Alkylation

Alkylation of fluoroolefins is probably the most studied reaction of that kind. Different aspects of this process, such as reaction with formaldehyde and hexamethylenetetramine in HF [3], or condensation of halomethanes with fluoroethylenes, catalyzed by $AlCl_3$ [27] have been reviewed. A recent development in this field mostly associated with the introduction of new catalysts, such as SbF_5 [7] or aluminum chorofluoride $AlCl_xF_y$ [12], was covered in two other review articles. Here a brief description of some of these reactions is given.

Solution of antimony pentafluoride in anhydrous HF was found to be an efficient medium for alkylation reactions of fluoroolefins. Compounds, known to give stable carbocations in superacidic media are excellent alkylating agents. For example, *t*-butyl chloride (**57**) reacts with TFE, giving the corresponding alkane **58**:

$$\underset{\mathbf{57}}{(CH_3)_3CCl} + TFE \xrightarrow[20^{\circ}C]{SbF_5/HF} \underset{\mathbf{58},\ 50\%}{(CH_3)_3CCF_2CF_3} + HCl \qquad (139)$$

This reaction proceeds as a conjugate process (see also Eq. 2, pathway B). It starts with ionization of **57** by action of super acid (HF/SbF_5) to give *t*-butyl cation. Its further attack on TFE with formation of another cation **59**, which is stabilized by adding fluoride anion, from either a counter anion or HF:

$$\mathbf{57} + H^+ \longrightarrow (CH_3)_3C^+ \xrightarrow{TFE} \underset{\mathbf{59}}{[(CH_3)_3CF_2CF_2^{+}]} \xrightarrow{F^-} \mathbf{58} \qquad (140)$$

Both 1,1- and 1,2-dichloroethane react with TFE under similar conditions, producing the same product, 2-chloropentafluorobutane **60**:

$$\left.\begin{matrix} ClCH_2CH_2Cl \\ CH_3CHCl_2 \end{matrix}\right] \xrightarrow[20^{\circ}C]{HF/SbF_5} \underset{\mathbf{60},\ 66\text{-}80\%}{CH_3CHClCF_2CF_3} \qquad (141)$$

Reaction of 1,1,1-trifluoroethane with fluorinated ethylenes requires pure antimony pentafluoride as a catalyst. Reaction with TFE gives a monoalkylation product in high yield; addition to $CH_2{=}CF_2$ results in formation of a branched alkane **61** [7]

$$CH_3CF_3 \xrightarrow[20\text{-}50^{\circ}C]{SbF_5} \begin{cases} \xrightarrow{TFE} \underset{90\%}{CH_3CF_2CF_2CF_3} \\ \xrightarrow{CH_2=CF_2} \underset{\mathbf{61},\ 40\%}{CH_3C(CH_2CF_3)_3} \end{cases} \qquad (142)$$

Antimony pentafluoride was also reported to catalyze condensation of CFC-113 with fluoroethylenes. Reaction proceeds under mild conditions, producing the corresponding butanes in high yield [7]:

$$\text{ClCF}_2\text{CCl}_2\text{F} \xrightarrow[\text{0-5°C}]{\text{SbF}_5} \begin{array}{l} \xrightarrow{\text{TFE}} \text{ClCF}_2\text{CCl}_2\text{CF}_2\text{CF}_3 \quad 80\% \\ \\ \xrightarrow{\text{CFCl=CFCl}} \text{ClCF}_2\text{CCl}_2\text{CFClCF}_2\text{Cl} \end{array} \qquad (143)$$

Recently discovered aluminum chlorofluoride AlCl_xF_y (ACF) is also an effective catalyst for condensation reactions. In some of them ACF is superior to both SbF_5 and AlCl_3. ACF was used for condensation of CHCl_2F [146], $\text{ClCF}_2\text{CCl}_2\text{F}$, $\text{CF}_3\text{CCl}_2\text{F}$ [147], and CF_2X_2 [148] with fluoroethylenes. For example, CF_2Cl_2 is able to react with two moles of TFE in the presence of ACF, giving pentane 62 in quite a high yield [148]:

$$\text{CF}_2\text{Cl}_2 + \text{CF}_2\text{=CF}_2 \xrightarrow[\text{30-40°C}]{\text{ACF}} \underset{17\%}{\text{CF}_3\text{CF}_2\text{CFCl}_2} + \underset{\textbf{62}, 52\%}{\text{CF}_3\text{CF}_2\text{CCl}_2\text{CF}_2\text{CF}_3} \qquad (144)$$

1,1,2-Trichlorotrifluoroacetone in the presence of SbF_5 reacts with trifluoroethylene or TFE, producing a mixture of approximately equal amounts of pentanone-2 and oxetane 63 [149], for instance:

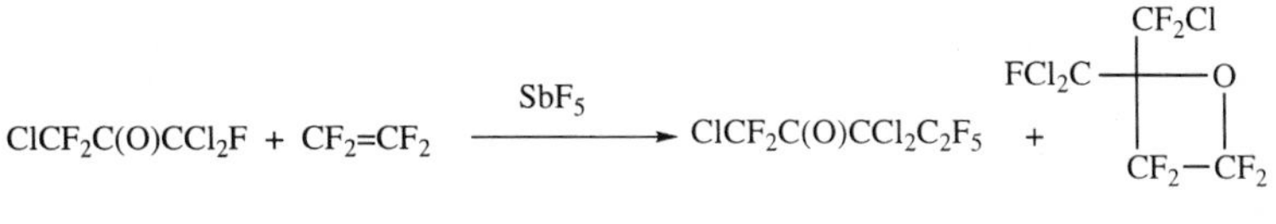

$$\text{ClCF}_2\text{C(O)CCl}_2\text{F} + \text{CF}_2\text{=CF}_2 \xrightarrow{\text{SbF}_5} \text{ClCF}_2\text{C(O)CCl}_2\text{C}_2\text{F}_5 + \underset{\substack{\textbf{63}}}{\begin{array}{c} \text{CF}_2\text{Cl} \\ \text{FCl}_2\text{C}\!-\!\!\!-\!\!\!-\text{O} \\ | \qquad | \\ \text{CF}_2\!-\!\text{CF}_2 \end{array}}$$
$$(145)$$

HFA and chloropentafluoroacetone are not active in this reaction [7]. Formation of ketone is a result of condensation similar to that of CFC-113 with TFE (see Eq. 143), while the second product is derived from competitive cycloaddition reaction of a zwitterionic species 64 and fluoroolefin:

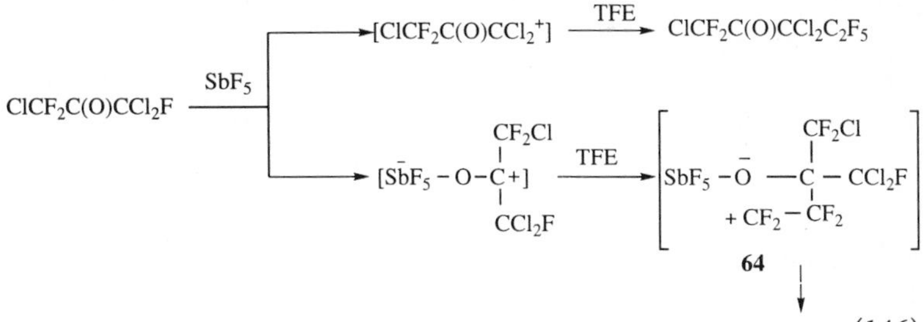

$$\text{ClCF}_2\text{C(O)CCl}_2\text{F} \xrightarrow{\text{SbF}_5} \begin{array}{l} \longrightarrow [\text{ClCF}_2\text{C(O)CCl}_2{}^+] \xrightarrow{\text{TFE}} \text{ClCF}_2\text{C(O)CCl}_2\text{C}_2\text{F}_5 \\ \\ \longrightarrow [\overset{-}{\text{SbF}_5}-\text{O}-\overset{\text{CF}_2\text{Cl}}{\underset{\text{CCl}_2\text{F}}{\overset{|}{\text{C}}{}^+}}] \xrightarrow{\text{TFE}} \left[\overset{-}{\text{SbF}_5}-\overset{-}{\text{O}}-\overset{\text{CF}_2\text{Cl}}{\underset{+\text{CF}_2-\text{CF}_2}{\overset{|}{\text{C}}-\text{CCl}_2\text{F}}} \right] \\ \hspace{9cm} \textbf{64} \\ \hspace{9.5cm} \downarrow \\ \hspace{9.3cm} \textbf{63} \quad (146) \end{array}$$

Recently it has been found that ACF is able to activate hexafluoroacetone (HFA) and that reaction of fluoroethylenes proceeds exclusively as electrophilic $2+2$ cycloaddition with formation of the corresponding oxetanes [49] (see also Eq. 35):

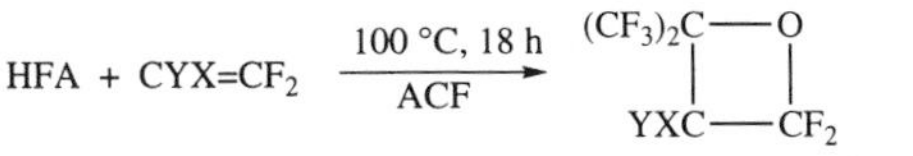

$$\text{HFA} + \text{CYX=CF}_2 \xrightarrow[\text{ACF}]{\text{100 °C, 18 h}} \begin{array}{c} (\text{CF}_3)_2\text{C}\!-\!\!\!-\!\!\!-\text{O} \\ | \qquad | \\ \text{YXC}\!-\!\!\!-\!\!\!-\text{CF}_2 \end{array} \qquad (147)$$
$$\begin{array}{l} \text{X =Y= F, 66\%} \\ \text{X = H, Y=F, 98\%} \\ \text{X = H, Y=Cl, 98\%} \\ \text{X = H, Y=Br, 91\%} \end{array}$$

This reaction and its mechanism are described in more detail in [12, 49].

6.1.10
Alkenylation

The first example of electrophilic condensation of polyfluorinated olefins with fluoroethylenes was reported in 1975 [150]. Reaction of HFP and TFE in the presence of SbF_5 catalyst produces *F*-pentene-2, mostly the *trans*-isomer:

$$CF_3CF{=}CF_2 + CF_2{=}CF_2 \xrightarrow[50°C]{SbF_5} \underset{70\text{-}95\%}{CF_3CF{=}CFCF_2CF_3} \qquad (148)$$

The mechanism includes formation of *F*-allyl carbocation **1** as a result of abstraction of activated allylic fluorine from HFP by a Lewis acid and addition **1** to TFE followed by isomerization of terminal olefin into product **11**:

$$CF_2{=}CFCF_3 + SbF_5 \;\rightleftharpoons\; \mathbf{1} \xrightarrow{\text{TFE}} [CF_2{=}CFCF_2CF_2CF_3]$$

$$\downarrow SbF_5 \qquad\qquad (149)$$

$$\mathbf{11}$$

$$\mathbf{11} + CF_2{=}CF_2 \xrightarrow[50°C]{\text{ACF}} CF_3CF{=}CF(CF_2)_3CF_3 + CF_3CF_2CF{=}CF(CF_2)_2CF_3 +$$

$$+ C_9F_{18}$$

yield C_7F_{14} 80%

Different fluoroolefins able to give relatively stable allylic cations under the action of SbF_5 (see Sect. 5) were used as alkenylating agents. Examples include $CF_3CH{=}CF_2$ [150], $(CF_3)_2C{=}CF_2$ [151], $c\text{-}C_4F_6$ [152], $(CF_3)_2C{=}CCl_2$, $CF_3CCl{=}CCl_2$ [153], $CF_3CF{=}CFH$, $C_2F_5CF{=}CFH$ [154], and $CF_3C(OR_f){=}CF_2$ [155] as alkenylating agents, and HFP and $CF_3CH{=}CF_2$ (electrophilic dimerization), TFE, $CFCl{=}CF_2$, $CFCl{=}CFCl$ and $CFH{=}CF_2$ as substrates. Internal fluoroolefins, such as *F*-pentene-2, $(CF_3)_2CFCF{=}CFCF_3$, or $c\text{-}C_5F_8$, are not active in this reaction when SbF_5 is used as a catalyst [7]. However, as shown recently, employment of ACF as a catalyst makes it possible to expand this reaction significantly. It catalyzes not only reaction of HFP/TFE but also condensation of *F*-pentene-2 with TFE, resulting in the mixture of *F*-heptene-2 and *F*-heptene-3 along with a small amount of C_9F_{18} [22, 156]. Both *F*-cyclopentene and $C_4F_9CH{=}CHC_4F_9$ react with TFE in the presence of ACF:

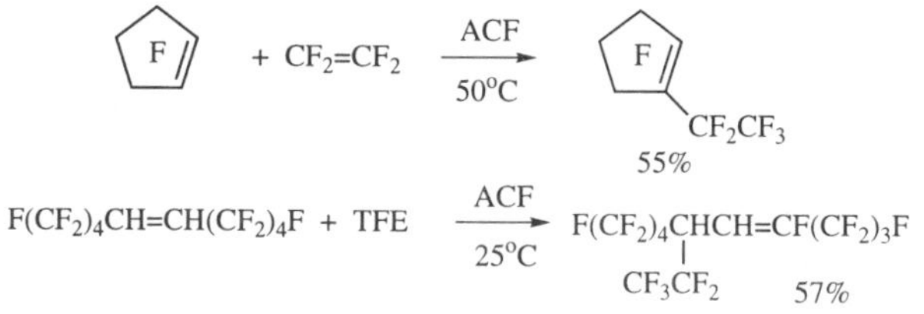

$$
\begin{array}{c}
\left\langle\!\!\!\bigcirc\!\!F\right\rangle + CF_2{=}CF_2 \xrightarrow[50°C]{\text{ACF}} \underset{55\%}{\left\langle\!\!\!\bigcirc\!\!F\right\rangle\!{-}CF_2CF_3} \\[4pt]
\end{array} \qquad (150)
$$

$$F(CF_2)_4CH{=}CH(CF_2)_4F + TFE \xrightarrow[25°C]{\text{ACF}} \underset{\substack{|\\CF_3CF_2 \quad 57\%}}{F(CF_2)_4CHCH{=}CF(CF_2)_3F}$$

Benzylic fluorides can be also condensed with fluoroolefins. *F*-Toluene was shown to form a 1:1 adduct with HFP [157]:

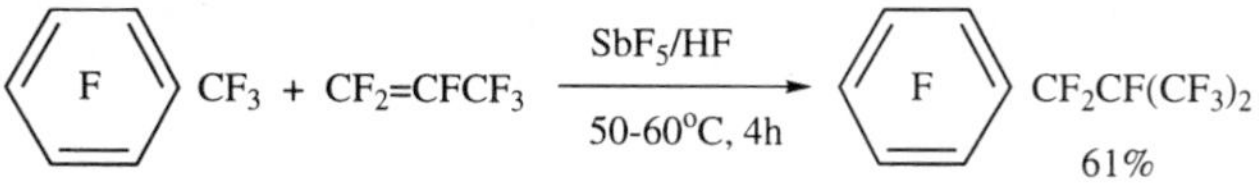

$$\text{(benzene ring with } F) \; CF_3 + CF_2{=}CFCF_3 \xrightarrow[50\text{-}60^\circ C, \, 4h]{SbF_5/HF} \text{(benzene ring with } F) \; CF_2CF(CF_3)_2 \qquad (151)$$

61%

Perfluorinated benzocycloalkanes also react with fluoroethylenes. Reaction proceeds under mild conditions in the presence of SbF_5 catalyst, producing a mixture of 1:1 and 1:2 adducts [158]:

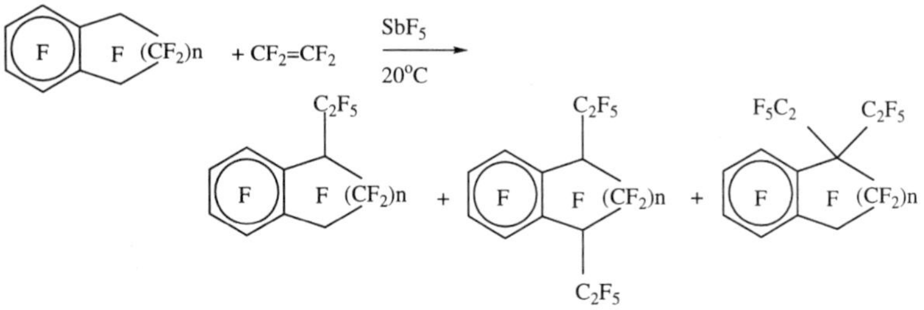

ratio %:

n = 0	89	6	5
n = 1	54	-	46
n = 2	90	10	-

$$(152)$$

A more detailed description of the reaction can be found in [12, 158].

Reaction of polyfluorinated dienes with TFE resembles the reaction of polyfluoropropynes with fluoroethylenes. Polyfluorinated dienes react with TFE at ambient temperature in the presence of SbF_5 catalyst [159]. F-Heptadiene-2,4 (65) was obtained in 77% yield in reaction of F-pentadiene-1,3 and TFE along with small amounts of F-nonadiene-3,5. F-Hexadiene-2,4 reacts with TFE at 50 °C to give a mixture of isomeric octadienes 66 a, b:

$$CF_2{=}CFCF{=}CFCF_3 + TFE \xrightarrow[20^\circ C]{SbF_5} CF_3CF{=}CFCF{=}CFC_2F_5$$

65, 77%

$$+ \; C_3F_7CF{=}CFCF{=}CFCF_2CF_3$$

$$CF_3CF{=}CFCF{=}CFCF_3 + TFE \xrightarrow[50^\circ C, \, 6\,h]{SbF_5} CF_3CF{=}CFCF{=}CFC_3F_7$$

66a

$$+ \; C_2F_5CF{=}CFCF{=}CFC_2F_5$$

66b

50%

$$65 \xrightarrow[20^\circ C, \, 24\,h]{SbF_5} \text{(cyclopentene ring with } F, \, CF_3, \, CF_3)$$

63% $\qquad (153)$

Polyfluorinated dienes undergo intramolecular cyclization in the presence of SbF_5. Diene 65 readily forms F-1,2-dimethylcyclopentene when it is dissolved in an excess of SbF_5 [159]. However, cyclization of its higher homologues proceeds at elevated temperature and results in the formation of isomeric 1,2-dialkyl-cyclopentenes [160]:

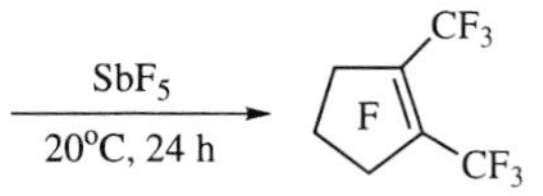

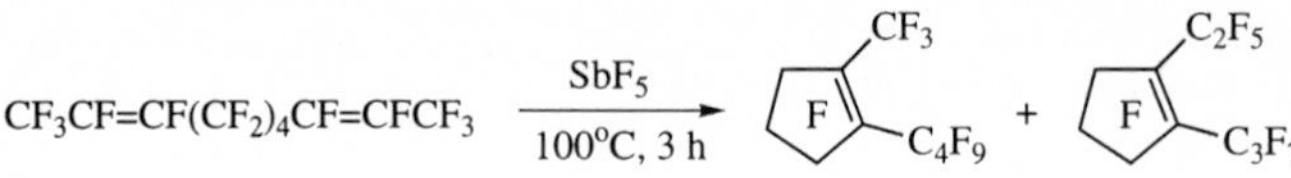

$$CF_3CF=CF(CF_2)_4CF=CFCF_3 \xrightarrow[100°C,\ 3\ h]{SbF_5} \quad \text{(154)}$$

3 : 7

84%

The mechanism of this reaction includes formation of the corresponding pentadienyl cation followed by a cyclization step which could be viewed either as a concerted ring-closure reaction of cation or an intramolecular version of alkenylation reaction:

65 + SbF$_5$

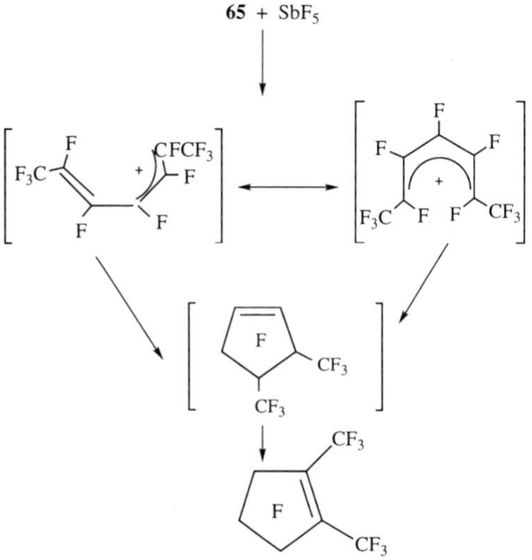

(155)

A similar reaction was found for some perfluorinated styrenes. Compound **67** at 130 °C was reported to undergo cyclization with formation of *F*-1,2-dimethylindane **68**:

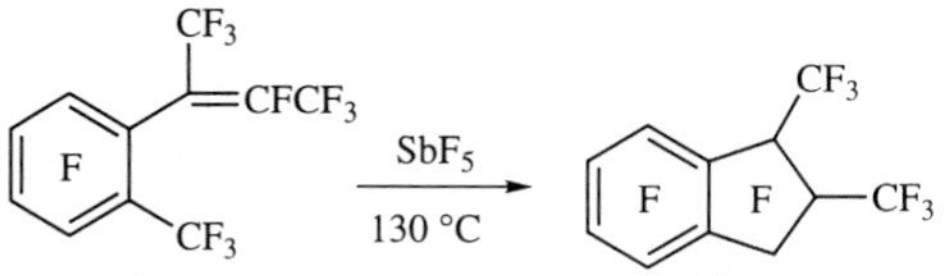

(156)

67   68

Formation of small amounts of a by-product – *F*-methylcyclobutane **69** – (in reaction of HFP and TFE) catalyzed by ACF was rationalized as a result of similar intramolecular alkylation reaction of the C=C bond in intermediate carbocation **70**, and it is probably the first example of electrophilic 2+2 cycloaddition reaction of fluoroolefin:

$$HFP + TFE \xrightarrow{ACF} CF_2=CFCF_2CF_2CF_2^+ \longrightarrow \quad \text{(157)}$$

70

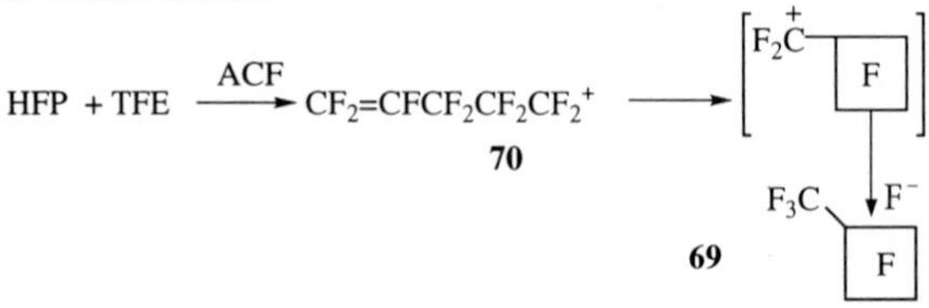

Formation of compound **69** is very important from a mechanistic standpoint, since it is evidence that such an elusive species as cation **70** is a real independent intermediate which has a lifetime sufficient for cyclization to occur. More detailed discussion of cyclization reaction of *F*-dienes, along with additional information on electrophilic 2+2 cycloaddition reaction of olefins and HFA are given in a review [12].

6.1.11
Acylation

Until recently, no successful attempts to carry out electrophilic acylation of polyfluoroalkenes using haloanhydrides of perfluorocarboxylic acids have been reported, although acylation of $CFH{=}CF_2$ catalyzed by SbF_5 and of $CH_2{=}CF_2$ catalyzed by acetyl fluoride was seen long ago [7, 161]. Unlike fluorides of perfluorinated saturated acids, acyl fluorides of perfluorinated α,β-unsaturated acids **71 a, b** were found to be active in reaction with TFE, CFCl=CFCl and HFP [162], for example:

$$\underset{\mathbf{71a,b}}{R_fCF{=}CF\overset{\overset{O}{\|}}{C}F} + TFE \xrightarrow[30\text{-}40\,^{\circ}C]{SbF_5} R_fCF{=}CF\overset{\overset{O}{\|}}{C}C_2F_5 \qquad (158)$$

$R_f = (CF_3)_2CF{-}$ 92%
$R_f = C_2F_5{-}$ 61%

It is interesting that *F*-ketene in this reaction also behaves as an acylating agent, giving a mixture of ketone **72** and acyl fluoride **73** in reaction with TFE:

$$(CF_3)_2C{=}C{=}O + TFE \xrightarrow[30\text{-}40\,^{\circ}C]{SbF_5} \underset{\mathbf{72}}{CF_2{=}C(CF_3)\overset{\overset{O}{\|}}{C}C_2F_5} + \underset{\mathbf{73}}{C_2F_5CF{=}C(CF_3)\overset{\overset{O}{\|}}{C}F}$$

yield of mixture 58% (159)

6.2
Reactions Involving Activated C-F Bonds of Fluoroolefins

6.2.1
Electrophilic Isomerization of Fluoroolefins

Under the action of strong Lewis acids, polyfluorinated olefins undergo isomerisation with migration of the double bond, which proceeds either through intermediate allylic carbocation, if the substrate has groups able to stabilize carbocation (C_6F_5 for example, **74**), or in most cases through a highly organized transition state **75**.

$$R_fCF_2CF{=}CF_2 \xrightarrow[\text{exothermic}]{SbF_5} R_fCF{=}CFCF_3$$

$R_f = CF_3$ - C_8F_{17}, 95-100% (160)

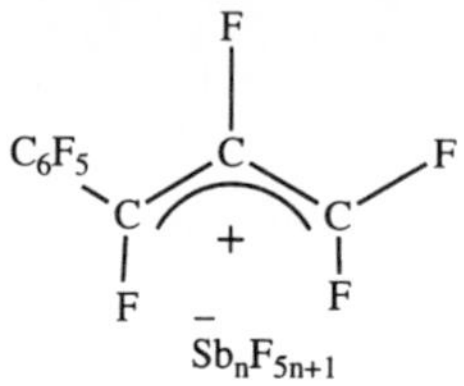

74

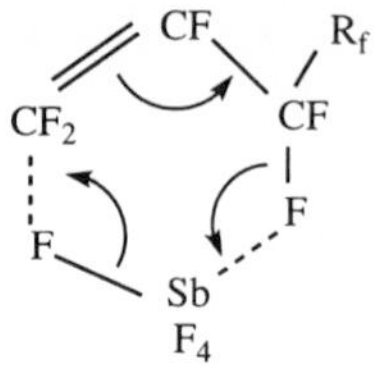

75

Scheme 3

The general concept of these processes was formulated in a review [7] and was later discussed in [12]. Linear *F*-alkenes-1 react exothermally with SbF_5 to give perfluoro-2-alkenes, with significant domination of the *trans*-isomer (up to 90%) [163]. Deeper migration of the double bond inside the carbon chain under the action of SbF_5 takes place only at elevated temperature and results in equilibrium between isomers containing an internal C=C bond:

$$CF_3CF{=}CFCF_2R_f + \text{cat.} \rightleftharpoons C_2F_5CF{=}CFR_f \quad (161)$$

R_f	temp.(°C)	cat.	ratio olefin-2/olefin-3
C_2F_5	70	SbF_5	25 / 75
n-C_3F_7	70	SbF_5	20 / 80
	25	ACF	5 / 95

It is interesting that the position of equilibrium depends on the type of catalyst: in ACF- catalyzed isomerization of *F*-heptene-1 the equilibrium ratio of olefin-2/olefin-3 is different (5:95 vs 25:75 for SbF_5-catalyzed process). It should be noted that this difference could also be a reflection of difference in reaction temperature, since isomerization catalyzed by more active ACF catalyst proceeds even at 25 °C [23]. When H, Cl, or *F*-alkyl group is located inside the carbon chain of polyfluoroolefin, the double bond always migrates to these groups, following the rule of a minimum number of vinylic fluorine substituents [7]:

$$CF_2{=}CFCF_2CFHCF(CF_3)_2 \xrightarrow{SbF_5} CF_3CF_2CF{=}CHCF(CF_3)_2 \quad (162)$$

Cyclic fluoroalkenes have the same reaction pattern, as illustrated by isomerization of cyclopentene **76** [159]:

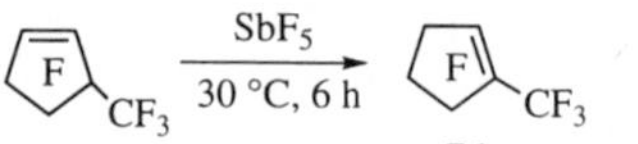

76 quant. (163)

The result of the SbF_5-catalyzed isomerisation of ω-H-*F*-alkenes-1 depends on the chain length of the olefin. This affects the position of equilibrium between terminal and internal isomers [164]:

$$CF_2{=}CF(CF_2)_nCF_2CF_2H \xrightarrow[50\text{-}100^\circ C]{SbF_5} F(CF_2)_nCF_2CF{=}CFH + F(CF_2)_nCF{=}CFCF_2H \qquad (164)$$

n	ratio olefin-1/olefin-2
1	100/0
2	75/25

The result of acid-catalyzed isomerisation of *F*-dienes depends on several factors: structure of substrate, catalyst, and temperature. Action of SbF_5 on terminal dienes under mild conditions causes a 1,3 fluorine shift occurring stereoselectively to give *trans*-, *trans*-, and *cis*-, *trans*- isomers of the corresponding internal dienes [160]:

$$CF_2{=}CFCF_2(CF_2)_nCF_2CF{=}CF_2 \xrightarrow[0\text{-}20\ ^\circ C]{SbF_5} CF_3CF{=}CF(CF_2)_nCF{=}CFCF_3 \qquad (165)$$

$n = 0,2,4,6,8,10$; 90-100%

At elevated temperature, the reaction is more complex and, depending on reaction conditions and the ratio of SbF_5 to diene, a mixture of perfluorinated cyclobutenes and cyclopentenes is formed [12, 160, 165].

Action of Lewis acids on conjugated dienes might result in isomerization into allene or acetylene. *F*-4-Methylbutadiene-1,3 isomerizes into allene **78** through intermediate formation of diene **77** [166]:

$$(CF_3)_2C{=}CFCF{=}CF_2 \xrightarrow[0^\circ C]{SbF_5} \underset{\mathbf{77}}{CF_2{=}C(CF_3)CF{=}CFCF_3} \xrightarrow[20\ ^\circ C]{SbF_5} \underset{\mathbf{78},\ 90\%}{(CF_3)_2C{=}C{=}CFCF_3} \qquad (166)$$

On the other hand, treatment of *F*-butadiene-1,3 by ACF may be a useful preparative route to *F*-butyne [23], since reaction proceeds under mild conditions and produces acetylene in a quantitative yield:

$$CF_2{=}CF\text{-}CF{=}CF_2 \xrightarrow[25^\circ C,\ 1\ h]{ACF} \underset{100\%}{CF_3C{\equiv}CCF_3}$$

$$CF_2{=}CFCF_2I \xrightarrow[\text{exotherm.}]{ACF} CF_3CF{=}CFI \qquad (167)$$

90% (50:50 *trans:cis*)

ACF may also be used for isomerisation of *F*-olefins [33]. *F*-Allyl iodide and olefin **79** were readily isomerized under the action of ACF and, as mentioned above, this catalyst is more active in isomerization of *F*-olefins [23]:

$$C_6F_5CF_2CF{=}CF_2 \xrightarrow[25^\circ C,\ 2h]{ACF} C_6F_5CF{=}CFCF_3 \qquad (168)$$

100% (54:46 *trans:cis*)

The reactivity of polyfluorinated cyclohexadienes toward antimony or niobium pentafluorides has some peculiarities. The equilibrium mixture of 1,4- and 1,3-cyclohexadienes is formed from pure isomers of each and a catalytic amount of MF_5 at room temperature where the *non-conjugated* diene predominates [167]. With an excess of SbF_5 both isomers disproportionate to give a mixture of aromatic compounds and cyclohexene polyfluoroaromatics and the corresponding cyclohexene [167, 168]:

X = F	92	8	
X = H	85	15	
X = Cl	85	15	
$X = Xe^+AsF_6^-$	100	0 (**80**)	

(169)

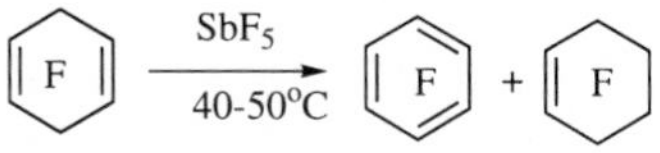

Interestingly, diene **80**, containing Xe^+ substituent, does not isomerize into 1,3-isomer under the action of SbF_5 and it is not ionized by the same Lewis acid into the corresponding arenonium cation [169].

Isomerization of bicyclic triene **81** proceeds fast even at low temperature in the presence of a catalytic amount of SbF_5; an attempt to observe the corresponding intermediate carbocation by NMR spectroscopy was unsuccessful [167]:

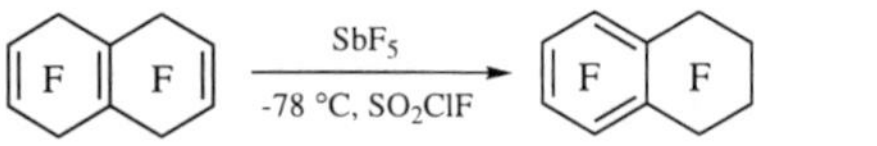

81 quant. (170)

6.2.2
Cleavage of Vinyl Ethers

Partially fluorinated vinyl ethers of fluoroolefins are quite susceptible to the action of Lewis acids. Reaction usually proceeds with ionization of the allylic C-F bond and results in formation of C=O group and elimination of alkyl halide. Indeed, 3-chloro-2-methoxyhexafluoro-2-butene **82** reacts with $AlCl_3$ with formation of trichlorovinyl ketone **83**, and cyclic alkoxyfluoroalkenes demonstrate similar behavior in reaction with aluminum or tin(IV) halides [170]:

$$CF_3C(OCH_3)=CClCF_3 \ (\mathbf{82}) \xrightarrow[\text{reflux}]{AlCl_3,\ CS_2} CF_3C(=O)CCl=CCl_2 \ (\mathbf{83}, 58\%) \tag{171}$$

1-Methoxy-*F*-propylene **84** reacts exothermally with SbF_5 to give perfluoroacroyl fluoride **85** [171]:

$$CH_3OCF=CFCF_3 \ (\mathbf{84}) \xrightarrow{SbF_5} CF_2=CFC(=O)F \ (\mathbf{85}, 47\%) \tag{172}$$

Interestingly, the perfluorinated version of vinyl ether **84** is even less stable to the action of SbF_5, and decomposition of this material with formation of **85** and CF_4 rapidly proceeds, even at –50 °C [155]. A similar reaction was reported for 1-methoxy-*F*-isobutene, but in this case it produces a mixture of methacroyl fluoride and *F*-ketene:

$$CH_3OCF=C(CF_3)_2 \xrightarrow[25\,°C]{SbF_5} O=C=C(CF_3)_2 + \overset{O}{\overset{\|}{F}}CC(CF_3)=CF_2 \qquad (173)$$

total yield 72%

When OR group is located inside the fluorocarbon chain, the corresponding vinyl ketone is the only product [172], and the CF(OCH3)-moiety is the preferential site for demethylation [171]:

$$(CF_3)_2CF\,CF=C(CF_3)\underset{OCH_3}{CFC_2F_5} \xrightarrow[0\,°C,\,2\,h]{SbF_5} (CF_3)_2CFCF=C(CF_3)\overset{O}{\overset{\|}{C}}C_2F_5$$

72%

$$CF_3C(OCH_3)=CF_2CF_3 \xrightarrow[\text{exotherm.}]{SbF_5} CF_3C(O)CF=CFCF_3$$

84%

$$CH_3OCF=C(CF_3)CF(OCH_3)CF(OCH_3)CF_3 \xrightarrow[25°C]{SbF_5} CH_3OCF=C(CF_3)C(O)C(O)CF_3$$

37%

(174)

6.2.3
Reactions with Sulfur Trioxide and Boron Triflate

Sulfur trioxide is a strong Lewis acid comparable in strength with such acids as $GaCl_3$, BCl_3 and, according to some data, even $SbCl_5$ [4]. Non-catalysed reaction of SO_3 with terminal polyfluoroalkenes results in formation of β-sultones, which is currently viewed as a concerted electrophilic 2 + 2 cycloaddition reaction [4]. This process has been intensively studied for over 30 years and data on this reaction has been reviewed [173]).

First indication that the course of the reaction of SO_3 with fluoroolefins could be different was found in 1976. The reaction of polyfluorocyclobutenes with stabilized sufur trioxide results in insertion reaction and produces the corresponding fluorosufates **86a – c** [174]:

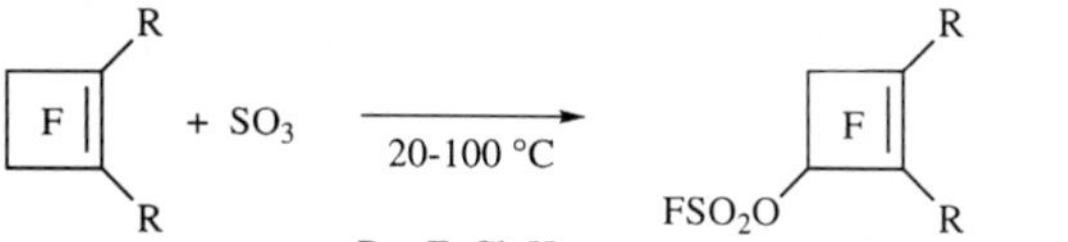

$$R = F, Cl, H \qquad \textbf{86a-c}, 44-74\% \qquad (175)$$

Later it was found that the result of reaction of unsaturated compounds with SO_3 strongly depends on the structure of fluoroolefin. While HFP gives β-sultone in a high yield under the action of pure SO_3 [173], perfluoroallylbenzene reacts with SO_3 exothermally to give a mixture of β-sultone and F-phenyl-propenyl fluorosulfate, but 3,3,3-trifluorotrichloropropene-1 or 2,3-dichloro-F-butene-2 reacts with SO_3, forming allyl fluorosulfates **87** or **88** [175, 176]:

$$C_6F_5CF_2CF=CF_2 + SO_3 \xrightarrow[\text{exotherm.}]{} \qquad + C_6F_5CF=CFCF_2O\text{S}$$

26 : 74 (176)

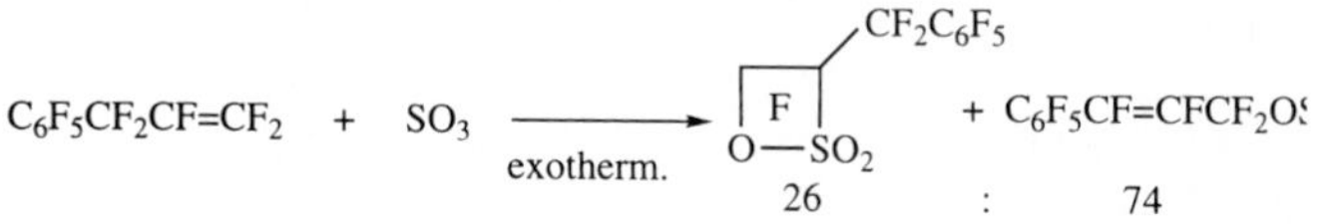

$$CCl_2{=}CClCF_3 + SO_3 \xrightarrow[100^{\circ}C,\ 20\ h]{} \underset{\mathbf{87}}{CCl_2{=}CClCF_2OSO_2F}$$
(continued 176)
$$CF_3CCl{=}CClCF_3 + SO_3 \xrightarrow[100^{\circ}C,\ 15\ h]{} \underset{\mathbf{88},\ 37\%}{CF_3CCl{=}CClCF_2OSO_2F}$$

A real breakthrough in this area came from the discovery that the direction of reaction could be totally changed by the use of a Lewis acid catalyst [5]. For instance, the reaction of a 30% excess of HFP with SO_3 in the presence of $B(OMe)_3$ produces *F*-allyl fluorosulfate **2** instead of β-sultone:

$$CF_2{=}CFCF_3 + 0.7SO_3 \xrightarrow[35\ ^{\circ}C,\ 6h]{B(OCH_3)_3} \underset{\mathbf{2},\ 52\%}{CF_2{=}CFCF_2OSO_2F}$$

$$CF_2{=}C(CF_3)_2 + SO_3 \xrightarrow[25\ ^{\circ}C,\ 16\ h]{BF_3} \underset{71\%}{CF_2{=}C(CF_3)CF_2OSO_2F} \qquad (177)$$

PFIB reacts in a similar manner [177]. This reaction was carried out at 25–35 °C in a closed system. Terminal *F*-olefins were converted into the corresponding fluorosulfates in the presence of B_2O_3 or $B(OMe)_3$ catalyst at higher temperature [175, 176]:

$$CF_3(CF_2)_nCF_2\,CF{=}CF_2 + SO_3 \xrightarrow[100\ ^{\circ}C]{} CF_3(CF_2)_nCF_2\,CF{=}CFCF_2OSO_2F \qquad (178)$$

n	cat.	time	yield%
2	B_2O_3	3h	37-42
3	$B(OCH_3)_3$	4d	29

Internal polyfluoroalkenes are less reactive toward SO_3. Indeed, fluorosulfate was obtained in low yield in reaction of *F*-2-butene and SO_3 and BF_3 at 150 °C [175]. However, much better results were achieved when a stronger Lewis acid was employed. Interaction of internal *F*-olefins with SO_3 stabilized by antimony pentafluoride proceeds at 75–80 °C resulting in the corresponding fluorosulfates **89a–e** [178, 179]:

$$R_fCF{=}CFCF_3 + SO_3 \xrightarrow[75\text{-}80\ ^{\circ}C,\ 2\ h]{SbF_5} R_fCF{=}CFCF_2OSO_2F \qquad (179)$$

R_f =	
C_2F_5	**89a**, 87%
n-C_3F_7	**89b**, 93%
i-C_3F_7	**89c**, 93%
t-C_4F_9	**89d**, 89%
n-C_5F_{11}	**89e**, 87%

Reaction of fluosulfates **89a–e** with an excess of SO_3 leads to formation of the corresponding acylfluorosulfates which were not isolated but converted into unsaturated acids **90** [179, 180]:

$$\mathbf{89a\text{-}e} + SO_3 \xrightarrow[100^{\circ}C,\ 4h]{B_2O_3} R_fCF{=}CFC(O)OSO_2F \xrightarrow{NaOH} R_fCF{=}CFC(O)ONa \xrightarrow{H_2SO_4} \underset{\mathbf{90},\ 70\text{-}92\%}{R_fCF{=}CFC(O)OH} \qquad (180)$$

The interaction of sulfur trioxide with polyfluorinated vinyl ethers proceeds without a catalyst at low temperature to give β-sultones. At higher temperature, sultones readily rearrange into the more stable acyl fluoride **91** or alkyl *F*-alkyl-β-ketosulfonates. **92, 93 a – b** [181]:

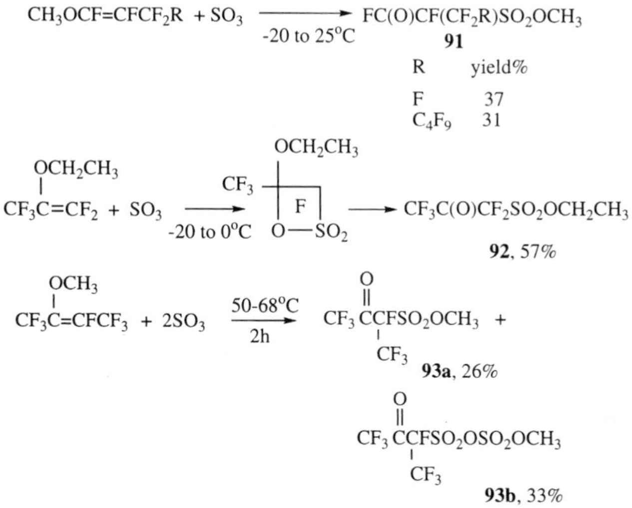

$$CH_3OCF{=}CFCF_2R + SO_3 \xrightarrow[\text{-20 to 25°C}]{} FC(O)CF(CF_2R)SO_2OCH_3$$

91

R	yield%
F	37
C_4F_9	31

92, 57%

(181)

93a, 26%

93b, 33%

Exothermic reaction of SO_3 with 2,3-dimethoxyhexafluorobutene-2 leads to a different product, cyclic sulfate **94**, which was converted into hexafluorobiacetyl by heating with H_2SO_4 [181]:

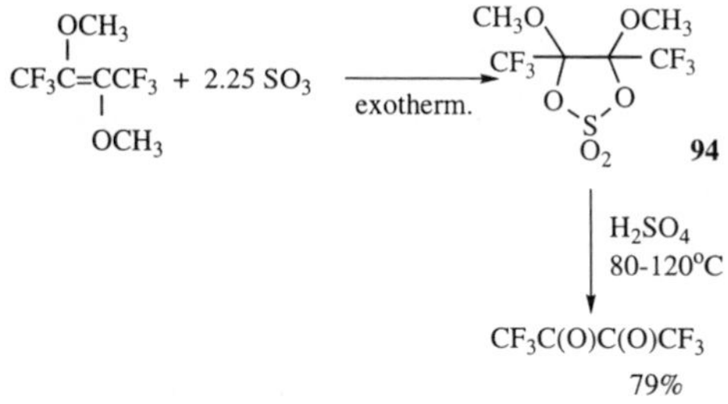

(182)

$$CF_3C(O)C(O)CF_3$$
79%

Cyclic vinyl ethers have different reactivity patterns. Reaction of cyclobutene and cyclopentene derivatives with SO_3 leads to the formation of cycloalkenones [182]:

(183)

X = F	2 d	84 %
Cl	12 h	77 %
OMe	12 h	61 %

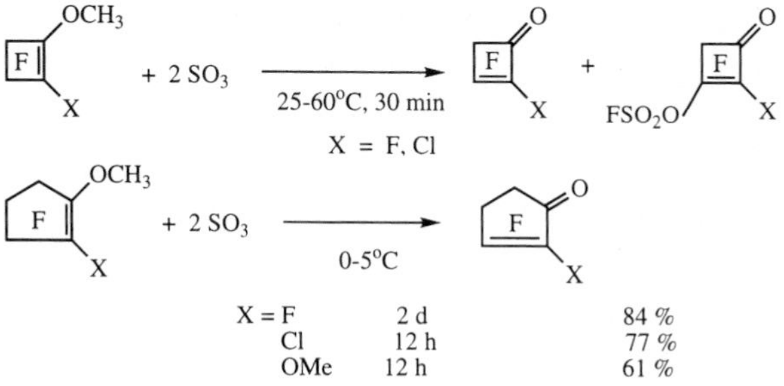

However, 1-methoxytrifluorocyclopropene behaves differently, and γ-sultone **95** is formed even at low temperature [182]:

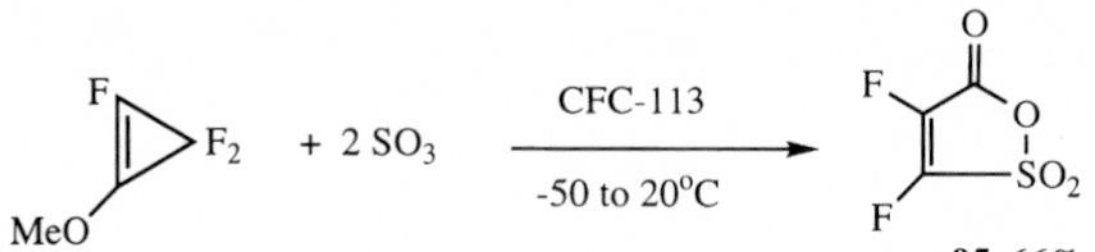

95, 66% (184)

A convenient route to difluoromaleic anhydride is based on the reaction of *F*-2,5-dihydrofuran with SO_3 [183]. Oxodefluorination proceeds at elevated. temperature and is catalyzed by trimethyl borate. *F*-2,5-Dihydrothiophene undergoes a similar conversion; when excess of SO_3 is employed, the primary thioanhydride **96** is oxidized to give **97** in high yield:

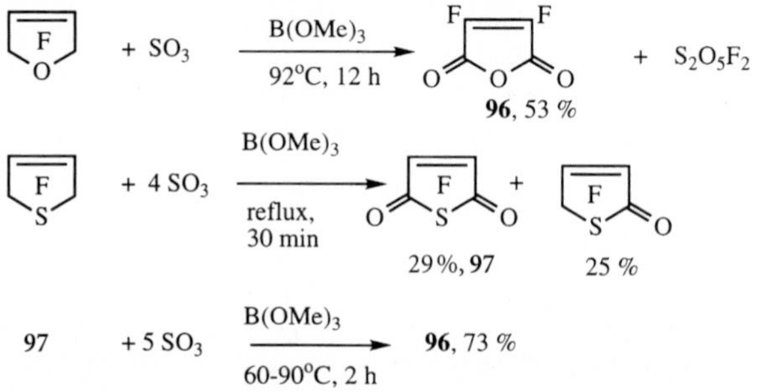

(185)

Recently, another strong Lewis acid, boron triflate $B(OSO_2CF_3)_3$ has also found application in fluoroorganic synthesis [25]. As demonstrated, this reagent can be used for the replacement of activated fluorine in fluorocarbons by OSO_2CF_3 group. Boron triflate readily reacts with CFC-112 and CFC-113 and *F*-toluene, producing corresponding triflates. Reaction of HFP with $B(OSO_2CF_3)_3$ rapidly proceeds at 0 °C to give *F*-allyl triflate, and BF_3 as a by-product:

$$B(OSO_2CF_3)_3 + 3CF_2{=}CFCF_3 \xrightarrow[0\,^{\circ}C,\ 5\ h]{} 3CF_2{=}CFCF_2OSO_2CF_3 + BF_3 \quad 68\%$$

$$B(OSO_2CF_3)_3 + CCl_2{=}CClCF_3 \xrightarrow[25\,^{\circ}C,\ 2\ h]{} CCl_2{=}CClCF_2OSO_2CF_3 \quad 63\% \qquad (186)$$

Trichloropropene reacts similarly to HFP. While internal *F*-olefins are not active enough for reaction with $B(OSO_2CF_3)_3$, more active 2,3-dichloro-*F*-butene-2 interacts with this Lewis acid at ambient temperature to give monotriflate **98** in a reasonable yield:

$$B(OSO_2CF_3)_3 + CF_3CCl{=}CClCF_3 \xrightarrow[25\,^{\circ}C,\ 4\ h]{} CF_3CCl{=}CClCF_2OSO_2CF_3 \quad \mathbf{98},\ 55\% \qquad (187)$$

7 Conclusions

The last 20 years or so have been a period of intensive development in the area of electrophilic reactions of fluoroolefins. Fluorine substituents impart unique reactivity to polyfluorinated olefins, and the knowledge of their behavior in reactions with electrophiles is important for synthetic chemists working in this area.

Advances in NMR spectroscopy have made this technique very effective for identification and structural characterization of reactive intermediates in electrophilic reactions. Nowadays it is a powerful tool which can be used for prediction and evaluation of reactivity of polyfluorinated materials.

The breakthrough in the area of inorganic chemistry results in the development of a number of powerful new electrophilic reagents (halogen fluorosulfates and triflates are examples). Application of these reagents in fluoroorganic synthesis has made it possible to increase significantly the number of substrates involved in these reactions. At the same time, the development of new catalysts such as aluminum chlorofluoride and catalytic systems (HF and solutions based on it) creates the condition for better control over reaction conditions and activity of the reagents.

Although this work probably does not cover all the details of the developments in electrophilic chemistry of fluoroolefins, it has nevertheless attempted to provide a strategic overview in this area, hoping that chemists working on related subjects should be able to find answers to basic questions associated with reaction conditions, types of electrophiles, reactivity of olefins and orientation of addition. We also wish to acknowledge the tremendous work accomplished in the last 30 years in the area of electrophilic reactions of fluoroolefins that has made this review possible.

Acknowledgements. PVA wishes to thank Dr. C. G. Krespan for stimulating discussions, Dr. A. C. Sievert for a number valuable suggestions, and Boris Vekker for help in preparation of the manuscript.

References

1. Olah GA, Prakash S, Sommer J (1985) Superacids. John Wiley & Sons, New York, p 210
2. Galakhov MV, Petrov VA, Belen'kii GG, Bakhmutov VI, German LS, Fedin EI (1986) Izv AN SSSR Ser Khim 1057
3. German LS, Knunyants IL (1969) Angew Chemie Int Ed 8:349
4. Zyk NV, Beloglazkina EK, Zefirov NS (1995) Zh Org Khim 31:1283
5. Krespan CG, England DC (1981) J Am Chem Soc 103:5598
6. Dyatkin BL, Mochalina EP, Knunyants IL (1969) Reactions of Fluoroolefins with Electrophilic Reagents. In: Tarrant P (ed) Fluorine Chemistry Reviews. Marcel Dekker Inc, New York and Basel, vol 3, p 45
7. Belen'kii GG, German LS (1984) New Reactions of Electrophilic Additions to Fluoroolefins. In: Volpin ME (ed) Soviet Scientific Rev/Sec B. Chemistry Reviews. HarwoodAcademic Publishers GmbH, New York, vol 5, p183
8. Fokin AV, Studnev Yu N (1984) Some Aspects of the Use of Peroxydisulfuryl Difluoride and Halogen Fluorosulfates in Organic Synthesis. In: ibid p 47

9. Katsuhara Y, DesMarteau DD (1980) J Am Chem Soc 45:2441
10. Bogouslavskaya LS (1989) Usp Khim 54:2024
11. Belen'kii GG (1996) J Fluorine Chem 77:105
12. Krespan CG, Petrov VA (1996) Chem Rev 96:3269
13. Hudlicky M (1976) Chemistry of Organic Fluorine Compounds, 2nd edn. Ellis Horwood, PTR Prentice Hall, New York
14. Hudlicky M, Pavlath A (1995) Chemistry of Organic Fluorine Compounds II A Critical Review ACS Monograph 187, Washington DC
15. Smart BE (1983) Fluorocarbons. In: Patai S, Rappoport Z (eds) The Chemistry of Halides, Pseudo-Halides and Azides Part 1. John Wiley & Sons, New York, p 603
16. Smart BE (1994) Characteristics of C-F Systems. In: Banks RE, Smart BE, Tatlow JC (eds) Organofluorine Chemistry. Principals and Commercial Applications. Plenum Press, New York, p 57.
17. Chambers RD (1973) Fluorine in Organic Chemistry. Durham, England. p64 – 92, 171 – 173
18. Bogouslavskaya LS Chuvatkin NN (1989) Halogen Fluorides in Organic Synthesis.In: German LS, Zemskov S (eds) New Fluorinating Agents for Organic Synthesis. Springer, New York, p 140
19. Furin GG, Bardin VV (1989) Higher Fluorides of Group V and VI Elements as Fluorinating Agents in Organic Synthesis. In: ibid, p 117
20. Miller WT, Freednan MB, Fried JH, Kock HF (1961) J Am Chem Soc 83:4105
21. Olah GA, Mo YK (1972) Fluorinated Carbocations. In: Tatlow JC, Peacock RD, Hynman HH, Stacey M (eds) Advances in Fluorine Chemistry. Butterworth, London, vol 7, p 69
22. Krespan CG, Dixon DA (1996) J Fluorine Chem 77:117
23. Petrov VA, Krespan CG, Smart BE (1996) ibid 77:139
24. Yakobson GG, Furin GG (1980) Synthesis 5:345
25. Petrov VA (1995) J Fluorine Chem 73:17
26. Miller WT (1940) J Am Chem Soc 62:993
27. Paleta O (1977) in Fluorine Chem Rev Tarrant P (Ed), Marcel Dekker, New York 8:39
28. Olah GA, Prakash S, Sommer J (1985) Superacids. John Wiley & Sons, New York, p 27
29. Bardin VV, Avramenko AA, Furin GG, Krasil'nikov VA, Karelin AI, Tushin PP, Petrov VA (1990) J Fluorine Chem 49:385
30. Olah GA, Prakash S, Sommer J (1985) Superacids. John Wiley & Sons, New York, pp 126 – 229
31. Gillespie RJ, Malhotra KC (1969) Inorg Chem 8:1751
32. Chambers RD, Clark DT, Holmes TF, Masgrave WK (1974) J Chem Soc Perkin 1 114
33. Avramenko AA, Bardin VV, Furin GG, Krasil'nikov VA, Karelin AI, Tushin PP (1988) Zh Org Khim 24:1443
34. Zhang D, Wang E, Mistry F, Powell B, Aubke F (1996) J Fluorine Chem 76:83
35. DesMarteau DD (1978) J Am Chem Soc 100:340
36. Katzuhara Y, Hammaker RM, DesMarteau DD (1980) Inorg Chem 19:607
37. Gould DE, Anderson LR, Young DE, Fox WB (1969) J Am Chem Soc 91:1310
38. Young DE, Anderson LR, Fox WB (1970) Inorg Chem 9:2602
39. Anderson JD, DesMarteau DD (1996) J Fluorine Chem 77:147
40. Smart BE, Reddy GS (1976) J Am Chem Soc 98:5593
41. Olah GA, Kuhn SJ (1967) Organic Synthesis 47:56
42. Davis TK, Moss RC (1970) J Chem Soc 1054
43. Wolf AA, Greenwood (1950) ibid 2200
44. Kopaevich YuL, Belen'keii GG, Mysov EI, German LS (1972) Zh Vses Khim O-va 17:236
45. Krespan CG (1992) US Pat 5,101,058; (1991) Chem Abstr 115:182630
46. Zeifman YuV, Ter-Garbielyan EG, Gambarayn NP, Knunayants IL (1984) Usp Khim 53: 256
47. Paddon-Dow MN, Santiago C Houk KN (1980) J Am Chem Soc 102:6561
48. Petrov VA Belen'kii GG, German LS (1980) Izv AN SSSR Ser Khim 2117
49. Petrov VA (1995) J Org Chem 60:3419
50. Petrov VA Krespan CG (1996) J Org Chem 61:9605
51. Lo ES, Readio D, Iserson H (1970) J Org Chem 35:2051

52. Kurykin MA, Krotovich IN, Studnev YuN, German LS, Fokin AV(1982) Izv AN SSSR Ser Khim 1861
53. Banks RE, Haszeldine RN, Taylor DR (1965) J Chem Soc 987
54. Wilson RD, Maya V, Philipowitch D, Christe KO (1983) Inorg Chem 22:1355
55. Chepick SD, Petrov VA, Galakhov MV, Belen'kii GG, Mysov EI, German LS (1990) Izv AN SSSR Ser Khim 1861
56. Olah GA, Chambers RD, Komisarov MB (1967) J Am Chem Soc 89:1268
57. Olah GA, Mo YK (1972) J Org Chem 37:1028
58. Olah GA, Heiliger L, Prakash GKS (1989) J Am Chem Soc 111:8020
59. Olah GA, Mo YK (1976) Halogenated Carbocations.In: Olah GA, Schleyer (eds) Carbonium Ions. Wiley Interscience, New York , 5:2135
60. Sargeant PB, Krespan CG (1969) J Am Chem Soc 91:415
61. Pozdnaykovich YuV, Shteingarts FD (1970) Zh Org Khim 6:1753
62. Chambers RD, Porkin A, Matthews RS (1976) J Chem Soc Perkin Trans I 2107
63. Bakhmutov VI, Galakhov MV (1988) Usp Khim 57:1467
64. Knunayats IL, Rokhlin EM, Abduganiev YG, Okulevich PO, Karpushina NI (1973) Tetrahedron 29:595
65. Galakhov MV, Petrov VA, Belen'kii GG, German LS, Fedin EI, Snegirev VF, Bakhmutov VI (1986) Izv AN SSSR Ser Khim 1063
66. Snegirev VF, Galakhov MV, Petrov VA, Makarov KN, Bakhmutov VI (1986) Izv AN SSSR Ser Khim 1318
67. Ramsey BG, Taft RW (1966) J Am Chem Soc 88:3058
68. Snegirev VF, Galakhov MV, Makarov KN, Bakhmutov VI (1985) Izv AN SSSR Ser Khim 2302
69. Galakhov MV, Petrov VA, Chepick SD, Belen'kii GG, Bakhmutov VI, German LS (1989) Izv AN SSSR Ser Khim 1773
70. Karpov VM, Mezhenkova TV, Platonov VE, Yakobson GG (1985) J Fluorine Chem 28:121
71. Karpov VM, Platonov VE (1994) Zh Org Khim 30:789.
72. West R, Kwitowskii PT (1966) J Am Chem Soc 88:5280
73. Petrov VA, Belen'kii GG, German LS (1984) Izv AN SSSR Ser Khim 438
74. Petrov VA, Belen'kii GG, Galakhov MV, Bakhmutov VI, German LS, Fedin EI (1984) Izv AN SSSR Ser Khim 2811
75. Petrov VA, Belen'kii GG, Galakhov MV, Bakhmutov VI, Kvasov BA, German LS, Fedin EI (1985) Izv AN SSSR Ser Khim 306
76. Chambers RD, Sailsbury MJ, Apsey GS, Moggi GJ (1988) J Chem Soc Chem Comm 680
77. Elsheimer SR (1985) Halogenation. In: ref 14, 364–387
78. Naae DJ (1980) J Org Chem 45:1394
79. Rondarev DS, Sass VP, Sokolov SV (1975) Zh Org Khim 11:937
80. Chambers RD, Musgrave WKR, Savory J (1961) J Chem Soc 3779
81. Probst A, Raab K, Ulm K, von Werner K (1988) J Fluorine Chem 37:223
82. Fokin AV, Studnev YuN, Rapkin AI, Krotovich IN, Tamarinov AS, Verinikin OV (1985) Izv AN SSSR Ser Khim 2298
83. Chepik SD, Belen'kii GG, Petrov VA, German LS (1993) J Fluorine Chem 65:223
84. BastockTW, Harley ME, Pedler AE, Tatlow JC (1975) J Fluorine Chem 6:331
85. Bardin VV, Furin GG, Yakobson GG (1981) Zh Org Khim 17:999
86. Bardin VV, Furin GG, Yakobson GG (1982) J Fluorine Chem 23:67
87. Karpov VM, Mezhenkova TV, Platonov VE, Yakobson GG (1986) Bull Soc Chim France 980
88. Karpov VM, Mezhenkova TV, Platonov VE (1996) J Fluorine Chem 77:133
89. Rao VNM (1994) Alternatives to Chlorofluorocarbons (CFC's). In ref 15, 159–175
90. Polishchuk VR, Leites LA, Mysov EI, Aerov AF, Kagramanova EM (1992) Izv AN SSSR Ser Khim 450
91. Polishchuk VR, Mysov EI, Stankevich IV, Chistyakov AL, Potechin KA, Struchkov YuT (1993) J Fluorine Chem 65:233
92. Mukhametshin FM (1980) Usp Khim 49:1260
93. Walker N, DesMarteau DD (1975) J Am Chem Soc 97:13

94. Hopkinson MJ, Walker N, DesMarteau DD (1976) J Org Chem 41:1407
95. Maya W, Shack CJ, Wilson RD, Murhead JS (1969) Tetrahedron Lett 3247
96. Anderson LR, Young DE, Gould DE, Juurik-Hogan R, Nuechterlein D, Fox WB (1970) J Org Chem 35:3730
97. Moldavskii DD, Temchenko VG, Slesareva VI, Antipenko GL (1973) Zh Org Khim 9:673
98. Johri KK, DesMarteau DD (1983) J Org Chem 48:242
99. Shack CJ, Criste KO (1978) J Fluorine Chem 12:325
100. Tari I, DesMarteau DD, (1983) J Org Chem 45:1214
101. Fokin AV, Studnev YuN, Rapkin AI, Pasevina KI, Potarina TM, Verinikin OV (1980) Izv AN SSSR Ser Khim 2369
102. Ref 27 pp73–118
103. Earl BL, Hill BK, Shreeve JM (1966) Inorg Chem 5:2184
104. Fokin AV, Studnev YuN, Kuznetsova LD, Krotovich IN(1978) Izv AN SSSR Ser Khim 649
105. Fokin AV, Studnev YuN, Kuznetsova LD, Rud VL (1974) ibid 471
106. Fokin AV, Studnev YuN, Rapkin AI, Chilikin VG, Verinikin OV (1983) ibid 659
107. Fokin AV, Studnev YuN, Rapkin AI, Potarina TM, Verinikin OV (1981) ibid 2376
108. German LS, Savicheva GI (1984) ibid 478
109. German LS, private communication
110. Olah GA, Wong Q, Sanford G, Prakash GKS (1993) J Org Chem 58:3194
111. Yamabe M, Munekata S, Samejima S (1982) U.S. 4,362,672; (1982) Chem Abstr 97:55318
112. Knunayants IL, Pervova EYa, Tuleneva VV (1956) Izv AN SSSR Ser Khim 843
113. Coudures C, Pastor R, Cambon A (1984) J Fluorine Chem 24:93
114. Karpov VM, Mezhenkova TV, Platonov VE (1974) Zh Org Khim 10:663
115. Chuikov IP, Karpov VM, Platonov VE, Beregovaya IV, Shegoleva LN (1993) J Fluorine Chem 65:29
116. Fokin AV, Studnev YuN, Rapkin AI, Chilikin VG, Verinikin OV (1983) Izv AN SSSR Ser Khim 1437
117. Fokin AV, Studnev YuN, Rapkin AI, Chilikin VG (1984) ibid 473
118. Fokin AV, Rapkin AI, Krylov II, Kutepov AP, Studnev YuN (1986) ibid 2364
119. Fokin AV, Studnev YuN, Rapkin AI, Chilikin VG (1982) ibid 2838
120. Fokin AV, Rapkin AI, Chilikin VG, Verinikin OV, Studnev YuN, (1986) ibid 891
121. Savostin VS, Krylov II, Kutepov AP, Rapkin AI, Fokin AV (1990) ibid 1891
122. Kopaevich Yu L, Belen'kii GG, German LS, Knunayants IL (1971) ibid 1224
123. Kopaevich Yu L, Belen'kii GG, German LS, Knunayants IL (1972) ibid 213
124. Belen'kii GG, Kopaevich Yu L, German LS, Knunayants IL (1972) ibid 973
125. Kolenko IP, Filayakova TI, Zapevalov AYa, Mochalina EP, German LS, Polischuck VR (1979) ibid 667
126. Chuikov IP, Karpov VM, Platonov VE (1992) ibid 1412
127. Frohn HJ, Bardin VV (1996) Z Naturforsch 51B:1011
128. Frohn HJ, Bardin VV (1996) Z Naturforsch 51B:1015
129. Belen'kii GG, Kopaevich Yu L, German LS, Knunayants IL (1972) Dokl AN SSSR 201:603
130. Desjodins CD, Passmore J (1977) Can J Chem 55:3136
131. Kopaevich Yu L, Belen'kii GG, Mysov EI, German LS, Knunayants IL (1972) Zh Vses Khim O-va 17:226
132. Furin GG, Shchegoleva LN, Yakobson GG (1975) Zh Org Khim 11:1290
133. Belen'kii GG, German LS, Knunayants IL, Furin GG, Yakobson GG (1976) Zh Org Khim 12:1183
134. Freeman F (1975) Chem Rev 75:439
135. Fokin AV, Kolomiets AF (1982) Izv AN SSSR Ser Khim 1820
136. Fokin AV, Kolomiets AF, Shkurak SN, Mukhametshin FM (1985) ibid 1835
137. Sizov AYu, Kolomiets AF, Fokin AV (1992) Usp. Khim 61:940
138. Il'in GF, Shkurak SN, Kolomiets AF, Sokol'skii GA (1983) Zh Vses Khim O-va 28:235
139. Dolbier WR (1996) Chem Rev 96:1557
140. Shkurak SN, Kolomiets AF, Fokin AV (1982) Izv AN SSSR Ser Khim 959
141. Shkurak SN, Ezhov VV, Kolomiets AF, Fokin AV (1984) ibid 1371

142. Shkurak SN, Kolomiets AF, Fokin AV (1981) Zh Org Khim 18:1549
143. Burmakov AI, Kunshenko BV, Alexeeva LA, Yagupol'skii LM. New Uses of sulfur Tetrafluoride in Organic Synthesis. In: ref 10,197–253
144. Muratov MN, Mokhamed M, Kunshenko BV, Alexeeva LA, Yagupol'skii LM (1986) Zh Org Khim 22:964
145. Fokin AV, Rapkin AI, Matveenko VI, Verinikin OV (1988) Izv AN SSSR Ser Khim 2367
146. Sievert AC, Weigert FG, Krespan CG (1992) US Pat 5,157,171; (1991) Chem Abstr 115: 70904
147. Sievert AC, Nappa MJ PCT WO 95/16655; (1995) Chem Abstr 123:339129
148. Sievert AC, Nappa MJ PCT WO 95/16656; (1995) Chem Abstr 123:339128
149. Belen'kii GG, Savicheva GI, Lur'e EP, German LS, Knunayants IL (1978) Izv AN SSSR Ser Khim 1430
150. Belen'kii GG, Lur'e EP, German LS (1975) ibid 2728
151. Petrov VA, Belen'kii GG, Mysov EI, German LS (1981) ibid 2098
152. Belen'kii GG, Lur'e EP, German LS (1976) ibid 2365
153. Petrov VA, Belen'kii GG, Kurbakova AP, Leites LA, German LS (1982) ibid 170
154. Petrov VA, Belen'kii GG, German LS (1982) ibid 1591
155. Chepik SD, Belen'kii GG, German LS (1991) ibid 1926
156. Krespan CG (1992) US Pat 5,162,594; (1992) Chem Abstr 117:69439
157. Knunyants IL, Yakobson GG (eds) (1977) Sintesy Ftororganitcheskikh Soedinenii. Monomery i Promezutochnye Producty (Russ) Khimiya Moscow, p 141
158. Karpov VM, Mezhenkova TV, Platonov VE, Yakobson GG (1985) Izv AN SSSR Ser Khim 2315
159. Petrov VA, Belen'kii GG, German LS (1990) ibid 920
160. Petrov VA, Belen'kii GG, German LS (1989) ibid 385
161. Belen'kii GG, German LS (1974) ibid 942
162. Chepik SD, Belen'kii GG, Cherstkov VF, Sterlin SR, German LS (1991) ibid 513
163. Belen'kii GG, Savicheva GI, Lur'e EP, German LS (1978) Izv AN SSSR Ser Khim 1640
164. Filyakova TI, Zapevalov AYa, Kodess MI, Kurykin MA, German LS (1994) ibid 1614
165. Petrov VA, Belen'kii GG, German LS (1982) ibid 2411
166. Filyakova TI, Belen'kii GG, Lur'e EP, Zapevalov AYu, Kolenko IP, German LS (1991) ibid 513
167. Avramenko AA, Bardin VV, Furin GG, Karelin AI, Tushin PP, Krasilnikov VA (1988) Zh. Org. Khim 24:1443
168. Shteigatrs VD (1992) Synthetic Aspects of Electrophilic ipso Reactions of Polyfluoroarenes. In: Olah GA, Chambers RD, Prakash GKS (eds) Synthetic Fluorine Chemistry, John Wiley & Sons New York, 259
169. Frohn HJ, Bardin VV (1995) Mendeleev Commun 114
170. Sherer O, Hoerlein J, Millauer H (1966) Chem Ber 99:1966
171. England DC (1984) J Org Chem 49:4007
172. Snegirev VF, Gervits LL, Makarov KN (1983) Izv AN SSSR Ser Khim 2765
173. Sokol'skii GA, Knunyants IL (1972) Angew Chem Int Ed Engl 11:583
174. Smart BE (1976) J Org Chem 41:2353
175. Krespan CG, Dixon DA (1986) J Org Chem 51:4460
176. Cherstkov VF, Sterlin SR, German LS, Knunyants IL (1982) Izv AN SSSR Ser Khim 2791
177. Cherstkov VF, Sterlin SR, German LS, Knunyants IL (1984) ibid 2152
178. Cherstkov VF, Sterlin SR, German LS, Knunyants IL (1982) ibid 1917
179. Cherstkov VF, Sterlin SR, German LS, Knunyants IL (1985) ibid 1864
180. Cherstkov VF, Sterlin SR, German LS (1992) ibid 2341
181. Krespan CG, Smart BE, Howard EG (1977) J Am Chem Soc 99:1214
182. Smart BE, Krespan CG (1977) ibid 99:1218
183. Krespan CG (1992) J Fluorine Chem 48:339

Fluorinated Free Radicals

William R. Dolbier, Jr.

Department of Chemistry, University of Florida, Gainesville, FL 32611-7200, USA
E-mail: wrd@server.chem.ufl.edu

All aspects of the structure, reactivity and chemistry of fluorine-containing, carbon-based free radicals in solution are presented. The influence of fluorine substituents on the structure, the stability and the electronegativity of free radicals is discussed. The methods of generation of fluorinated radicals are summarized. A critical analysis of the reactivities of perfluoro-*n*-alkyl, branched chain perfluoroalkyl and partially-fluorinated free radicals towards alkene addition, H-atom abstraction, and towards intramolecular rearrangement reactions is presented. Lastly, a summary of the synthetically-useful chemistry of fluorinated radicals is presented.

Keywords: Radicals, fluorinated radicals, radical structure, radical reactivity, radical rearrangements

Topics in Current Chemistry, Vol. 192

1 Introduction

Free radical reactions comprise an important part of the chemistry of organofluorine compounds. The purpose of this review will be to present a current perspective of what is known about the structure, reactivity and chemistry of carbon-based, fluorine-containing free radicals in solution, with an emphasis, in the case of the chemistry, being given to the progress made in the field during the last decade.

In contrast to closed shell molecules, free radicals are species which have an odd number of electrons. Simply speaking, all electrons in free radical species are considered to be paired up, except for one orbital which contains the single electron. The molecular orbital which describes the distribution of this odd electron is called the SOMO (singly occupied MO). In the ground state of the radical, the SOMO is also the HOMO. In a carbon-based free radical the SOMO is generally strongly localized to a trigonal carbon atom.

Free radicals were for a long time believed to be too reactive and indiscriminate in their reactivity to be harnessed usefully for synthesis. However, because of increased understanding of the nature of their reactivity, synthetic chemists have learned to tame these highly reactive intermediates to such an extent that reactions involving free radical intermediates are now considered to be quite useful in synthesis, particularly with respect to carbon-carbon bond forming and carbocyclic ring forming processes.

Fluorinated radicals have played a significant role in the history and development of the field of free radical chemistry, and it was recognized quite early that they have natures which are quite different from those of their hydrocarbon counterparts. As a result, there has been much effort directed towards defining and understanding these differences with respect to their structure, reactivity and chemistry.

1.1 Influence of Fluorine as a Substituent

Substituents give rise to a perturbation of any "standard" system, whether it be a reactive intermediate, such as a radical, or a valence-satisfied molecule, and to a first approximation, the character of a substituent is considered to remain basically unaltered from one molecular environment to another. Substituent effects can be broadly divided into steric effects and polar (or electronic) effects, with electronic effects being further divided into σ inductive effects and π conjugative (resonance) effects. Although it is not a rigid rule, because of the small size of a fluorine substituent and the relatively non-sterically-demanding nature of the transition states for most types of radical reactions, the influence of fluorine substituents upon structure and reactivity of radicals is usually considered to derive largely from the electronic nature of fluorine [1]. Fluorine is the most electronegative atom, and it thus exhibits a potent σ inductive electron withdrawing effect in all situations. It is also a potentially strong π electron donor to carbon π-systems, including the semi-occupied molecular orbital

(SOMO) of a carbon radical, because of the good match up in size of the lone pair 2*p* orbitals of fluorine with those of carbon. The effectiveness of this conjugative interaction is a function of the energetic separation of the interacting orbitals, as well as of the degree of their overlap, both of which are significantly influenced by the strong inductive withdrawing nature of the fluorine substituent.

Thus, the net impact of a substituent such as fluorine, which is inductively withdrawing and π-donating, will result from a complex interplay of these disparate interactions. To make matters even more complicated, the combined influences of *multiple* fluorine substituents is not additive, and cannot be readily derived from an understanding of the effect of a single fluorine substituent.

2 Structure

Fluorine substituents have a dramatic impact upon the structure of alkyl radicals. The methyl radical itself is planar; UV, IR, PES and ESR spectroscopy, as well as the highest level of theoretical analysis, all indicate that its conformational properties are best defined as deriving from a single minimum [2]. Fluoromethyl radicals, on the other hand, are increasingly pyramidal [3], with the trifluoromethyl radical being essentially tetrahedral [3–7], with a significant barrier to inversion [8, 9].

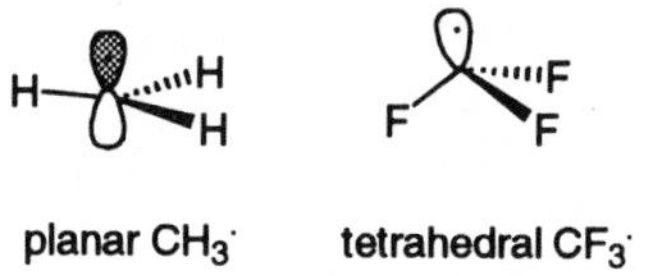

Scheme 1

ESR spectroscopy is perhaps the best method for the unequivocal detection and observation of free radicals, and ESR ^{13}C hyperfine splitting (hfs) constants are considered to be a very useful indicator of a radical's geometry because non-planarity introduces *s* character into the orbital that contains the unpaired electron. The methyl radical's $^{13}C_\alpha$ value of 38 G is consistent with a planar structure. Fluoromethyl radicals exhibit increased $^{13}C_\alpha$ values, as shown in Table 1, thus indicating increasing non-planarity, with trifluoromethyl radical's value of 272 G lying close to that expected for its sp^3 hybridization [4].

As can also be seen from Table 1, the α-F hfs interactions exhibited by CH_2F, CHF_2 and CF_3 are also consistent with their increasingly pyramidal nature, and Table 2 provides data for other α-fluorinated radicals which indicate their degree of bending. Direct fluorine substitution at the radical site also gives rise to large increases in the radical's barrier to inversion, with barriers of ~1,7 and 25 kcal/mol being calculated for CH_2F, CHF_2, and CF_3, respectively [9].

Such a strong influence of fluorine substituents on the geometry of a radical can be understood largely in terms of the effect of the σ inductive influence of

Table 1. ESR hyperfine splitting constants for fluorinated methyl radicals [4]

Radical	CH_3	CH_2F	CHF_2	CF_3
$a\,^{13}C$	38.5	54.8	148.8	272
$a\,^{19}F_\alpha$	–	64.3	84.2	143.7
$a\,H_\alpha$	– 23.0	– 21.1	22.2	–
g	2.0026	2.0045	2.0041	2.0026

Table 2. ESR hyperfine splitting constants for 1°, 2° and 3° alkyl and fluorinated alkyl radicals [4, 10–13]

Radical	CH_3CH_2	$(CH_3)_2CH$	$(CH_3)_3C$	CH_3CF_2	CH_3CF_2	$(CF_3)_2CF$	$(CF_3)_3C$
$a\,^{13}C_\alpha$	39.1	41.3	49.5	–	–	–	44.3
$a\,^{19}F_\alpha$	–	–	–	94.0	87.6	70.3	–

the fluorine substituent on the thermodynamics of bonding. There is a thermodynamic advantage for the carbon orbitals used in bonding to fluorine to be relatively high in *p*-character, as they would be in an increasingly bent radical. In such a case, the orbital containing the unpaired electron in a fluoromethyl radical would have increasing *s*-character as the number of fluorines is increased. It has also been suggested that conjugative effects can contribute significantly to pyramidalization, but Bernardi and coworkers concede that in the case of fluorine substituents, the inductive effect is primarily responsible for the observed conformational trends [9].

From a simple MO perturbational perspective, pyramidalization of a radical $\cdot CH_{(3-n)}X_n$ occurs when it can lead to mixing of the SOMO with the lowest occupied σ MO (LUMO). (This would lead to charge transfer, and increased ionic character to the C-X bonds.) The more electronegative the substituent X is, the lower the LUMO energy, hence the lower the SOMO-LUMO gap, which results in more mixing. Being the most electronegative, fluorine substituents have the strongest influence on non-planarity [9].

Methyl substituents also induce some bending, with ethyl, isopropyl and *tert*-butyl radicals becoming increasingly pyramidal [2], but in contrast to the influence of fluoro-substituents, such radicals have shallow potential energy functions with very small (<1 kcal/mol) barriers to inversion [14]. (Table 2 provides the hyperfine splitting constant data for 1°, 2°, and 3° alkyl and fluoroalkyl radicals.) Placing fluoro-substituents at the β-position, as in the 2-fluoroethyl radical, gives rise to conformational preferences which appear to be of minor structural consequence at the radical site [15–17]. Even the strongly electronegative trifluoromethyl substituent would appear to induce less pyramidalization than a methyl substituent. For example, the smaller $^{13}C_\alpha$ hfs constant exhibited by the perfluoro-*tert*-butyl radical, combined with its normal temperature dependence properties, indicate a more planar geometry than that of *tert*-butyl radical [10, 14].

With the demonstrated strong inducement by fluorine substituents to pyramidalize a radical site, the question arises whether benzylic or allylic conjugation would be sufficient to make the radical site planar. Indeed, that seems to be the case, but as one might expect, such radicals appear to have less resonance stabilization than their hydrocarbon analogues.

In an ESR study of 1,1,3,3-difluoroallyl radicals, Krusic and coworkers were able to demonstrate that the barrier to rotation of such apparently planar radicals is substantially reduced [18]. Although allyl itself has a rotational barrier of 15 kcal/mol [19, 20], 1,1,3,3-tetrafluoroallyl, **1**, had a barrier of but 7.2 kcal/mol. The observed $^{19}F_{\alpha}$ hfs constants (42.6 and 39.7 G) were consistent with **1** being a planar system. It is likely that the lowering of the rotational barrier of **1** derives from a destabilizing interaction between the fluorine lone pairs and the doubly-occupied allyl π-MO which diminishes the net allylic resonance energy, as well as from stabilization of the transition state due to pyramidalization.

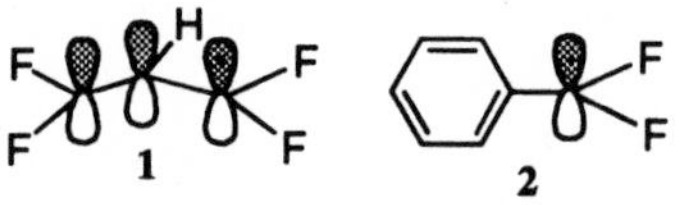

Scheme 2

Likewise, Pittman has examined the α,α-difluorobenzylradical (**2**) by ESR [21]. This radical also exhibited small $^{19}F_{\alpha}$ hfs (51.4 G) which is consistent with a planar or near planar radical. Because of the symmetry of the system, he could obtain no information on the radical's rotational barrier. Yoshida et al. have recently calculated a 20° distortion from planarity for this radical and have rationalized its relatively high reactivity on the basis of its non-planar nature [22].

To summarize the considerable available structural data with respect to fluorine substitution, one can conclude that non-conjugated carbon radicals bearing at least two fluorine substituents will be strongly pyramidal, σ-radicals, while β-fluorine substituents appear to have little influence on the geometry of a radical. The strong σ-character of CF_3, CHF_2, and perfluoro-*n*-alkyl radicals has a considerable influence on their *reactivity*.

3 Thermochemical Properties of Fluorinated Radicals

3.1 Radical Stabilities

The influence of fluorine substituents on the stability of alkyl radicals derives from the same complex interplay of inductive and resonance effects that affects their structure. Simple orbital interaction theory predicts that substituents of the –X: type (that is, electronegative substituents bearing lone pairs) should destabilize inductively by virtue of their group electronegativities, and stabilize by resonance to the extent of their ability to delocalize the odd electron.

The effect of fluorine substitution on the thermodynamic stability of a radical has been difficult to assess experimentally. Homolytic C-H BDEs of hydrocarbons bearing heteroatom substituents have long been considered to provide good estimates of the stabilities of the corresponding alkyl radicals [23]. For example, Bordwell and Pasto have devoted considerable attention to the prediction of BDEs of mono- and disubstituted methanes, and they have also presented cogent interpretations of the validity and significance of the derived radical stabilization energies (RSEs), which are defined as the change in the total energy for the isodesmic reaction shown in Eq. (1), and as such are recognized not to be identical in definition to "resonance" energies [24, 25].

$$X_nCH_{3-n}^{\bullet} + CH_4 \longrightarrow X_nCH_{4-n} + CH_3^{\bullet} \qquad (1)$$

Thus, on the basis of relative C-H BDEs, $(CH_3)_3C$-H (96.4), $(CH_3)_2CH$-H (98.6), CH_3CH_2-H (101.1) and CH_3-H (104.8 kcal/mol) [23, 26], one reaches the conclusion that the stability of hydrocarbon radicals decreases: 3° > 2° > 1° > CH_3. It is, of course, recognized that the variable degree of steric strain of the molecules within this series limits the quantitative impact of these numbers vis-a-vis radical stability.

Using the experimental bond dissociation energies for the fluorinated methanes [BDEs: CH_3-H (104.8 ± 0.2), CH_2F-H (101.2 ± 2), CHF_2-H (103.2 ± 2), CF_3-H (106.7 ± 1 kcal/mol)] [26–29] indicates that, by equation iii, the impact of fluorine substitution on RSEs is positive for the first fluorine but then increasingly negative along the series. Whereas a single fluorine substituent is stabilizing by 3.6 kcal/mol, two stabilize by a mere 1.6 (which can be compared to the 2.6 kcal/mol stabilization calculated for the $CH_3CF_2^{\bullet}$ radical, compared to $CH_3CH_2^{\bullet}$), and three destabilize by 1.9 kcal/mol. This trend has been explained by Epiotis and Bordwell as deriving from the fact that substituents of the –X: type (that is electronegative substituents bearing lone pairs) should destabilize a radical inductively and stabilize the radical to the extent of their ability to delocalize the odd electron [9, 24]. Because the energy of its lone pairs (compared to those on nitrogen or oxygen) is not conducive to their interaction with the SOMO and with such delocalization further diminishing with increasing pyramidalization, such interaction actually becomes destabilizing for $CF_3^{\bullet}$, which leads to the considerable negative RSE exhibited by $CF_3^{\bullet}$ and hence to the very high CF_3-H BDE. Indeed, because of the strong impact of fluorine substitution on the ground state energies of polyfluoromethanes, one must question the use of RSE values derived from Eq. (1) for evaluating the degree of "radical stabilization" in the fluoromethyl radical series.

On the basis of meager available BDE data for fluorinated ethanes, β-fluorine substitution would appear to give rise to "radical destabilization" relative to the ethyl radical, as defined by an isodesmic equation analogous to Eq. (1). Indeed, calculations of some missing members of the fluorinated ethyl series {BDEs: CH_3CH_2-H, 97.7 (101.1, exptl); CH_2FCH_2-H, 99.6; CHF_2CH_2-H, 101.3; CF_3CH_2-H, 102.0 kcal/mol (106.7, exptl) [30, 23, 26]} indicate that the inductive effect of even a *single* β-fluorine substituent is sufficient to "destabilize" an ethyl radical.

What one is really saying however, in the ethane series, is that a β-CF_2 group serves to *strengthen* β C-H (and C-C) bonds, probably because of strong electrostatic attractions in these C-H and C-C bonds [31]. The relevance of such BDE data to intrinsic radical stabilizations is at this point highly debatable.

Experimental support for the given order of stability in the fluoromethyl series has been provided in a study by Jiang et al. of the radical fragmentation of the respective series of fluorinated *tert*-butoxy radicals [32]:

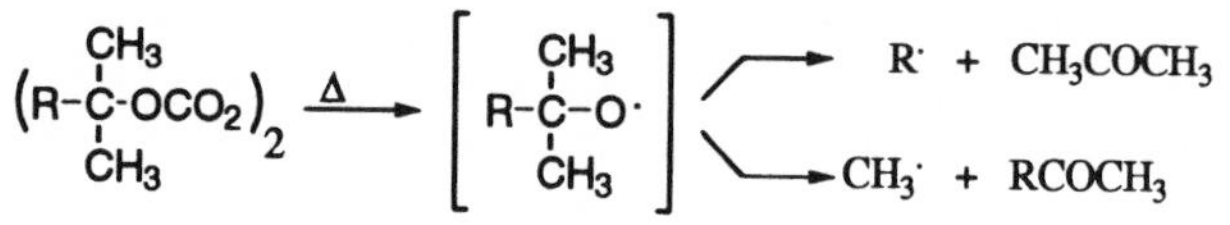

$$\left(R\text{-}\underset{\underset{CH_3}{|}}{\overset{\overset{CH_3}{|}}{C}}\text{-}OCO_2\right)_2 \xrightarrow{\Delta} \left[R\text{-}\underset{\underset{CH_3}{|}}{\overset{\overset{CH_3}{|}}{C}}\text{-}O\cdot\right] \begin{array}{c} \longrightarrow R\cdot + CH_3COCH_3 \\ \\ \longrightarrow CH_3\cdot + RCOCH_3 \end{array}$$

$$R = \quad CF_3 < CH_3 < CHF_2 \sim CH_2F$$

$$k_{rel} = \quad 0.08 : 1 \quad : 10.2 \quad : 9.0$$

Scheme 3

One would expect that any radical stabilization or destabilization effects deriving from fluorine substituents ought to be reflected in the experimental rates of thermal rearrangement via homolytic processes. However, in the few such systems which have been studied, one or two fluorine substituents do not seem to have a significant impact upon such rates [33, 34].

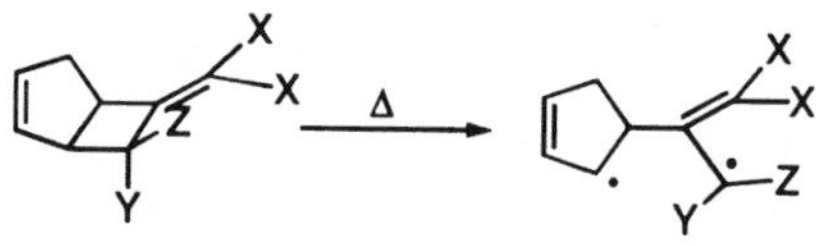

	Log A	E_a
For X=H, Y=Z =D	13.6	39.9
For X=Z=H, Y=F	13.7	39.6
For X=F, Y=Z=H	13.6	38.4

Scheme 4

In kinetic studies where there have been trifluoromethyl groups on a C-C bond undergoing homolytic cleavage [35–38], steric effects sometimes led to results which were difficult to interpret [36], and other substituents were observed to influence the effect of the CF_3 group [35, 36]. However, the trifluoromethyl substituent has generally been found to have minimal or even a negative effect on the rates of the rearrangements. A study of the vinylcyclopropane rearrangement of *trans*-2-(trifluoromethyl)-vinylcyclopropane exemplifies the lack of significant influence of a CF_3 group on C-C bond dissociation energies and hence on carbon radical stabilities [38]. In this study Dolbier and McClinton found that the C_1-C_2 and C_1-C_3 bonds were cleaved competitively, and the over-

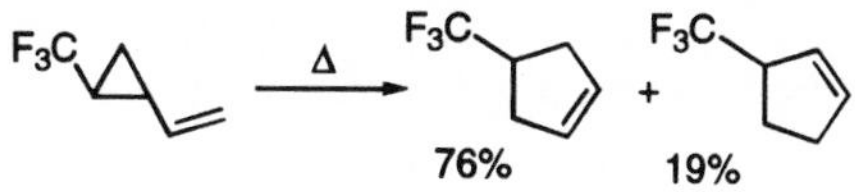

Scheme 5

all rate of rearrangement was, within experimental error, identical to that of unsubstituted vinylcyclopropane.

Within our discussion of the stabilities of fluorinated radicals we have, of course, been referring to thermodynamic stabilities. In fact, most fluorinated radicals will be seen to have enhanced *kinetic* reactivity in reactions with closed shell molecules. However, appropriate fluorine substitution can also give rise to long-lived, or persistent radicals, the most dramatic example being Scherer's radical, **3**, which persists at room temperature, even in the presence of molecular oxygen [39]:

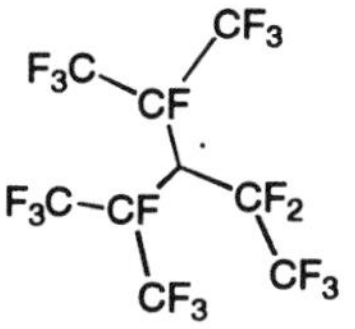

3

Scheme 6

The incredible kinetic stability of Scherer's radical most likely derives from steric isolation of the site of free valence. Models indicate that the radical site is essentially buried within a protective shield of surrounding fluorine substituents.

Recently, there have been a number of other examples of stabilized perfluoro radicals reported in the literature [40–46]. Included in this work are reports of the first isolable, functionalized radicals, and the first isolable perfluoro*vinyl* radical.

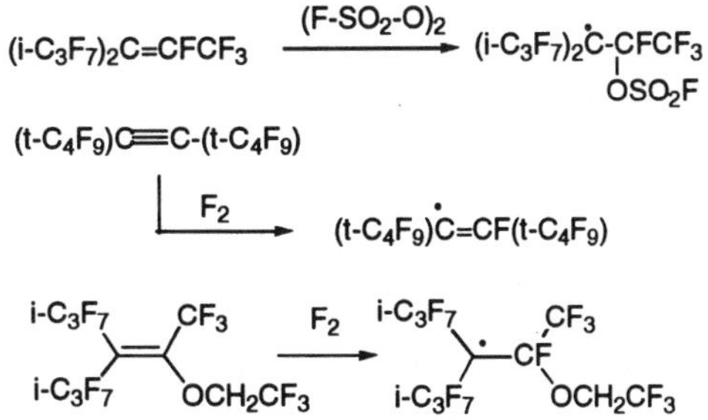

Scheme 7

3.2
Electronegativities, Ionization Energies, and Electron Affinities

In order to assess the contribution of polar factors to the reactivities of fluorinated radicals, one needs a measure of their electronegativities. Absolute electronegativities (χ) may be derived if one knows both the IPs and the EAs of radicals [47], as shown below in Eq. (2).

$$\chi = \frac{IP + EA}{2} \tag{2}$$

Unfortunately, few experimental ionization potentials or electron affinities of fluorinated radicals have been reported, and the calculation of such molecular properties is fraught with difficulties, although reasonable trends can be predicted [48–49]. Table 3 provides what numbers are presently available [50–55].

Table 3. Experimental ionization potentials, electron affinities and absolute electronegativities of alkyl and fluorinated alkyl radicals [50–55]

Radical	CH_3	CH_3CH_2	$(CH_3)_2CH$	$(CH_3)_3C$	CH_2F	CHF_2	CF_3
IP (eV)	9.84	8.12	7.37	6.70	9.04	8.73	9.25
EA (eV)	0.08	– 0.26	– 0.32	– 0.16		1.3	1.84
χ	4.96	3.93	3.53	3.27		5.0	5.55
Radical	CH_3CF_2	$CH_3CF_2CF_2$	$(CH_3)_2CF$	$(CF_3)_3C$	CH_3CF_2	HCF_2CF_2	CF_3CHF
IP (eV)	9.98	10.06	10.50		7.92	9.29	9.60
EA (eV)	1.81	> 2.65	> 2.65	3.4			
χ	5.90	> 6.36	> 6.58				

Pearson has observed that the reactivity of various organic substrates, including radicals, can be correlated with their absolute electronegativies [56]. It can be seen that although trifluoromethyl and pentafluoroethyl radicals are much more electronegative than the more nucleophilic alkyl radicals, such as *tert*-butyl, methyl itself should not be much more nucleophilic than trifluoromethyl. *Nucleophilicities* of alkyl radicals increase: $CH_3 < 1° < 2° < 3°$. Although there are not sufficient IP or EA data available to substantiate the issue, the reactivity studies which are described in Sect. 5 demonstrate that the *electrophilicities* of perfluoroalkyl radicals increase: CF_3 $1° < 2° < 3°$.

4
Methods of Generation of Fluorinated Radicals

Consistent with the growing appreciation of the importance of free radical chemistry to the field of organofluorine chemistry, both mechanistically and synthetically, over the years there have been numerous methods developed for the purpose of generating perfluoro and partially fluorinated alkyl radicals. Such methods include thermal and photochemical homolysis, radical initiation, and electron-transfer processes. Whereas the most common type of precursor is a perfluoroalkyl iodide or bromide, fluorinated free radicals can also be generated from fluoroalkyl sulfonyl halides, carboxylic acids, diacyl peroxides, and azoalkanes, as well as from a number of other less common precursors, including perfluorocarbons themselves.

In this section, each method of generating carbon-based fluorinated radicals will be introduced and discussed in terms of mechanism, and it will be seen that virtually all of the useful methods for generating perfluoro and partially fluorinated radicals in a practical manner involve well defined and controlled free radical chain reactions. Some representative examples which demonstrate the preparative use of these methodologies will be presented in the final section of this review.

4.1
From Perfluoroalkyl Iodides

Perfluoroalkyl iodides comprise the most important and commonly-used source of perfluoroalkyl radicals [57–59], and for the most part, the methods which have been developed for such perfluoro systems also work well for generation of partially-fluorinated alkyl radicals from their respective iodides.

4.1.1
Thermal and Photochemically-Induced Homolysis

Simple homolysis of the C-I bond by heating or by light is the most straightforward approach and was the first used for adding perfluoroalkyl iodides to olefins. One presumes that both the thermal and the photochemically induced addition reactions of perfluoroalkyl radicals proceed via free radical chain reactions as depicted in the Scheme below. However, the conditions of these reactions are rarely ideal for preparative purposes because high temperatures are required for the thermolytic process and long photolysis times are required for the photolytic method [60].

$$R_FI \xrightarrow{\Delta \text{ or } h\nu} R_F{}^\cdot + I^\cdot \quad \text{(initiation)}$$

$$R_F{}^\cdot + \text{\Large$>$\!\!=\!\!\Large$<$} \longrightarrow R_F\text{—}\overset{\cdot}{\text{—}} \xrightarrow{R_FI} R_F\text{—}\text{—}I + R_F{}^\cdot$$

$$\text{(propagation)}$$

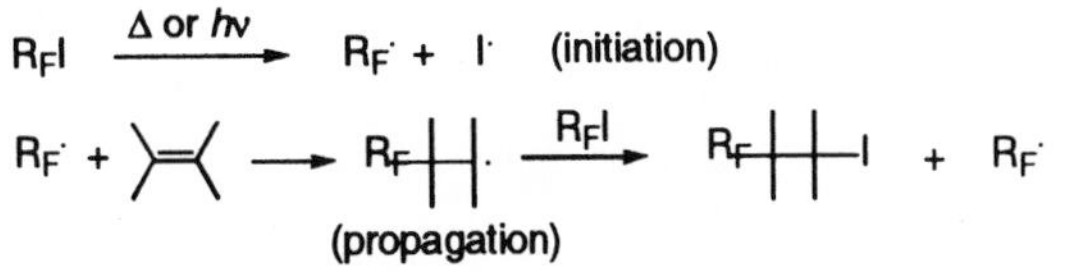

Scheme 8

4.1.2
Use of Free Radical Initiators

The use of free radical initiators in such reactions can be very useful. They allow the reactions to be run at much lower temperatures and generally make them more efficient [60].

$$In_2 \xrightarrow{\Delta,\ 60^{\circ}C} 2\ In^{\cdot} \xrightarrow{R_FI} R_F^{\cdot} \xrightarrow{Substrate} etc$$

$$In\text{-}I$$

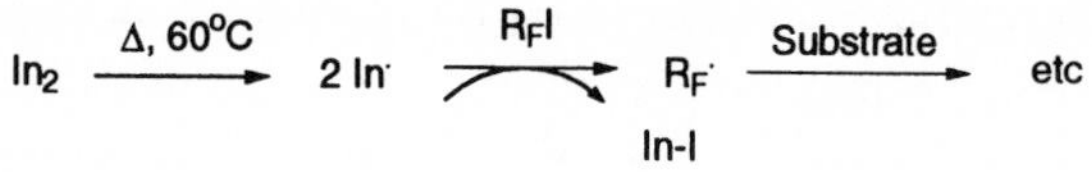

Scheme 9

4.1.3
Chemical Reduction (SET)

The most important recent development in this area has been the use of various single-electron reductants to initiate the free radical chain process. Such reductants have been most commonly metals or anionic species, and such processes have been used either to initiate addition processes or substitution ($S_{RN}1$) processes

<u>Addition:</u>

$$M\ (or\ Nu^{-}) + R_FI \longrightarrow M^{\cdot+}\ (or\ Nu^{\cdot}) + R_FI^{\cdot-} \xrightarrow{-I^{-}} R_F^{\cdot}$$

$$R_F^{\cdot} + \hspace{-0.3em}\rangle\hspace{-0.4em}=\hspace{-0.4em}\langle \longrightarrow R_F\hspace{-0.2em}+\hspace{-0.3em}+\hspace{-0.3em}\cdot \longrightarrow etc$$

<u>Substitution:</u>

$$Nu^{-} + R_FI \longrightarrow Nu^{\cdot} + R_FI^{\cdot-} \xrightarrow{-I^{-}} R_F^{\cdot}$$

$$R_F^{\cdot} + Nu^{-} \longrightarrow R_FNu^{\cdot-} \xrightarrow{R_FI} R_FNu + R_FI^{\cdot-} \longrightarrow etc$$

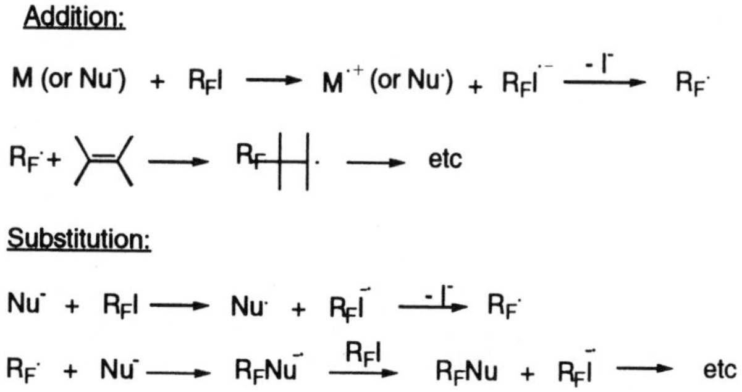

Scheme 10

4.1.4
Electrochemical Methods

Perfluoroalkyl radicals can be produced electrochemically from perfluoroalkyl iodides by cathodic reduction, and although there can be problems in controlling side reactions such as dimerization, such methodology can have considerable synthetic advantage.

$$R_FI + e^{-} \xrightarrow{redn} R_FI^{\cdot-} \xrightarrow{-I^{-}} R_F^{\cdot} \longrightarrow etc$$

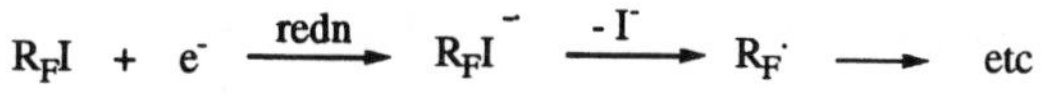

Scheme 11

4.1.5
Fenton Methodology

Perfluoroalkyl radicals have also been generated from perfluoroalkyl iodides by use of Fenton conditions in DMSO, with the final step being abstraction of iodine from R_FI by methyl radical [61].

$$H_2O_2 + Fe(II) \longrightarrow \cdot OH + {}^-OH + Fe(III)$$

$$\cdot OH + CH_3SOCH_3 \longrightarrow \underset{H_3C}{\overset{\cdot O}{\underset{}{\overset{}{S}}}}\overset{OH}{\underset{CH_3}{}} \longrightarrow CH_3\cdot + CH_3SO_2H$$

$$CH_3\cdot + R_FI \longrightarrow CH_3I + R_F\cdot$$

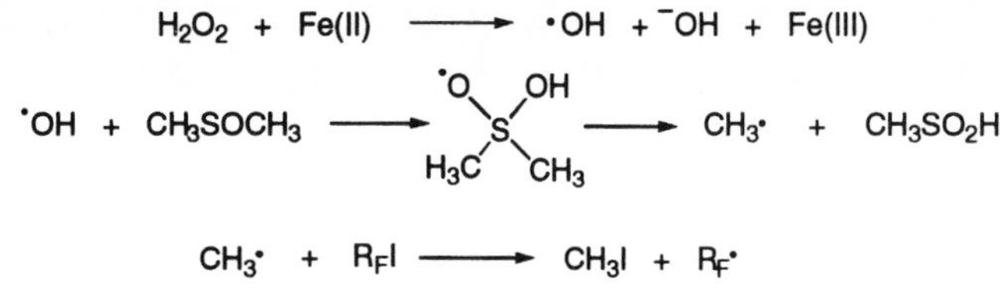

Scheme 12

4.2
From Perfluoroalkyl Sulfonyl Halides

Perfluoroalkyl sulfonyl halides are also good photochemical sources of perfluoroalkyl radicals, and they may also be used under thermal-induction, with a radical initiator, to form $R_{\bar{F}}$ in a synthetically useful manner [62].

$$R_FSO_2Br \xrightarrow{h\nu} R_FSO_2\cdot + Br \xrightarrow{-SO_2} R_F\cdot \longrightarrow etc$$

Scheme 13

Chain transfer of a Br atom from the sulfonyl bromide seems to be more efficient than that from perfluoroalkyl iodides or bromides, but problems can be encountered with sulfonyl bromides because SO_2 expulsion is somewhat slow and sometimes competes with alkene addition of $RSO_2\cdot$.

4.3
From Perfluoroalkanoic Acids

Electrochemical oxidation of perfluoroalkanoic acids can lead to radical derived products, although the same problems apply to such oxidations as applied to the electrochemical reduction processes of perfluoroalkyl iodides.

$$R_FCO_2{}^- - e^- \xrightarrow{oxidn} R_FCO_2\cdot \xrightarrow{-CO_2} R_F\cdot \longrightarrow etc$$

Scheme 14

Electron transfer from perfluoroalkanoic acids to *xenon difluoride* was also reported to give perfluoroalkyl radicals which were found to add to benzene derivatives [63].

$$2 R_FCO_2H + XeF_2 \longrightarrow 2 R_FCO_2\cdot + 2 H^+ + Xe + 2 F$$

$$R_FCO_2\cdot \xrightarrow{-CO_2} R_F\cdot \xrightarrow{Ar-H} R_FAr$$

Scheme 15

Perfluoroalkanoic acids also undergo *Hunsdiecker reactions,* with the greatest utility for such methodology being the preparation of perfluoroalkyl iodides, bromides and chlorides [64].

$$CF_3CO_2^- Ag^+ \xrightarrow{\ I_2\ } CF_3CO_2\text{-}I \longrightarrow CF_3CO_2^{\cdot} \ + \ I^{\cdot}$$

$$CF_3CO_2^{\cdot} \xrightarrow{\ -CO_2\ } CF_3^{\cdot} \xrightarrow{\ I_2\ } CF_3I$$

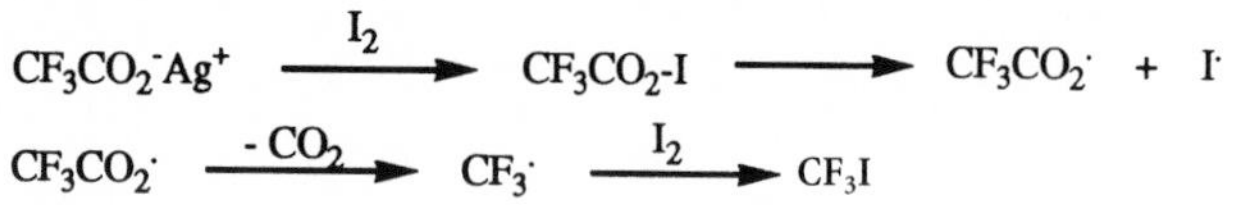

Scheme 16

Another way that has potential for the generation of perfluoroalkyl radicals from carboxylic acids is the use of *Barton esters.* However, unlike the situation for their hydrocarbon analogs, fluorinated thiohydroxamate esters have thus far only been able to be prepared in situ [65].

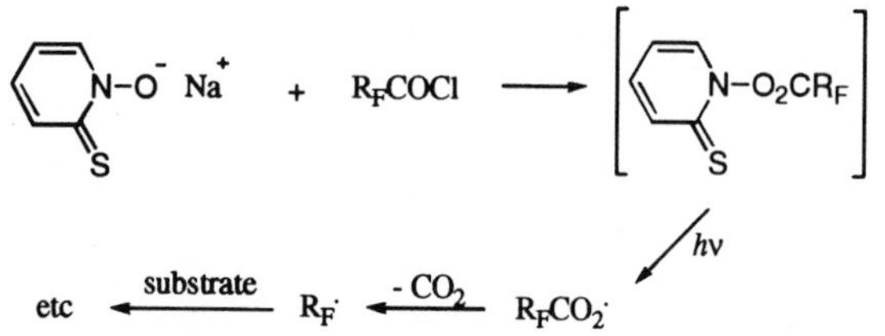

Scheme 17

4.4
From Fluorodiacyl Peroxides

4.4.1
Thermal and Photochemical Homolysis

Perfluoro-*n*-alkyl diacyl peroxides decompose homolytically to give perfluoro-alkyl radicals under mild conditions, and radicals formed in such a manner have been used synthetically or as radical initiators for polymerizations [66, 67].

$$(R_FCO_2)_2 \xrightarrow{\ \Delta\ \text{or}\ h\nu\ } 2\ R_FCO_2^{\cdot} \xrightarrow{\ CO_2\ } R_F^{\cdot} \longrightarrow \text{etc}$$

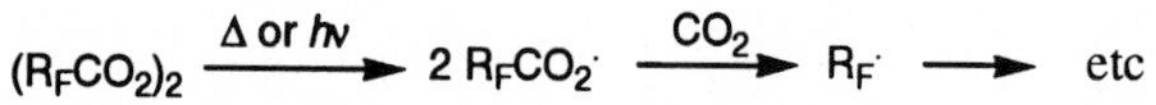

Scheme 18

Such reactions are, of course, *unimolecular* decompositions, not free radical chain processes. This fact made perfluorodiacyl peroxides ideal precursors for the laser flash photolysis studies which will be described in Sect. 4. Kinetics for the thermal decomposition of a number of perfluorodiacyl peroxides have been measured and their $\Delta H^{\ddagger}$ values were approximately 24 kcal/mol, about 5 kcal/mol lower than for analogous hydrocarbon diacyl peroxides [68, 69]. Their typical half-life is ~ 5 hours at 20 °C, while $(HCF_2CF_2CO_2)_2$ is anomolously reactive and has a half life of only 81 minutes. Although other partially-fluorinated diacyl peroxides have also been prepared, for the purpose of LFP studies [70], other than for the case of 2,2-difluoropropionyl peroxide [71], their thermal kinetic parameters have not yet been determined.

4.4.2
Electron Transfer Processes

It appears that in the presence of electron-rich π-systems, either olefinic or aromatic, these electron-deficient diacyl peroxides undergo electron-transfer, decarboxyation, and cage-recombination to give adducts in good yield [68].

$$(R_FCO_2)_2 \ + \ ArH \ \longrightarrow \ (R_FCO_2)_2^{-\bullet} \ + \ ArH^{+\bullet} \ \longrightarrow$$

$$R_F^{\bullet} \ + \ ArH^{+\bullet} \ + \ CO_2 \ + \ R_FCO_2^{-} \ \longrightarrow \ R_FArH^{+} \ \xrightarrow{-H^{+}} \ R_FAr$$

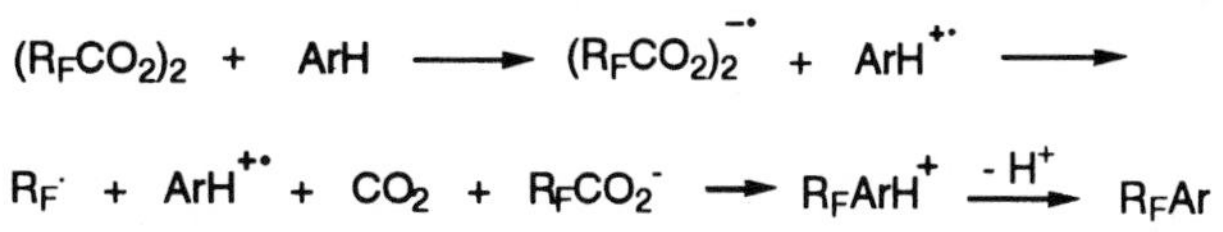

Scheme 19

4.5
From Perfluoroazoalkanes

Perfluoroazoalkanes [72] have also been utilized as thermal or photochemical precursors of perfluoroalkyl radicals.

$$R_F\text{-}N{=}N\text{-}R_F \ \xrightarrow{h\nu} \ 2\,R_F^{\bullet} \ + \ N_2 \ \longrightarrow \ \text{etc}$$

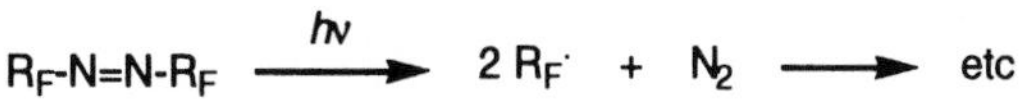

Scheme 20

Although this method has been generally limited by the low efficiency of photodeazetation and the high temperatures needed for thermal deazetation (half-life $\approx$ 1 h at 332 °C) [73], it has recently been found that use of 185 nm light provides reasonable quantum yield of fluorinated radicals ($\Phi = 0.15$), via a two photon process involving *trans* to *cis* conversion followed by photofragmentation [74]. Nevertheless, the significant intervention of cage-recombination in these reactions limits the synthetic utilization of the radicals which are generated.

4.6
Some Other Sources

There are other sources of perfluoroalkyl radicals which have found occasional use, such as photolysis of hexafluoroacetone to generate trifluoromethyl radicals [75, 76], photolysis of perfluoroacyl halides [77], Umemoto's photolysis of

N-nitroso-*N*-trifluoromethyl trifluoromethanesulfonamide [78], photolysis of bis-(trifluoromethyl)tellurium [79], thermal AIBN-induced decomposition of bis-(trifluoromethyl)mercury [80], thermolysis of highly branched perfluoro-carbons [81], and even thermolysis of persistent perfluoroalkyl radicals such as that discovered by Scherer [39, 45]. Recently it has even been found that per-fluorocarbons can be a source of perfluoroalkyl radicals when they undergo photoinduced reduction by NH_3 or by Cp_2TiF_2 [82, 83].

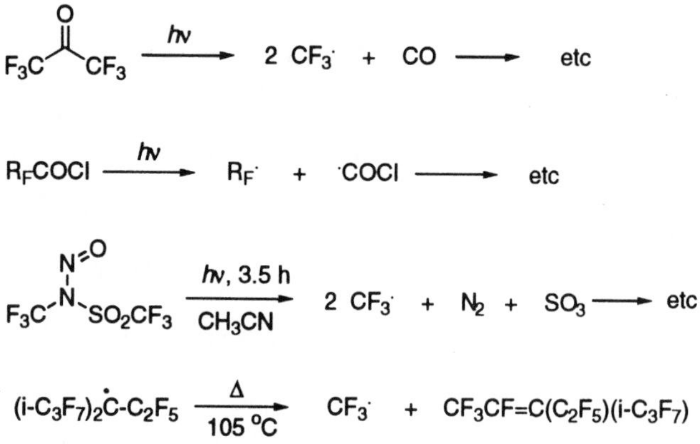

$$F_3C \overset{\overset{\displaystyle O}{\|}}{\underset{}{C}} CF_3 \xrightarrow{h\nu} 2\ CF_3\cdot\ +\ CO \longrightarrow etc$$

$$R_FCOCl \xrightarrow{h\nu} R_F\cdot\ +\ \cdot COCl \longrightarrow etc$$

$$F_3C\overset{\overset{\displaystyle N^{\nearrow O}}{|}}{\underset{}{N}}{}_{SO_2CF_3} \xrightarrow[CH_3CN]{h\nu,\ 3.5\ h} 2\ CF_3\cdot\ +\ N_2\ +\ SO_3 \longrightarrow etc$$

$$(i\text{-}C_3F_7)_2\dot{C}\text{-}C_2F_5 \xrightarrow[105\ ^{\circ}C]{\Delta} CF_3\cdot\ +\ CF_3CF{=}C(C_2F_5)(i\text{-}C_3F_7)$$

Scheme 21

$$\underset{F_3C}{\overset{F_3C}{{>}}}CF{-}\underset{F_2C{-}CF_3}{\overset{CF_2}{|}} \xrightarrow[(+e^-,\ \text{-}F)]{Hg/h\nu/NH_3} \underset{F_3C}{\overset{F_3C}{{>}}}CF{-}\underset{F_2C{-}CF_3}{\overset{\dot{C}F}{|}}$$

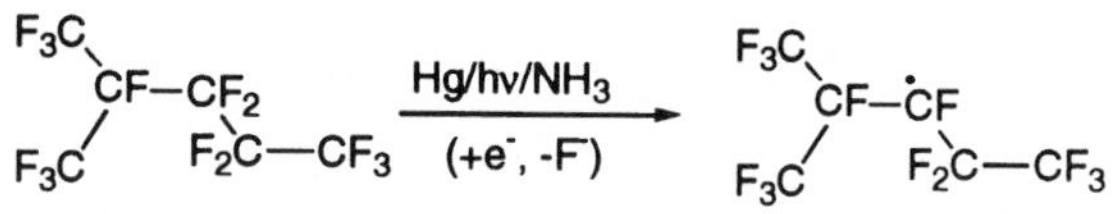

Scheme 22

4.7
From Radical Addition to Perfluoroolefins

In order to be complete in our discussion of methods for generation of fluori-nated radicals, it must be mentioned that perfluoroalkyl radical intermediates are also formed in every reaction in which radical species such as halogen atoms, thiyl radicals, or other carbon radicals add to fluoroolefins. As will be seen in Sect. 6.3.2, such processes are especially important in the telomerization or polymerization of fluorinated olefins

5
Reactivity of Fluorinated Radicals

Discussions about reactivity must be carried out within the context of some reaction. For free radicals, the fundamentally most important types of reactions are those involving their addition to π-bonds, particularly their additions to alkenes, and their hydrogen-abstraction reactions. Therefore, virtually all assessments of the reactivity of radicals involve studies of such reactions.

5.1
Alkene Addition Reactions

5.1.1
Early Kinetic Studies

There has been considerable effort directed towards obtaining a fundamental understanding of the factors that govern the reactivities of carbon-centered radicals in bimolecular reactions, particularly with respect to their addition to alkenes [84]. From early liquid and gas phase studies, reactivity in such addition reactions was concluded to derive from a "complex interplay of polar, steric, and bond-strength terms [85]," which is much influenced by the nature and position of substituents on both the radical and the alkene.

Competition studies from Szwarc's group provided excellent quantitative insights into the relative affinities of methyl and trifluoromethyl radicals for a host of alkenes [86–88], and from this work came the first general recognition that substituted alkyl radicals could exhibit polar characteristics ranging from nucleophilic to electrophilic. On the basis of such early work, methyl and trifluoromethyl were taken to be the prototypical nucleophilic and electrophilic radicals, respectively, characterizations which it turns out are somewhat exaggerated in both cases.

In a series of papers in the early 1980s, Sokolov's group reported relative rate studies which were similar in nature to those of the early Szwarc studies. Sokolov generated various perfluoroalkyl radicals via thermal decomposition of the respective perfluoro diacyl peroxides in heptane containing various olefins [89] or arenes [90]. Determination of the ratio of olefin addition products to hydrogen abstraction products provided the relative rate data given in Table 4 [89].

The relative rates measured in these early studies generally correlate quite well with absolute rate data which have recently been derived from direct measurements (see below).

Table 4. Relative rates of addition of perfluoroalkly radicals to olefins vs their rates of hydrogen-atom abstraction from heptane at 50 °C [89]

Olefin	$CF_3^{\cdot}$	$C_2F_5^{\cdot}$	$C_3F_7^{\cdot}$	Olefin	$CF_3^{\cdot}$	$C_2F_5^{\cdot}$	$C_3F_7^{\cdot}$
$CH_2{=}CH_2$	132	340	290	$CF_2{=}CF_2$	8	7	<0.3
$CH_2{=}CHF$	30	108	40	$CF_2{=}CFCF_3$		0.33	
$CH_2{=}CF_2$	9	13		$CF_2{=}CFOCF_3$		1.1	
$CHF{=}CF_2$	6	9					

5.1.2
Early Regiochemical Studies

The regiochemical data deriving from the early Szwarc studies on the addition of $CF_3^{\cdot}$ to unsymmetrical alkenes provided useful insight, and Haszeldine's studies which defined the preferred mode of addition of $CF_3^{\cdot}$ to $CH_2{=}CHF$, $CH_2{=}CF_2$, $CHF{=}CF_2$, $CH_2{=}CHCH_3$, $CH_2{=}CHCF_3$, $CF_2{=}CHCH_3$, $CF_2{=}CHCF_3$, $CF_2{=}CFCF_3$, etc., provided the foundation for our present understanding of the regiochemical behavior of fluorinated radicals [91]. These data and additional, more precise data obtained in their own labs has been reviewed critically by Tedder and Walton, and is partially summarized in Table 5 [85, 92, 93].

Table 5. Regiochemistry of trifluoromethyl radical additions to olefins [85, 92, 93]

Olefin	$CH_2{=}CHF$	$CH_2{=}CF_2$	$CHF{=}CF_2$	$CH_2{=}CHCl$	$CHCl{=}CF_2$
Ratio	1:0.09	1:0.05	1:0.50	1:0.02	11.5
Olefin	$CH_2{=}CHCH_3$	$CH_2{=}C(CH_3)_2$	$CH_2{=}CHCH{=}CH_2$	$CH_2{=}CHCF_3$	
Ratio	1:0.1	1:0.08	1:< 0.01	1:0.01	
Olefin	$CHF{=}CHCF_3$	$CF_2{=}CHCH_3$	$CF_2{=}CHCF_3$	$CF_2{=}CFCF_3$	
Ratio	1:0.33	1:50	1:1.5	1:0.25	

As concluded by Tedder, it would appear that a combination of polar and steric effects on the part of both the olefin and the trifluoromethyl radical are sufficient to determine the observed regioselectivities [93].

Recently, a quantitative study of the regiochemistry of addition of a number of different fluoroalkyl radicals to $CHF{=}CF_2$, summarized in Table 6, indicated that the observed selectivity could be correlated with the postulated relative electrophilicity of the radicals, with the conclusion being reached that the secondary n-$C_5F_{11}(CF_3)CF^{\cdot}$ radical was the most electrophilic [94].

Table 6. Regioselectivities of addition of some fluoroalkyl radicals to trifluoroethylene [94]

$$R_F^{\cdot} + CHF{=}CF_2 \rightarrow R_FCHF\text{-}CF_2^{\cdot}\ (\alpha \text{ or } {}^{\cdot}CHF\text{-}CF_2R_F\ (\beta)$$

Radical	n-$C_5F_{11}(CF_3)CF^{\cdot}$	$(CF_3)_2CF^{\cdot}$	$CF_3(CF_2)_2CF_2^{\cdot}$	$CF_3(CF_2)_3CH_2CF_2^{\cdot}$
$\alpha:\beta$	93:7	90:10	75:25	60:40

5.1.3
Factors which Affect Radical Reactivity

Additions of carbon-centered radicals to alkenes are generally strongly exothermic since a σ-bond is formed at the expense of a π-bond (e.g., addition of methyl radical to styrene has a $\Delta H^0 = -38.5$ kcal/mol). Thus, according to the Hammond postulate, such reactions should have early transition states with little bond-making or bond-breaking being involved. This is supported by the measured activation energies for such additions, which generally lie between 3 and 8 kcal/mol [95], as well as by theoretical calculations which indicate the involvement of unsymmetrical addition transition states that are relatively independent of the electrophilic or nucleophilic nature of the adding radical species. The calculations also indicate an approach to bonding at the α-carbon in which the radical is far removed from the β-carbon [96–99].

Correlation of the effect of substituents on the rates of reactions with early transition states often is best accomplished in terms of perturbational molecular orbital theory, and polar effects can play a major role for such reactions [100, 101]. Essentially this theory states that energy differences between the highest occupied molecular orbital (HOMO) of one reactant and the lowest unoccupied molecular orbital (LUMO) of the other reactant are decisive in determining the reaction rate: the smaller the difference in energy, the faster the predicted rate of reaction [102, 103]. Since the HOMO of a free radical is the SOMO, the energy difference between the SOMO and the alkene HOMO and/or LUMO is of considerable importance in determining the rates of radical additions to alkenes [84].

5.1.4
Absolute Rate Data

In recent years, direct, time-resolved methods have been extensively employed to obtain absolute kinetic data for a wide variety of alkyl radical reactions in the liquid phase, and there is presently a considerable body of data available for alkene addition reactions of a wide variety of radical types [104]. For example, rates of alkene addition reactions of the nucleophilic *tert*-butyl radical (with its high-lying SOMO) have been found to correlate with alkene electron affinities (EAs), which provide a measure of the alkene's LUMO energies [105, 106]. The data indicate that the reactivity of such nucleophilic radicals is best understood as deriving from a dominant SOMO-LUMO interaction, leading to charge transfer interactions which stabilize the early transition state and lower both the enthalpic and entropic barriers to reaction, with consequent rate increase. A similar recent study of the methyl radical indicated that it also had nucleophilic character, but its nucleophilic behavior is weaker than that expressed by other alkyl radicals [107].

Data is also available for addition reactions of "electrophilic" radicals, $^{\cdot}CH(CN)_2$, and so-called ambiphilic radicals, $^{\cdot}CH(CO_2Et)_2$ and $^{\cdot}CH_2CO_2$-*t*-Bu, which derive their electrophilic character from π-delocalization of the carbon-centered radical onto electron-attracting substituents [108–111], and for which the enthalpy of the addition process rather than the polar nature of the radicals may be the primary rate-determining factor [112, 113].

Fluorinated radicals, in contrast, would be expected to derive their electrophilicities virtually entirely from fluorine's inductive effect. One would expect the reactivity of perfluoro-*n*-alkyl radicals to differ significantly from that of their hydrocarbon counterparts, since the latter are electron-rich, planar π-radicals, whereas the former are electron poor, nonplanar σ-radicals.

5.1.4.1
Perfluoro-n-Alkyl Radicals

Recent laser flash photolysis (LFP) studies have provided absolute rates of addition of perfluoro-n-alkyl radicals to a variety of alkenes in solution [114,115]. In these studies, $C_2F_5^{\cdot}$, $C_3F_7^{\cdot}$, and n-$C_7F_{15}^{\cdot}$ were generated "instantaneously" by photolysis of the respective diacyl peroxides. The initially-formed perfluoroacyloxyl radicals decarboxylated rapidly to yield the perfluoroalkyl radicals, after which the additions of these radicals to styrene, α-methylstyrene, etc. were monitored directly via observation of the growth of UV absorption due to the transient benzylic radicals.

$$(R_F\text{-}CO_2)_2 \xrightarrow{hv} 2\ R_FCO_2^{\cdot} \xrightarrow[\text{fast}]{-CO_2} R_F^{\cdot} \xrightarrow[k_{add}]{CH_2=CHPh} R_FCH_2\overset{\cdot}{C}HPh \quad 320\ nm$$

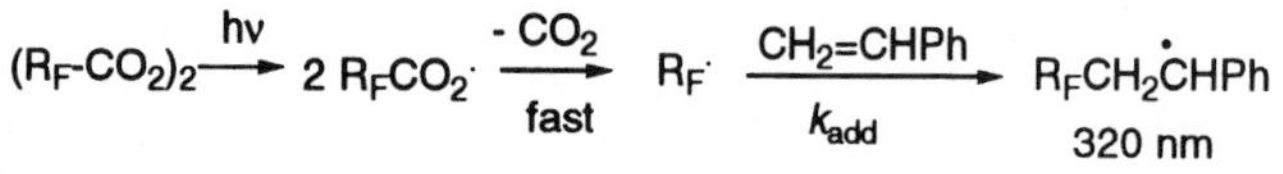

Scheme 23

The rate constants, k_{add}, obtained from the LFP experiments for addition of the perfluoro-n-alkyl radicals to the various alkenes in 1,2,2-trichloro-1,1,2-trifluoroethane (F113) are given in Table 7. It can be seen that such radicals are *much* more reactive than their hydrocarbon counterparts, particularly in additions to electron-rich alkenes, with n-$C_3F_7^{\cdot}$ adding to 1-hexene 30 000 times faster, and to styrene 350 times faster than an n-alkyl radical [114,115].

Table 7. Absolute rate constants for the reaction of perfluoro-n-alkly radicals with various unsaturated substrates at $298 \pm 2\ °K$, as measured by LFP in F113 [114,115]

Alkenes (IP's)[a]	$k_{add}/10^6 M^{-1}s^{-1}$						
	n-C_3F_7	n-C_7F_{15}	n-C_8F_{17}	C_2F_5	CF_3	RCH_2	$(CH_3)_3C$
α-methylstyrene (8.19)	78	89		94	87[b]	0.059[c]	
β-methylstyrene (8.10)	3.8	3.7		7.0	18[d]		
styrene (8.43)	43	46	46	79[d]	53[b]	0.12[f]	0.13[f]
pentafluorostyrene (9.20)	13			23[d]	26[b]	0.31[b]	
4-methylstyrene	130		61[h]		140		
4-methoxystyrene			65[h]				
4-chlorostyrene			36[h]				
4-(CF_3)styrene	29	24	25[h]				
1,4-dimethylene-cyclohexane (9.12)	41					1.3×10^{-4g}	
1-hexene (9.14)	6.2	7.9		16		2×10^{-4g}	
$nC_4F_9CH_2CH=CH_2$				2.5			
cyclohexene (8.94)	1.3						
$H\text{-}C{\equiv}CCMe_2OH$ (10.18)	0.9						
$CH_2=CCl_2$ (9.79)	5.2						0.35[h]
$CH_2=C(CH_3)COOCH_3$ (9.70)	19						
$CH_2=C(CH_3)CN$			3.2[j]				
$CH_2=CHCN$ (10.91)	2.2	1.6	2.0[j]	3.2	4.4[e]		2.4[i]

[a][50] [b][70] [c][116] [d][117] [e][118] [f][119] [g][120] [h][121] [i][122] [j] Rates obtained from competition study [115].

Although the high electrophilicity of perfluoroalkyl radicals is probably the dominant factor giving rise to their high reactivities, there are a number of other factors which undoubtedly also contribute.

Steric effects can*not* be contributing to the observed enhanced rates since, although fluorine is a small substituent, it is certainly larger than a hydrogen atom. No doubt of some relevance is the σ-nature of perfluoro-*n*-alkyl radicals. Since substantial bending (14–15 °C from planarity) is apparently required in the transition state for alkyl radical addition to alkenes, non-planar perfluoroalkyl radicals would be expected to have an inherent energetic advantage over a (planar) alkyl radical in addition reactions [96, 123]. The greater reactivity of "bent" σ-radicals relative to planar π-radicals has been noted earlier in studies on aryl radicals [124], and the influence of pyramidality on radical reactivities has been previously suggested in a study of α-hydroxyalkyl radicals [125]. The energy required to bend the methyl and *tert*-butyl radicals to the same extent as in their respective transition state structures for addition to ethylene has been calculated to be 1.6 and 1.5 kcal/mol, respectively [123]. In contrast, the ESR parameters for perfluoro-*n*-alkyl radicals, as discussed earlier, implies that their configuration at the radical center should not require further bending in order to reach their transition states for addition to alkenes.

Also of relevance is the significantly stronger (ca. 10 kcal/mole) C-C bond that forms when $R_f^{\cdot}$ vs $R^{\cdot}$ adds to an alkene (CH_3-CH_3, BDE=91 vs CF_3-CH_3, BDE=101 kcal/mol) [23]. Although this greater exothermicity of the perfluoroalkyl radical addition reactions must be to some degree relevant, the relatively small, seven-fold difference in the rates of addition of n-$C_3F_7^{\cdot}$ to styrene vs 1-hexene, two processes which differ in exothermicity by ~14 kcal/mol, indicates that the rates of such early transition state processes *cannot* be greatly affected by differences in ΔH^0.

The dominant factor that gives rise to the observed high reactivities of perfluoro-*n*-alkyl radicals, particularly in their additions to electron-rich alkenes, would appear to be the high electrophilicities of these very electron-deficient radicals [114]. A perfluoro-*n*-alkyl radical, which one can assume to have a low-lying SOMO, should exhibit a dominant SOMO-HOMO interaction in its additions to alkenes, and polarization of the type shown in Fig. 1 will stabilize the early transition state in which little radical character has been transferred to the substrate alkene. Therefore, if steric hindrance is equivalent for a series of alkenes, the rates of addition of $R_F^{\cdot}$ should correlate with the alkene IPs (which should reflect HOMO energies). As Fig. 2 indicates, there is indeed a respectable correlation between log k_{add} for typical perfluoro-*n*-alkyl radicals and terminal

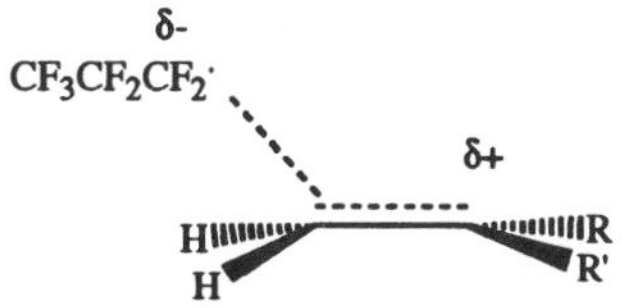

Fig. 1. Typical polar transition state for addition of perfluoroalkyl radical to an electron-rich olefin

alkene IPs. The styrenes all appear to be slightly more reactive than would be expected based upon their IPs, a result which implicates the slight intervention of enthalpy effects. However, the correlation expressed by Fig. 2 is consistent with the electrophilic character of perfluoro-*n*-alkyl radicals. For steric reasons, the non-terminal olefins, *β*-methylstyrene and cyclohexene, are noticeably less reactive than might have been anticipated from their IPs.

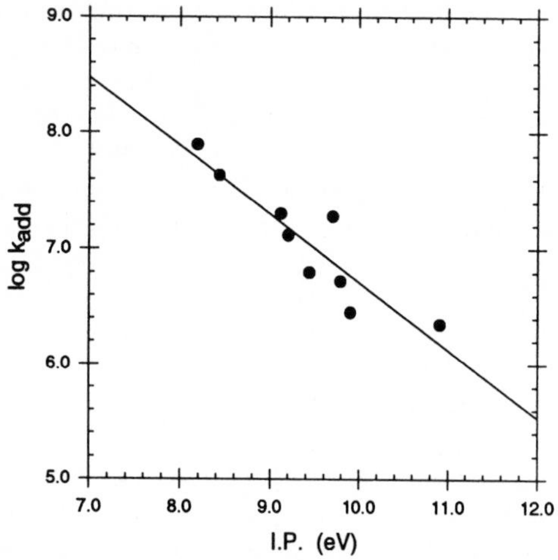

Fig. 2. Plot of the log of the rate constants for addition of heptafluoro-*n*-propyl radical to some alkenes the ionization potentials of the alkenes

The electrophilic character of *n*-perfluoroalkyl radicals was confirmed by a correlation of the rates of addition of the *n*-$C_8F_{17}\cdot$ radical to a series of para-substituted styrenes with Hammett σ values, as shown in Fig. 3. The *ρ* value is negative (–0.53), as would be expected for an electrophilic reactant [115].

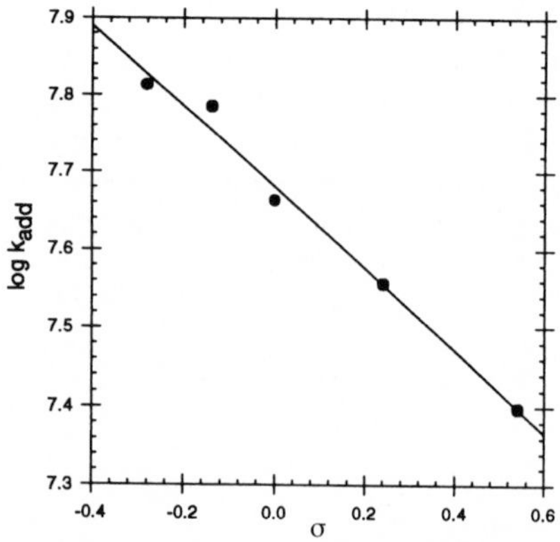

Fig. 3. Plot of the log of the rate constants for addition of perfluoro-*n*-octyl radicals to some para-substituted styrenes Hammett σ values of the substituents

In a similar competition study, Jiang et al. observed a value of –0.99 for the addition of $CF_3\cdot$ to p-substituted phenyl acetylenes. They found that applying a dual parameter equation ($\sigma + \sigma\cdot_{JJ}$) led to a much better correlation [126].

Upon examination of the data in Table 7 one notices slight differences in reactivity for n-$C_3F_7\cdot$ vs $C_2F_5\cdot$ and $CF_3\cdot$. Whereas n-$C_3F_7\cdot$, n-$C_7F_{15}\cdot$, and n-$C_8F_{17}\cdot$ have identical reactivities within experimental error, and, as such, the addition rates of these radicals can be considered to be of *generic* n-perfluoroalkyl radicals, $C_2F_5\cdot$ and $CF_3\cdot$ each exhibit incrementally greater reactivities than these radicals (on the average about 1.8 and 1.4 times more reactive, respectively). Such enhanced reactivities do not seem to derive from a greater electrophilicity, since the plots of log k_{add} vs alkene IPs for these two radicals lie virtually parallel to those of the generic n-perfluoroalkyl radicals. In all likelihood, the slightly enhanced reactivities of $C_2F_5\cdot$ and $CF_3\cdot$ derive from some combination of steric, pyramidalization, or enthalpic factors.

5.1.4.2
2° and 3° Perfluoroalkyl Radicals

A group of perfluoroalkyl radicals which do exhibit marked increases in reactivity due to enhanced electrophilicity are the branched, 2° and 3° perfluoroalkyl radicals, specifically the perfluoro-*iso*-propyl and *tert*-butyl radicals [118].

Table 8 provides the absolute rates of addition of $(CF_3)_3C$, $(CF_3)_2CF\cdot$, $CF_3CF_2\cdot$, and $CF_3\cdot$ to a group of alkenes of variable reactivity. It can be seen from the Table that both the perfluoro-*iso*-propyl and perfluoro-*tert*-butyl radicals give evidence of *much greater* electrophilicity in their alkene addition reactions. For example, the latter radical reacts significantly (6.8 times) *faster* than $CF_3\cdot$ with the nucleophilic α-methylstyrene (IP=8.9 eV), while reacting somewhat (1.6 times) *slower* than $CF_3\cdot$ with the more electrophilic pentafluorostyrene (IP=9.2 eV). Comparative plots of all of the available rate data for alkene additions of $CF_3\cdot$ and $(CF_3)_3C\cdot$ vs alkene IPs, as seen in Fig. 4, leave no doubt as to the relative electrophilicity of the two species. Moreover, the rates of addition of the very electrophilic, but *non*-σ, planar perfluoro-*tert*-butyl radical to the more nucleophilic

Table 8. Absolute rate constants for the addition of trifluoromethyl, pentafluoroethyl, heptafluoro-*iso*-propyl, and nonafluoro-*tert*-butyl radicals to various olefins at 290 °K in F113 [117, 118]

Olefin	$k_{add}/10^6 M^{-1} s^{-1}$			
	CF_3	C_2F_5	$(CF_3)_2CF$	$(CF_3)_3C$
Styrene	53	79	120	363
Pentafluorostyrene	26	23	81	16
α-methylstyrene	87	94		589
β-methylstyrene	17	7.0	1.9	2.5
$CH_2=CMeCO_2Me$				3.8
$CH_2=CHCN$	4.4	3.2		

alkenes, which approach diffusion-control, also leave no doubt as to the great importance of electrophilicity in giving rise to the extraordinary reactivity of *all* perfluoroalkyl radicals.

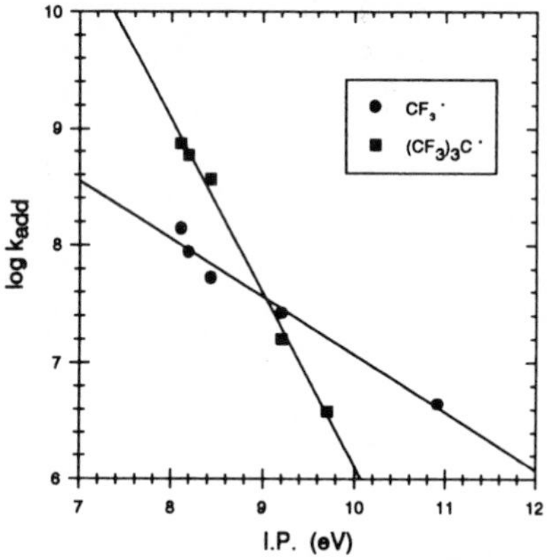

Fig. 4. Plot of the log of the rate constants for additions of tifluoromethyl and perfluoro-*tert*-butyl radicals to some alkenes the ionization potentials of the alkenes

As mentioned earlier, steric effects apparently inhibit addition of virtually any radical to alkenes that are substituted at both ends of the double bond, such as β-methylstyrene. That being the case, one would expect that the larger the attacking radical, the greater should be its steric impact upon rate of addition. The validity of this expectation is indicated from that data in Table 8, which compares the rates of addition of various perfluoroalkyl radicals to β-methylstyrene with those to styrene, with the relative rate being taken as a measure of steric impact of attacking radical. It can be seen from this data that only a small steric influence is observed for $CF_3\cdot$, while progressively greater impact is exhibited for C_2F_5, *n*-C_3F_7, *iso*-C_3F_7, and *tert*-C_4F_9, with the rate of addition for the perfluoro-*tert*-butyl radical to β-methylstyrene being 144 times slower than that for its addition to styrene, this in spite the greater overall nucleophilicity of the former substrate as indicated by its IP.

5.1.4.3
Partially-Fluorinated Radicals

In order to determine whether the enhanced reactivities of perfluoroalkyl radicals could be attributed to some linear combination of the individual contributions of fluorine atoms on the α-carbon (the radical center), the β-carbon and the γ-carbon atom, the absolute rates of addition of a number of *partially*-fluorinated alkyl radicals to α-methylstyrene, styrene, and pentafluorostyrene were determined by LFP [70]. The data in Table 9 clearly indicated that this is not the case, that γ- and β-fluorinated-*n*-alkyl radicals exhibit little enhancement, while α-difluoroalkyl radicals, although more reactive, remain very much less-reactive than analogous perfluoro species.

Table 9. Absolute rate constants for reactions of alkyl and fluorine-substituted alkyl radicals with three styrenes in Freon 113 at 298 ± 2° K, as measured by LFP [70]

	$k_{add}/10^6 M^{-1} s^{-1}$		
Radical	$C_6H_5CH=CH_2$	$C_6H_5C(CH_3)=CH_2$	$C_6F_5CH=CH_2$
$CH_3CH_2CH_2CH_2CH_2\cdot$	0.12[a]	0.06[a,b]	0.31
$CF_3(CF_2)_3CH_2CH_2\cdot$	0.13	0.34	0.23
$CH_3CH_2CH_2CF_2CH_2\cdot$	0.52	0.98	0.39
$CH_3CH_2CH_2CH_2CHF\cdot$	0.46		0.70
$CH_3CH_2CH_2CH_2CF_2\cdot$	2.7	3.3	3.1
$CF_3\cdot$	53	87	26
$CF_3CF_2CF_2\cdot$	43	78	13

[a] From [116].
[b] Probably too low by approximately a factor of five.

Looking at the rates of addition to styrene, it can be seen that $RCH_2CHF\cdot$ is 3.8 times more reactive than n-$C_4F_9CH_2CH_2$. However, the $RCH_2CF_2\cdot$ radical is not another 3.8 times more reactive than $RCH_2CHF\cdot$, but is, instead, about 6 times more reactive. That is, the second α-fluorine atom produces an extra (synergistic) enhancement in the radical's reactivity of roughly a factor of two. Furthermore, if we continue this series of α-substitutions all the way to the $CF_3\cdot$ we see that this trend of unexpectedly large rate enhancements continues with the $CF_3\cdot$ radical being about 20 times as reactive as the $RCH_2CF_2\cdot$ radical. Thus $CF_3\cdot$ with its 441-fold rate enhancement (compared to n-pentyl) is about 8 times more reactive than would be expected based upon the incremental impact of three single F substituents. Looking at $CF_3CF_2CF_2$, it can be seen to exhibit a rate enhancement for addition to styrene of 358 over that of the n-pentyl radical, whereas based upon a linear combination of the effects of α-, β- and γ-fluorine substitution, one would expect a rate enhancement of only 104. Thus, *per*fluorination of an n-alkyl radical system gives rise to a synergistic impact of about 3.5-fold increase in rate.

The limited data available for partially fluorinated methyl radicals is consistent with the above data in that $CH_2F\cdot$ and $CHF_2\cdot$ appear to have reactivities roughly comparable to those of $RCH_2CHF\cdot$ and RCH_2CF_2, respectively. For example, for their additions to pentafluorostyrene in acetonitrile: k_{add} $(CH_2F\cdot)=3.5\times10^5$ $M^{-1} s^{-1}$, and k_{add} $(CHF_2\cdot)=5.5\times10^6$ $M^{-1} s^{-1}$ [70].

The data in Table 9 also allows one to reach conclusions regarding the electrophilicity vs nucleophilicity of the partially-fluorinated radicals since the three styrene substrates have a considerable range in IP values. β-Fluorine, and to a lesser extent γ-fluorine, substitution would appear to have a small impact on electrophilicity, whereas a single α-fluorine substituent seems to impart slightly nucleophilic properties. In a recent study, Takeuchi et al. have examined both computationally and experimentally radicals which bear both an α-fluorine substituent and an electron withdrawing ester function. They found that the α-fluorine substituent makes such radicals more electrophilic, but that they still add more readily to styrene than they do to acrylonitrile [127].

α,α-Difluoro substitution appears to give rise to a radical which has neither electrophilic nor nucleophilic characteristics [128]. Two conclusions can be reached on the basis of the partially fluorinated rate studies: (1) polar effects on transition state energies are very much less important for partially fluorinated radicals than for perfluorinated radicals, and (2) the effect on radical reactivity of *per*fluorination is considerably greater than the sum of its parts.

5.1.4.4
Solvent Effects on Rates

As would be expected for reactions with polar transition states, additions of perfluoroalkyl radicals to alkenes are faster in CH_3CN than in Freon 113 with the observed solvent effects being greater for additions to alkenes which are more electron-rich [70, 117]. Table 10 provides comparisons of rates in the two solvents.

Table 10. Solvent effects on the rates of addition of perfluoroalkyl radicals to styrene and pentafluorostyrene at 298 ± 2° K [70, 117]

	$k_{add}/19^6 M^{-1}s^{-1}$			
	Styrene		Pentafluorostyrene	
	F113	CH_3CN	F113	CH_3CN
$CF_3\cdot$	53	171	26	33
$C_2F_5\cdot$	79	127	23	28
n-$C_3C_7\cdot$	43	108	13	27

For example, the rate accelerations in acetonitrile relative to F113 for additions to styrene by $CF_3\cdot$ and $CF_3CF_2CF_2\cdot$ are a factor of 3.2 and 2.5, respectively, but for additions of these two radicals to pentafluorostyrene the solvent effects are only 1.3 and 2.1. For comparison, Salikhov and Fischer have found that the rate of addition of the nucleophilic *tert*-butyl radical to (electron-deficient) acrylonitrile (IP=10.9 eV) is also somewhat accelerated in more polar solvents, e.g., $k_{add}(CH_3CN)/k_{add}(c\text{-}C_6H_{12})=2.8$ [129].

5.2
Hydrogen Atom Abstractions

An understanding of the factors that influence the rates of hydrogen atom abstraction processes is *very* important in order to maximize the utility of radical-based processes in carbon-carbon bond-forming reactions. This is because most such reactions are chain reactions in which one of the key propagation steps involves transfer of a hydrogen atom from some hydrogen atom transfer agent, such as tri-n-butyltin hydride.

The rates of hydrogen atom abstractions by radicals are subject to the same factors that control rates of alkene additions [130]. Both enthalpic and polar factors are very important in determining the rates of hydrogen abstraction. Enthalpic considerations are important in that an abstraction process will be faster (a) the lower the hydrogen BDE of the molecule from which the H is abstracted, and (b) the stronger the new C-H bond which is formed. For example, as indicated in Table 11, the rates of hydrogen abstraction by an alkyl radical are found to correlate very well with the BDEs of a series of related H-donors [131].

Table 11. Some rates hydrogen abstraction by a primary alkyl radical [132]

	Et_3SiH	$(TMS)_2SiMeH$	n-Bu_3GeH	$(TMS)_3SiH$	n-Bu_3SnH
$k_H/10^4 M^{-1}s^{-1}$	0.07	3.2	10	38	240
M-H BDE[a]	90.1[b]	85.3[b]	82.6[c]	79.0[b]	73.6[d]

[a] kcalmol [b] [133] [c] [134] [d] [135].

Polar factors will be important when the "polarity" of the abstracting radical is significantly different from that of the group which comprises the H-donor. A good example is given below, wherein it is seen that although hydrogen-abstraction from HCl by $CF_3\cdot$ is more exothermic by 2 kcal/mol than that of $CH_3\cdot$ [23], its E_a for H-abstraction is double that of CH_3, because $CH_3\cdot$ provides the better match-up of polarities in the abstraction transition state [130].

Reaction	E_a	ΔH^o
$CH_3^{\cdot} + HCl \longrightarrow CH_3\text{-}H + Cl^{\cdot}$	3.5	-2
$CF_3^{\cdot} + HCl \longrightarrow CF_3\text{-}H + Cl^{\cdot}$	8.0	-4

Scheme 24

As in the case for alkene additions, if the SOMO of the radical is relatively high in energy, such as is the case for alkyl radicals, the principal interaction with the abstractable X-H bond will be with its unoccupied σ^* MO (one-electron-two-orbital type), and such a radical would be considered nucleophilic. If the SOMO is relatively low in energy, such as is the case for perfluoroalkyl radicals, the principal interaction with the abstractable X-H bond will be with its occupied σ MO (three-electron-two-orbital type), and the radical is considered *electrophilic.* Either way, a good match-up in polarities in an H-atom transition state will give rise to beneficial transition state charge-transfer interaction [130, 136, 137].

5.2.1
Perfluoro-n-Alkyl Radicals

Early work by Brace indicated that perfluoroalkyl radicals were pretty good abstractors of hydrogen [138], and only five years ago, in a study of the electrochemistry of perfluoroalkyl halides, Saveant et al. made the comment that perfluoroalkyl radicals "are much better H-atom scavengers" than alkyl radicals [139]. Therefore, there was considerable indication that highly electronegative perfluoroalkyl radicals should exhibit significantly greater reactivity towards hydrogen abstraction than their hydrocarbon counterparts.

With the availability of LFP-determined rates for addition of per-fluoro-*n*-alkyl radicals to alkenes, it was possible to determine the rates of hydrogen abstraction using competition methods, of the type shown below [140, 141].

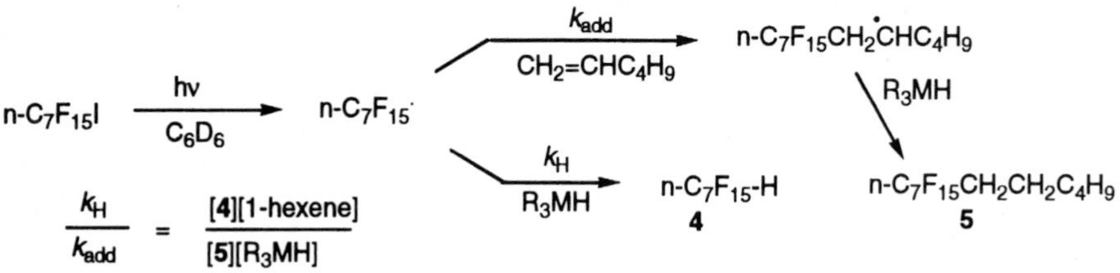

Scheme 25

From these experiments, and using the equation above, it was possible to determine the ratios of rate constants, k_H/k_{add}, and from these ratios it was possible to obtain values for k_H since the value for k_{add} was known. As can be seen from a comparison of Tables 11 and 12, all of the silane, stannane and germane reducing agents exhibit substantial rate enhancements in their transfer of a hydrogen atom to the perfluoro-*n*-alkyl radical in comparison to similar transfer to a hydrocarbon radical [140–142].

Table 12. Absolute rate constants for hydrogen abstraction by perfluoron-*n*-heptyl radical in C_6D_6 at 303 ± 2° K [140–142]

	Et_3SiH	$(TMS)_2SiMeH$	$n Bu_3GeH$	$(TMS)_3SiH$	n-Bu_3SnH
$k_H/10^6 M^{-1}s^{-1}$	0.75	16	15	51	203

Such rate enhancements range from a factor of 75 for the most reactive *n*-Bu_3SnH to 880 for the least reactive $(CH_3CH_2)_3SiH$. Like the rates of H-atom transfer to *n*-alkyl radicals, these rates also exhibit a good correlation with the H-BDEs of the respective reducing agents. Interestingly, such kinetic results indicate that triethylsilane, which reduces hydrocarbon radicals so slowly as to be virtually useless as an effective chain sustaining reducing agent in alkyl radical systems, reduces perfluoro-*n*-alkyl radicals efficiently and at a rate which should make it a very useful agent for relatively slow chain processes involving fluorinated radicals.

Why are perfluoroalkyl radicals so much more reactive with such H-atom donors? The observed hydrogen-atom abstractions by the perfluoro-*n*-octyl radical are a little more *exothermic* than those by an analogous *n*-alkyl radical (BDEs of CF_3CF_2-H and CH_3CH_2-H are 103 and 101 kcal/mol, respectively) [23], and greater rates for such processes were therefore to be expected on the basis of greater exothermicity. However, that this cannot be the entire explanation was evident from a study of the rates of hydrogen-atom abstraction from benzene thiol and its *p*-substituted derivatives. Benzene thiol is an excellent reducing agent for *alkyl* radicals ($k_H = 1.4 \times 10^8 M^{-1}s^{-1}$) [132], transferring a hydrogen atom at a rate >50-times that of *n*-Bu_3SnH.

In contrast, benzene thiol was found to be a relatively poor H-atom transfer agent to perfluoro-*n*-octyl radical, exhibiting a rate of $3.3 \times 10^5 M^{-1}s^{-1}$ which is ~420-times *slower* than its rate of reduction of *n*-alkyl radicals [140, 141]. Since the same relative exothermicities prevail for *this* reduction as for those of the other reducing agents, relative heats of reaction must not be the complete reason for observed differences in H-abstraction reactivity between R· and R_f· The contrasting relative reactivities of electropositive H-atom donors such as silanes, stannanes and germanes, and a relatively electronegative H-atom donor such as benzene thiol, undoubtedly derive from differences in how they facilitate polarity interactions in the transition states for their particular hydrogen transfers. The absolute electronegativies of *n*-alkyl and *n*-perfluoroalkyl radicals [4.00 and 5.9 (value given in Sect. 3.2 for C_2F_5· used)] reflect their respective nucleophilic and electrophilic characters, whereas C_6H_5-S· itself has a value of 5.5, and the electronegativities of the R_3Sn, R_3Ge, and R_3Si radicals, although unknown, should lie below the value for R_3C, which is 3.3 [56]. Therefore, one can see that for H-abstractions from the silane, stannane and germane hydrides, perfluoro-*n*-alkyl radicals should give rise to a particularly good match-up of electronegativities which should lead to more-highly polarized transition states for these H-transfers than for those to an alkyl radical (See Fig. 5).

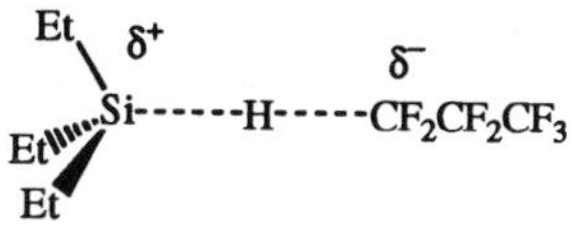

Fig. 5

In contrast, because of similar electronegativities, the transition state for hydrogen transfer from benzene thiol to a perfluoro-*n*-alkyl radical should have little polar character. In confirming the important role of polar effects in these hydrogen transfer processes, a good Hammett correlation was observed for the reduction of *n*-$C_7F_{15}I$ by a series of arene thiols [141, 143].

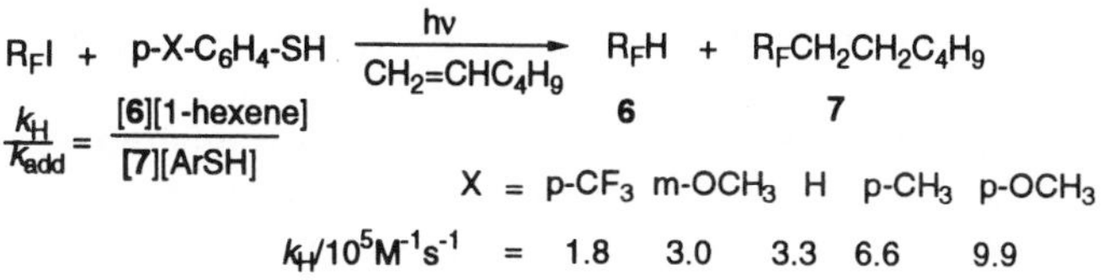

Scheme 26

In this study it was found that the best correlation was with σ^+ values ($\rho^+ = -0.56$), which may be compared to the value of -0.30 observed in the correlation of rates of H-abstraction by *tert*-butoxyl from arene thiols [144]. In fact, it would appear that perfluoro-*n*-alkyl radical reactivity in H-atom abstraction approaches the reactivity/selectivity characteristics of the highly electronegative *tert*-butoxyl radical. For example, both *n*-$C_7F_{15}\cdot$ and $(CH_3)_3C$-O· abstract H from *n*-Bu_3Sn-H at the same rate ($2.0 \times 10^8 M^{-1} s^{-1}$) [145]. However, as the data in Table 13 indicate, the differential in rates becomes quite considerable when it comes to abstraction from a C-H bond, perhaps because of the more sterically-demanding nature of such H-abstractions. Nevertheless, the rates of H-abstraction by *n*-$R_f\cdot$ are still $>10^3$ larger than those of analogous hydrocarbon radicals.

Table 13. Comparison of rates of H-abstraction by perfluoro-*n*-alkyl and *tert*-butoxyl radicals [141, 145]

	$k_H/10^7 M^{-1} s^{-1}$				
	n-Bu_3SnH	*n*-Bu_3GeH	Et_3SiH	THF	Et_3N
n-$C_7F_{15}\cdot$	20	1.4[a]	0.075	0.061[a]	5[a]
$(CH_3)CO\cdot$	20	9.2	0.57	8.3[b]	180[c]

[a] [146] [b] [147] [c] [148].

5.2.2
2° and 3° Perfluoroalkyl Radicals

Only one rate has been measured for hydrogen abstraction by a branched chain, perfluoroalkyl radical, and that was for H-abstraction from Et_3SiH by $(CF_3)_2CF\cdot$. This rate was obtained from a standard series of experiments involving competition between addition of $(CF_3)_2CF\cdot$ to pentafluorostyrene vs abstraction of hydrogen from Et_3SiH [149]. The observed rate, $3 \times 10^6 M^{-1} s^{-1}$ at 30 °C, was a factor of ~5 times faster than the analogous abstraction by *n*-$R_F\cdot$, this in spite of the fact that $(CF_3)_2CF\cdot$ is a planar π-radical. The enhanced reactivity exhibited by $(CF_3)_2CF\cdot$ can be attributed virtually entirely to its significantly greater electrophilicity than *n*-R_F, as was the case for the exceptional reactivities which were also exhibited by $(CF_3)_2CF\cdot$ and $(CF_3)_3C\cdot$ in their alkene addition reactions.

5.2.3
Partially-Fluorinated n-Alkyl Radicals

The trend of reactivities which is observed for hydrogen atom abstraction by partially fluorinated radicals is qualitatively similar to that for their addition to styrene. However, the absolute rates and the range of reactivities for each type of process can be seen to differ significantly. Thus, absolute rate constants for

Table 14. Rate constants for hydrogen abstraction from n-Bu_3SnH and for addition to styrene by various fluorinated radicals at 25 ± 3 °C [31]

Radical	$k_H/10^6 M^{-1} s^{-1}$	k_{rel}	$k_{add}/10^6 M^{-1} s^{-1}$	k_{rel}
$RCH_2CH_2\cdot$	2.4[(e)]	(1)	0.12[(f)]	(1)
$RCH_2CF_2\cdot$ [b]	9.1	3.8	2.7[(g)]	22.5
$RCF_2CH_2\cdot$ [b]	14	5.8	0.52[(g)]	4.3
$R_fCH_2CH_2\cdot$ [c]	2.1	0.9	0.13[(g)]	1.1
$RCF_2CF_2\cdot$ [d]	92	38	20	167
n-$C_7F_{15}\cdot$	200[(h)]	83	46[(i)]	383
$CF_3CF_2\cdot$	320	133	79	658

[b] For k_H experiment, R=n-C_4H_9; for k_{add} expt, R=n-C_3H_7; [c] for k_H expt, R_f =n-C_6F_{13}; for k_{add} expt, R_f =n-C_4F_9; [d] for k_H expt, R =n-C_4H_9; for k_{add} expt, R=C_2H_5; [e] [150]; [f] [113]; [g] [70]; [h] [140, 141]; [i] [114, 115].

hydrogen atom abstraction from tin hydride by the radicals listed in Table 14 are greater by roughly an order of magnitude (range 3.4–27) than the absolute rate constants for addition of these same radicals to styrene. Indeed, the rate constants for hydrogen atom abstraction by the two perfluorinated radicals are within an order of magnitude of the diffusion-controlled limit. As a natural consequence, the range of fluorine-induced changes in reactivity is smaller for the hydrogen abstraction reactions (range 0.9–133) than for the addition reactions (range 1–658).

5.2.3.1
α,α-Difluoroalkyl Radicals

If one considers the C-H bond-weakening effect of α,α-difluoro substitution, (CH_3CH_2-H BDE, 101.1 vs CH_3CF_2-H BDE, 99.5 kcal/mol) [26–28] along with the demonstrated lack of impact of α,α-difluoro substitution on radical electrophilicity [70], it is quite clear that the small (factor of 3.8) rate enhancement for hydrogen atom abstraction from n-Bu_3SnH by $RCH_2CH_2CF_2\cdot$ relative to $RCH_2CH_2CH_2\cdot$ must derive mainly from the pyramidal nature of the former radical [31, 70]. The opposite effect of α,α-difluoro substitution on C-C BDEs (CH_3-CH_3 BDE, 89.9 vs CH_3-CF_3 BDE, 101.2 kcal/mol) helps to explain why there is a much more significant rate enhancement (factor of 22.5) for addition to styrene by $RCH_2CH_2CF_2\cdot$ relative to $RCH_2CH_2CH_2$.

5.2.3.2
B,β-Difluoroalkyl Radicals

The β,β-difluorinated radicals are more interesting in that $RCH_2CF_2CH_2\cdot$ is ca. five times as reactive as $RCH_2CH_2CH_2\cdot$ in both hydrogen atom abstractions from n-Bu_3SnH and in additions to styrenes. Since the $RCH_2CF_2CH_2\cdot$ radicals are effectively planar, their enhanced reactivities must derive from either polar or enthalpic factors. The latter is probably the more important. That is, the C-H bond in $CH_3CF_2CH_2$-H is calculated to be 2.8 kcal/mol stronger than that in $CH_3CH_2CH_2$-H (BDEs 103.1 and 100.3 kcal/mol, respectively) [31] , and the C-C bond in $CH_3CF_2CH_2$-CH_3 is calculated to be 3.2 kcal/mol stronger than that in $CH_3CH_2CH_2$-CH_3 (BDEs 89.9 and 86.7kcal/mol, respectively) [31]. The greater exothermicities of the $RCH_2CF_2CH_2\cdot$ radical reactions appear to be quite sufficient to account for its very modest increase in reactivity relative to primary alkyl radicals [151].

5.2.3.3
γ,γ-Difluoroalkyl Radicals

The γ,γ-difluorinated radical $R_FCF_2CH_2CH_2\cdot$ has essentially the same reactivity as a primary alkyl radical and essentially (within 1.1 kcal/mol) the same C-C and C-H BDEs as in the corresponding alkane [31].

5.2.3.4
Polyfluorinated Radicals

All α,α-difluoroalkyl radicals are pyramidal, and furthermore, the degree of bending at the radical center would appear to be rather similar whether or not the radicals are more extensively fluorinated [31, 70]. The enhanced reactivity of the $RCH_2CF_2CF_2$, $R_FCF_2CF_2CF_2$, and $CF_3CF_2\cdot$ radicals relative to $RCH_2CH_2CF_2\cdot$ radicals must therefore arise from either the greater electronegativity of these polyfluorinated radicals, or to more favorable thermodynamic factors, or both. The trend in reactivities for these radicals correlates poorly with the thermodynamic driving force for the reactions in question [31]. Thus, for the hydrogen atom abstraction, the rate constants (relative to that for the *n*-alkyl radical) for $RCH_2CH_2CF_2$, $RCH_2CF_2CF_2$, and $CF_3CF_2\cdot$ are 3.8, 38 and 133, respectively, whereas the relevant C-H BDEs for the products which would be formed are 97.7, 100.1, and 99.5 kcal/mol, respectively [31]. Similarly, the rate constants (relative to *n*-alkyl) for the addition to styrene of these same radicals are 22.5, 167 and 658, respectively, whereas the relevant C-C-BDEs are 91.6, 95.4 and 95.5 kcal/mol, respectively [31]. Therefore, although there may be a small thermodynamically-induced rate enhancement for the abstraction and addition reactions on changing from the -$CH_2CF_2\cdot$ moiety to the -$CF_2CF_2\cdot$ and $CF_3CF_2\cdot$ moieties, it cannot fully account for the observed changes in rate constants. We therefore conclude that polar effects arising from the greater electronegativities of $RCH_2CF_2CF_2$, $R_FCF_2CF_2CF_2\cdot$ and $CF_3CF_2\cdot$ vs that of $RCH_2CH_2CF_2\cdot$ are responsible for the enhanced reactivities of these first three radicals relative to the last.

The synergistic impact of multiple fluorine substitution on radical reactivity has been discussed in the earlier section on alkene additions and will not be remarked upon again at this time [70].

5.3
Radical Rearrangements

In recent years, radical cyclization processes, particularly those of the 5-hexenyl system, have become very important tools within the synthetic repertoire of chemists who wish to construct five-membered rings either singly or in a tandem fashion [132]. Like other productive radical-based synthetic processes, these reactions are chain reactions in which one of the key propagation steps involves transfer of a hydrogen atom from some reducing agent, in the case of hydrocarbons usually n-Bu_3SnH or $[(CH_3)_3Si]_3SiH$. Such cyclizations have long been utilized for the purpose of gaining insight into reactivity factors which pertain to cyclization processes, but which also inevitably provide considerable insight into the chemistry of alkene addition processes. For example, our understanding of Baldwin's Rules [152], and their underlying factors [153], was enhanced by the study of such systems, and the proposition of the "Beckwith chair" transition state provided a breakthrough in providing understanding of the regio- and stereochemistry of such cyclizations [154–157]. Much insight has also been obtained related to the influence of substituents at or near the radical site (of Thorpe-Ingold, or gem-dimethyl, nature [158, 159], or of electronic origin) [160]. Until recently, however, there had been no quantitative studies of how fluorine substituents affected the rate or regiochemistry of the cyclization process.

Once a reasonable arsenal of reducing agents with accurately determined rate constants for H-atom transfer had been acquired for perfluoro-n-alkyl radicals, then it became possible to take a quantitative look at *cyclizations* of fluorinated radicals. As a result, a series of perfluoro- and partially-fluorinated alkenyl radical systems have been examined for the purpose of obtaining the rates and regiochemistries of their cyclizations [140, 161, 162]. All of these systems were studied by means of the competition method, where cyclization was allowed to compete with direct reduction of the acyclic fluorinated radical, usually using Et_3SiH as the hydrogen-transfer agent.

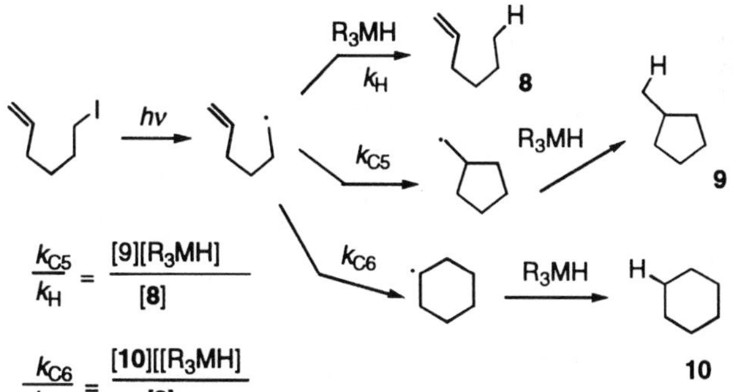

$$\frac{k_{C5}}{k_H} = \frac{[9][R_3MH]}{[8]}$$

$$\frac{k_{C6}}{k_H} = \frac{[10][[R_3MH]}{[8]}$$

Scheme 27

5.3.1
5-Hexenyl Radical Cyclizations

The kinetic study of the *perfluoro*-5-hexenyl radical system **11** led to remarkable results in that its rate constant for cyclization, k_{C5}, and its regiochemistry (i.e., dominant *exo-trig*) were only slightly different from those of the parent hydro-carbon system ($k_{rel} = 2.0$), with k_{C6} being negligible for both systems [140, 162].

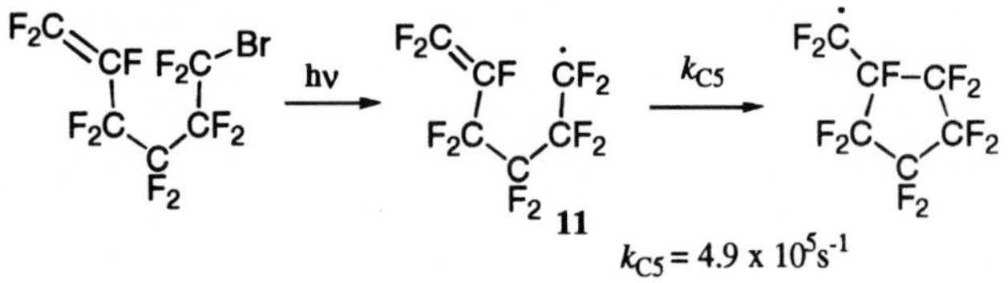

Scheme 28

This similarity in reactivities probably derives from a fortuitous cancellation of substituent effects in **11**. Fluorination increases chain stiffness and creates an unfavorable polarity mismatch between an electrophilic radical and an electron-poor double bond, but this is offset by the significant decrease of π-bond energy in **11**. The vinyl ether **12** analog cyclizes about seven times faster than **11**, which is consistent with the known lower π-bond energy and higher free-radical reactivity of perfluorovinyl ethers vs perfluoroalkenes [142].

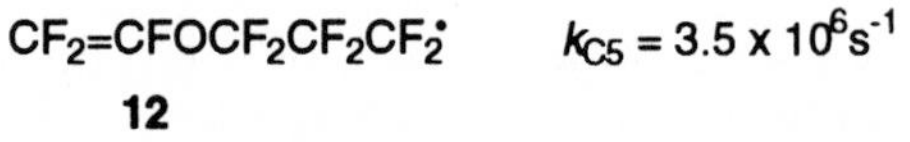

$$CF_2{=}CFOCF_2CF_2CF_2^{\bullet} \qquad k_{C5} = 3.5 \times 10^6 \, s^{-1}$$

Scheme 29   **12**

A kinetic study of *partially*-fluorinated 5-hexenyl radicals has demonstrated the kinetic importance of polarity factors while providing substantial insight into a number of factors which affect both the rate and the regiochemistry of 5-hexenyl radical cyclizations.

It was initially presumed that one would observe a polarity-driven kinetic advantage in 5-hexenyl radical cyclizations if one constructed such radicals with either a hydrocarbon radical site adding to a fluorinated alkene segment or, vice-versa, with a fluorinated radical site adding to a hydrocarbon alkene segment. As it turns out, only the latter combination led to a significant cycliza-tion rate enhancement.

5.3.1.1
Cyclizations Involving Hydrocarbon Radical/Fluorinated Alkene

Kinetic data for radicals **13–17**, all of which involve 5-hexenyl radical cycliza-tions of a 1° hydrocarbon radical site onto a fluorinated alkene segment, given in Table 15, indicate that the degree of fluorination of the double bond has little impact upon the rate of cyclization of a 5-hexenyl radical [163, 164].

Only the 5-fluoro-, **13**, and the tri- and pentafluoro systems, **17** and **18**, exhibit any significant deviation from the cyclization rate of the parent system, with the first being significantly slowed, and the latter two being slightly enhanced.

Table 15. Absolute rates for 5-hexenyl radical cyclizations in involving hydrocarbon radical adding to fluorinated alkene, at 30 °C in C_6D_6 [161, 162]

Radical	$k_{C5}/10^5 s^{-1}$
$CH_2{=}CH(CH_2)_3CH_2\cdot$	2.7[a]
$CH_2{=}CF(CH_2)_3CH_2\cdot$, **13**	0.25
$CHF{=}CH(CH_2)_3CH_2\cdot$, **14**	1.7
$CF_2{=}CH(CH_2)_3CH_2\cdot$, **15**	2.1
$CHF{=}CF(CH_2)_3CH_2\cdot$, **16**	2.0
$CF_2{=}CF(CH_2)_3CH_2\cdot$, **17**	6.1
$CF_2{=}CF(CF_2)_2CH_2CH_2\cdot$, **18**	4.3

[a] [116].

Such relative lack of kinetic impact of olefinic fluorine substituents on alkyl radical addition reactions is consistent with Tedder's early studies on methyl affinities, the results of which are shown in Table 16, where the range of reactivities observed for the addition of methyl radical to ethylenes with varying fluorine content is seen to be relatively small [93, 163].

Table 16. Relative rates of addition of methyl and trifluoromethyl radicals to some fluoroethylenes[a]

Radical	$CH_2{=}CH_2$	$CH_2{=}CHF$	$CHF{=}CH_2$	$CHF{=}CF_2$	$CF_2{=}CHF$
$CH_3\cdot$	(1)	0.9	0.2	1.9	3.9
$CF_3\cdot$	(1)	0.45	0.05	0.033	0.017

[a] [93, 163].

A single fluorine substituent at C-5 (as in radical **13**) leads to a significant, ~11-fold decrease in rate constant. This decrease no doubt derives largely from the steric effect which would be expected from any substituent at the 5-position. A methyl substituent, for example, gives rise to a 45-fold decrease in cyclization rate [164]. Interestingly, whereas the presence of a 5-methyl substituent causes endo-cyclization to become preferred (63%), the cyclization of 5-fluoro-5-hexenyl radical remains *exo*-specific within our NMR analytical methodology (±4%).

A slight overall enhancement in reactivity (2.2-fold) is observed in the cyclization of the 5,6,6-trifluoro-5-hexenyl radical (**17**). The π-bond of **17** thus is at least reactive enough (consider that the heat of hydrogenation of trifluoroethene (–45.7 kcal/mol) is 13 kcal/mol greater than that of ethylene) [165] to overcome the steric inhibition of its 5-fluoro substituent. Polar influences, although possibly of some minor importance in the cases of **17** and **18**, should not be playing a significant role in *any* of these cyclizations, since the reported electron affinities of ethylene (–1.78 eV), fluoroethene (–2.39 eV), *cis*- and *trans*-1,2-difluoroethene (–2.18 and –1.84 eV), and 1,2,2-trifluoroethene (–2.45 eV) encompass a total range of only 0.7 eV [166]. In the olefin addition reactions of the more nucleophilic tert-butyl radical, Fischer observed a rate variation of only ~5 got olefins with a 0.7 eV difference in EA [105]. Therefore one would not expect polar influences to be very significant for cyclizations of **13**–**17**.

The 3,3,4,4,5,6,6-heptafluoro-5-hexenyl radical, **18**, exhibits very little rate enhancement relative to the hydrocarbon parent ($k_{rel}=1.6$), and its rate constant is virtually the same as that of the perfluoro radical, **11**. A recent study of the reactivity of $R_FCH_2CH_2\cdot$ – type radicals demonstrated that they do not exhibit electrophilic character in their additions to alkenes [70]. They are π-radicals with a reactivity profile much like that of an *n*-alkyl radical.

The reactivities of vinyl fluorine-substituted radicals **14**–**16** can be effectively rationalized in terms of combinations of modest steric and enthalpic effects. The lack of significant influence of single or geminal fluorine substituents at the 6-position or of vicinal, 5,6-difluoro substituents likely derives from a canceling out of advantageous and disadvantageous effects in each case. The single 6-fluoro substituent should stabilize by approximately the same amount, both the olefin [167] and the radical which results from cyclization [31, 168]; hence, no resultant net effect. Geminal 6,6-difluoro substituents appear to stabilize the π-system slightly, based upon the 3.7 kcal/mol greater π-bond dissociation energy ($D_\pi{}^0$) of $CH_2{=}CF_2$ than that of ethylene [165]. With the stability of the resultant radical from cyclization being essentially unaffected by the presence of the geminal fluorine substituents, there should be little effect on the cyclization rate constant by 6,6-difluoro substitution. Thermodynamic data indicate that vicinal fluorination, such as in the 5,6-difluoro system, **16**, destabilizes the π-system by ~5 kcal/mol [165]. This, combined with the small stabilization of the cyclized radical, are apparently enough to offset the steric inhibition of the 5-fluoro substituent to give the observed kinetic result.

5.3.1.2
Cyclizations Involving Fluorinated Radical Adding to Hydrocarbon Alkene

In contrast, when the mode of substitution is reversed, that is when the radical is fluorinated and the alkene fragment is not, as is the case with radicals **19**, **21**–**23**, a much greater impact on reactivity is observed.

The overall reactivities of these radicals in their *uni*molecular 5-hexenyl cyclization processes reflects those same factors which affect the reactivity of partially-fluorinated radicals in their bimolecular addition reactions with alkenes, such as styrene. Table 17 indicates this clearly, and it also reflects the general leveling effect which would be expected for the more facile unimolecular cyclization processes which have log A's about 1–2 units larger than those for the bimolecular additions.

Our earlier studies of the bimolecular alkene addition reactivity of α,α-difluoro alkyl radicals indicated that they exhibited little "philicity", reacting with styrene and pentafluorostyrene (IPs 8.43 and 9.20, respectively) at virtually the same rate [70]. Their significantly greater reactivity in bimolecular additions, hydrogen-abstraction reactions and unimolecular cyclizations can therefore be largely attributed to their pyramidal nature, with some possible thermodynamic con-

Table 17. Absolute rate constants for 5-hexenyl radical cyclizations involving fluorinated radical addition to hydrocarbon alkens, at 30 °C in C_6D_6 [140, 162]

Radical	$k_{C5}/10^6 s^{-1}$	k_{C5}(rel)	$k_{C6}/10^6 s^{-1}$	k_{C6}(rel)
$CH_2=CH(CH_2)_3CH_2\cdot$ [a]	0.27	(1)	0.0055	(1)
$CH_2=CH(CH_2)_3CF_2\cdot$, **19**	3.5	13	–	–
$CH_2=CH(CH_2)_2CF_2CH_2\cdot$, **20**	1.1	4.1	0.11	20
$CH_2=CH(CH_2)_2CF_2CF_2\cdot$, **21**	8.8	32.6	1.9	345
$CH_2=CHCH_2(CF_2)_2CF_2\cdot$, **22**	44	163	5.2	945
$CH_2=CH(CF_2)_3CF_2\cdot$, **23**	11	40.7	3.5	636

[a] [116].

tribution, as has been discussed in the earlier section on bimolecular additions of such radicals.

The slight enhancement observed for cyclization of radical **20** is consistent with the slight electrophilicity of such radicals which was demonstrated earlier in the studies of their bimolecular olefin addition reactivity [70]. The similar reactivities of **20** and hydrocarbon parent are consistent with the similarity of the ESR parameters for these two types of radicals [169]. That is they are both effectively planar π-radicals.

The $\alpha,\alpha,\beta,\beta$-tetrafluoro-5-hexenyl radical, **21**, of course, retains the reactivity which comes from its σ-nature, but it also gets a significant boost in reactivity because of its substantial electrophilic character. Although LFP data is limited, the rate constant for addition of $CH_3CH_2CF_2CF_2\cdot$ to styrene ($k_{rel}=167$, relative to n-alkyl) indicates a substantial enhancement compared to an α,α-difluoroalkyl radical [70]. Its rate of H-abstraction from n-Bu_3SnH ($k_{rel}=38$, compared to n-alkyl) also reflects the very favorable matchup of transition state polarities which is characteristic of highly fluorinated radicals [137, 138]. Lastly, the sparse IP data which is available for such radicals (see Table 3) also reflects a significant electrophilicity, approaching but not equal to that of perfluoroalkyl radicals.

In the system with three CF_2 groups, i.e. **22**, the radical takes on perfluoroalkyl character and the impact on cyclization rate is magnified still further. The dominant factor which has been credited for giving rise to the high reactivities of perfluoro-n-alkyl radicals in their additions to alkenes, particularly to electron-rich alkenes, is their high electrophilicities. That is, charge transfer interactions, e.g. $[(CF_3CF_2CF_2)^{\delta-}\ (\text{alkene})^{\delta+}]^{\ddagger}$ stabilize an early transition state and lower both the enthalpic and entropic barriers to reaction.

The large rate enhancement observed for cyclization of **22** is consistent with the 30 000-fold polarity-driven rate ratio for n-$C_3F_7\cdot$ vs $RCH_2CH_2\cdot$ addition to 1-hexene [115]. Nevertheless, system **22** is still not ideal in terms of transition state polarity matchup because the proximity of the perfluoroalkyl group to the olefinic segment should serve to diminish its nucleophilicity significantly.

The reactivity of the octafluoro-5-hexenyl radical system **23** is seen to be diminished relative to that of the hexafluoro system **22**. This can be attributed, at least in part, to the impact of the perfluoroalkyl group on the nucleophilicity of the terminal alkene segment which will serve to further diminish the nucleophilicity of the terminal double bond of **23**, and hence make it less reactive with its highly fluorinated radical terminus.

5.3.1.3
Regiochemistry

Whereas cyclizations of parent hydrocarbon, perfluoro-, and most of the partially-fluorinated 5-hexenyl radical systems occur with the dominant exo-selectivity we have come to expect in such cyclizations, radicals **21**–**23** and even **20** exhibit an unexpected enhanced propensity to undergo a 6-*endo* cyclization [140, 162] as indicated in Table 18. The 25% endo cyclization exhibited by **23**, for example, means that this cyclization proceeds 700 times faster than the endo cyclization of the parent radical.

Table 18. Regiochemistry of some cyclizations of fluorinated 5-hexenyl radicals [170]

	Endo-Selectivity	
Radical	Observed	Predicted
$CH_2{=}CHCH_2CH_2CF_2CF_2\cdot$ (**21**)	18.8% endo	15.6%
$CH_2{=}CHCH_2CF_2CF_2CF_2\cdot$ (**22**)	10.6%	5.7%
$CH_2{=}CHCF_2CF_2CF_2CF_2\cdot$ (**23**)	25%	24.2%
$CH_2CHCH_2CH_2CF_2CH_2\cdot$ (**20**)	8.8%	8.9%

All others, < 4%.

At the present time there is not a good explanation for the regiochemical diversity exhibited by these radicals in their cyclizations. It should be noted, however, that such diversity was predicted computationally [170].

5.3.2
4-Pentenyl Radical Cyclizations

Cyclizations to form 4-membered rings are rare in hydrocarbon systems and only occur when there is a radical stabilizing group at the terminus [171, 172]. However, because fluorinated cyclobutanes appear to be less-strained than their

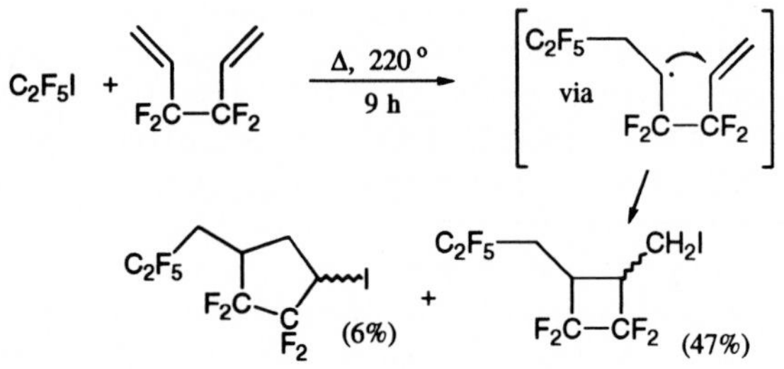

Scheme 30

hydrocarbon counterparts [165], cyclizations of fluorinated 4-pentenyl radicals appear to be both kinetically and thermodynamically feasible. Piccardi was the first to observe such a cyclization in the thermal addition reactions of C_2F_5I and CCl_4 to 3,3,4,4-tetrafluoro-1, 5-hexadiene, where 4-*exo-trig* 4-membered ring formation was found to be favored over 5-*endo-trig* 5-membered ring formation [173].

Nevertheless, in spite of this observation, it was found that the parent system, the perfluoro-4-pentenyl radical, **24**, failed to cyclize even when Et_3SiH was employed as the hydrogen atom transfer agent [140]. Either the equilibrium between **24** and the cyclized radical must be very unfavorable or the rate constant for **24**'s cyclization is less than 1×10^4 s^{-1} at 30 °C. Thus the lack of cyclization of **24** could be due to either thermodynamic or kinetic factors.

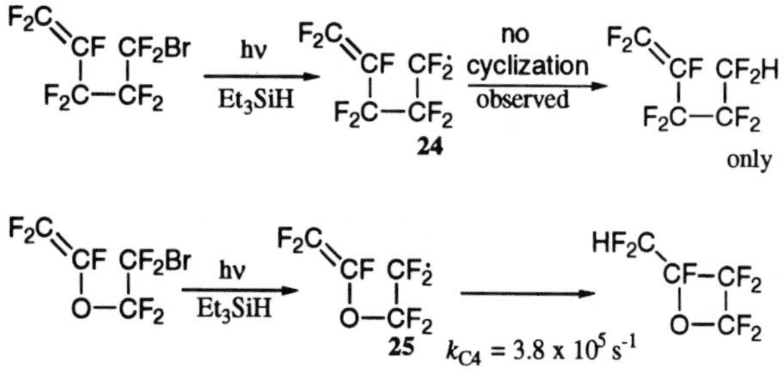

Scheme 31

Either way, ether analog **25** was expected to be more reactive, and indeed, **28** was found to cyclize quite efficiently with a rate constant for cyclization of 3.8 $(\pm 0.3)\times10^5\,s^{-1}$ [140]. Only the *exo*-mode of cyclization was observed, in contrast to Piccardi's results. Further studies of fluorinated 4-pentenyl and cyclobutylcarbinyl radical systems will hopefully provide eventual definitive insight into those factors which govern rates and equilibria in this system.

5.3.3
Cyclopropylcarbinyl Radical Ring-Openings

The cyclopropylcarbinyl ring opening comprises a very fast clock process in the hydrocarbon system, with a rate constant of $9.4\times10^7\,s^{-1}$ [132]. The ring opening of the 2,2-difluorocyclopropyl carbinyl radical, **26**, occurs with exclusive C_1-C_3 bond cleavage to produce the 2,2-difluorobut-3-enyl radical, **27** [174].

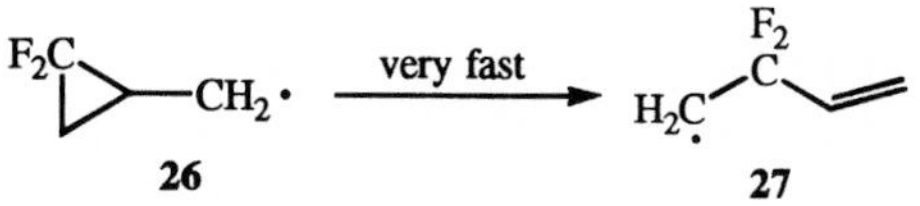

Scheme 32

Such a result is consistent with a wealth of data relating to the thermal isomerizations of gem-difluorocyclopropane systems which indicate that the cyclopropane C-C bond distal to the CF_2 group is substantially weaker towards homolytic cleavage than the proximal bonds [175]. The rate for this ring opening has not yet been determined, although it is certainly considerably faster than the already very fast ring-opening of the parent, hydrocarbon system. With only 3,3-difluorobutene being observed as a product in the reduction of 2,2-difluoro-1-bromomethylcyclopropane in neat n-Bu_3SnH, a minimum rate of 8×10^8 can be calculated for the cleavage of radical **26**.

There are just a few examples of such radical ring-openings in the literature (See Sect. 6.6.4), and all of them proceed with exclusive distal bond cleavage and with no observed trapping of the precursor cyclopropylcarbinyl radical.

5.4 Summary

In summary, perfluoroalkyl radicals exhibit extraordinary reactivity in both their alkene addition reactions and their hydrogen-abstraction processes, relative to their hydrocarbon counterparts. This reactivity can be attributed partially to the increased exothermicity of such reactions when compared to the analogous reactions of hydrocarbon radicals, and partially also to the fact that perfluoro-n-alkyl radicals are σ-radicals. However the major source of the reactivity of 1°, 2°, 3° perfluoroalkyl radicals must be their high electronegativity, which gives rise to stabilizing polarization of the transition states of these radicals' addition and hydrogen-abstraction processes.

6 Reactions Involving Fluorinated Alkyl Radicals

Interest in organic free radical reactions has increased in recent decades as radical-based methodology for organic synthesis has evolved, particularly with regard to carbon-carbon bond forming reactions [176–179]. The incorporation of fluorinated alkyl groups into organic compounds has also become an area of increasing interest as the beneficial effect of such substituents upon the pharmacological properties of molecules has become recognized [180, 181].

Because perfluoroalkyl radicals are easily generated by a variety of means, and because they exhibit high reactivity towards diverse types of organic substrates, the use of fluoroalkyl radicals has become a preferred method for the incorporation of fluoroalkyl groups into organic compounds.

6.1 Disproportionation and Coupling

Because of the very strong β-C-F bonds of perfluoroalkyl radicals, such species do not disproportionate. Thus the only combinatorial reaction of perfluoroalkyl radicals is that of *coupling*. When two $R_F\cdot$ radicals are generated within a solvent cage, as is the case with perfluoroazoalkane and perfluoroketone photolyses, there always results a significant amount of in-cage coupling (in the case of perfluoroazomethane this consumes about 25% of the radicals before they emerge from the cage) [86] which limits the use of such methodology in synthesis.

Nevertheless, Nakamura and Yabe found that the photolytic generation of *n*-$C_8F_{17}\cdot$ from perfluoroazooctane in the presence of naphthalene to be a useful method for synthesis of the isomeric perfluoro-*n*-octylnaphthalenes [74].

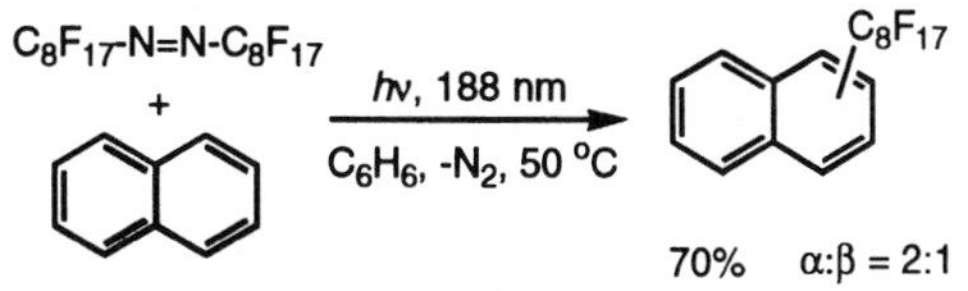

Scheme 33

Sometimes cage recombination can be used to advantage, such as in Golitz and de Meijere's method for coupling alkyl groups to trifluoromethyl groups in which he enhances the amount of cage recombination by use of a viscous, non-H-atom donating solvent [182].

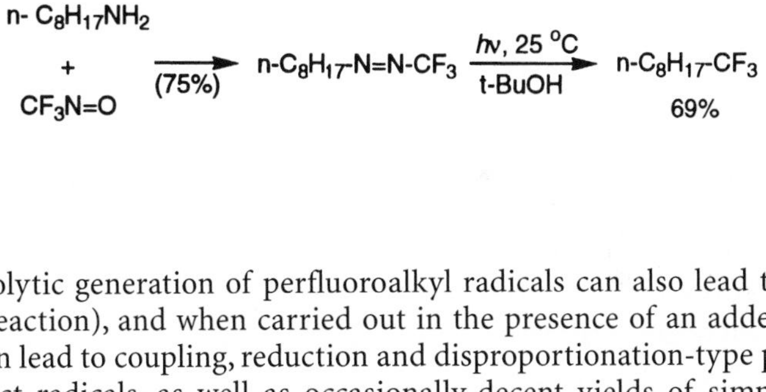

Scheme 34

Electrolytic generation of perfluoroalkyl radicals can also lead to coupling (Kolbe Reaction), and when carried out in the presence of an addend, like an olefin, can lead to coupling, reduction and disproportionation-type products of the adduct radicals, as well as occasionally-decent yields of simple adducts [183–188].

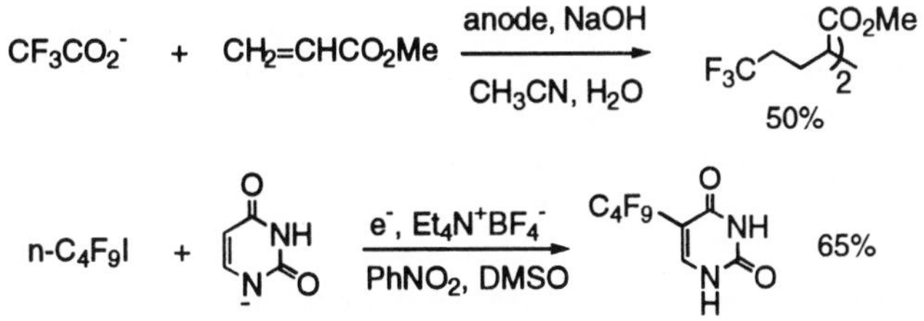

Scheme 35

Thermal unimolecular decomposition of perfluorodiacyl peroxides seems to be less prone to cage-recombination, with only 5% of coupling remaining when such decomposition is carried out in the presence of an excess of a radical scavenger such as CCl_3Br [68]. Of course, donor-induced decomposition of diacyl peroxides leads to clean chain processes with virtually no radical recombination being observed [66].

6.2
Thermolytic Cleavage of Persistent Perfluoroalkyl Radicals

Because perfluoroalkyl radicals cannot disproportionate, when they are constructed in such a way that they are also unable to couple (by means of steric isolation of the radical center), it becomes possible to study the thermal homolytic behavior of these persistent radicals. It has been found that they generally extrude CF_3· radicals when heated. As a matter of fact, the facility and temperature variability of such extrusions make these persistent radicals good free radical initiators.

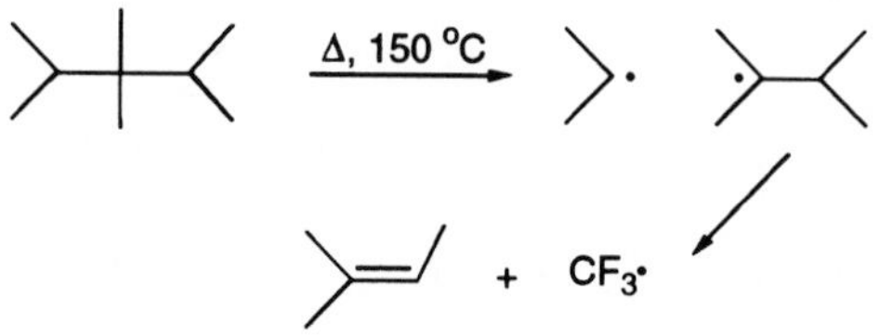

Scheme 36

The ease of CF_3· extrusion depends upon the degree of steric strain in the mother radical [81].

6.3
Addition to Unsaturated Systems

Most useful reactions of perfluoroalkyl radicals involve efficient chain processes, and the challenge has been to find conditions where efficient propagation of the chain via chain transfer can occur. The development of such methodology has derived largely from the huge amount of work which has been devoted to studies of the addition of perfluoroalkyl radicals to unsaturated systems, particularly olefins.

Perfluoroalkyl iodides serve as the primary source of the propagating radicals in such additions, although there are situations where other precursors such as perfluoroalkyl sulfonyl bromides and diacyl peroxides may be used effectively.

6.3.1
Thermal and Photochemical, Homolytically-Induced Additions

The discovery in the late 1940s by Emeleus and Haszeldine that perfluoroalkyl iodides could be cleaved by light or heat to give perfluoroalkyl radicals certainly has proved to be historically one of the most important advances in synthetic fluorine chemistry [189]. Haszeldine was the first to recognize the potential ability of perfluoroalkyl iodides to take part in free radical chain processes which involve the intermediacy of perfluoroalkyl radicals [91, 190, 191], and such fundamental thermal and photochemical methodologies continue to be used and continue to evolve as useful techniques for adding perfluoroalkyl radicals to olefins and alkynes [192].

$$CF_3I \ + \ CH_2{=}CH_2 \xrightarrow[\text{48 h}]{\Delta,\ 250\,°} CF_3CH_2CH_2I\ (75\%) \ + \ \text{higher telomers (25\%)}$$
$$\text{(excess)}$$

$$CF_3I \ + \ CH_2{=}CH_2 \xrightarrow[\text{46\% conv.}]{h\nu,\ 108\ h} CF_3CH_2CH_2I\ (82\%) \ + \ \text{higher telomers (18\%)}$$
$$\text{(excess)}$$

$$CF_3I \ + \ CF_2{=}CF_2 \xrightarrow[\text{10 h}]{\Delta,\ 200\,°} \text{mostly telomers}$$
$$1:1$$

$$CF_3I \ + \ CH_2{=}CHCH_3 \xrightarrow{h\nu} CF_3CH_2CHICH_3\ (98\%)$$

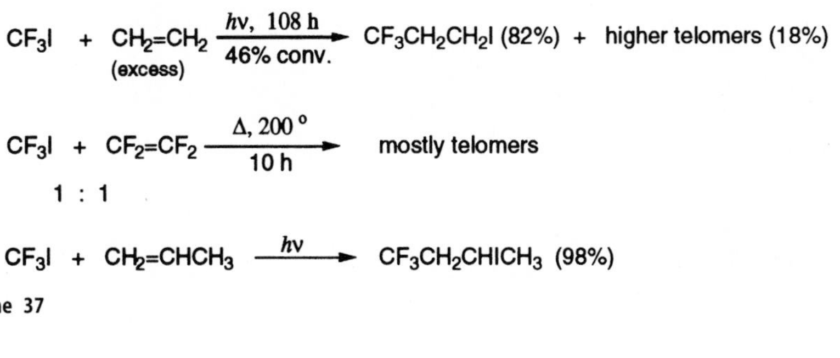

Scheme 37

An interesting recent example of a successful thermal process involves the thermal reaction of perfluoroalkyl iodides with perfluoroallyl chloride [193].

$$C_8F_{17}I \ + \ CF_2{=}CF\text{-}CF_2Cl \xrightarrow[74\%]{\Delta,\ 250\ °C} C_8F_{17}CF_2\text{-}CF{=}CF_2$$

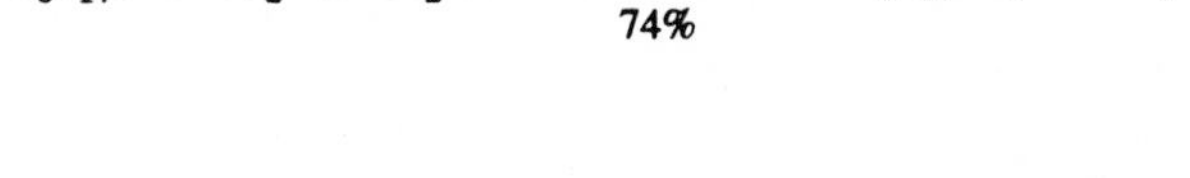

Scheme 38

The photo-initiated addition process appears to have general applicability, although it can require extensive photolysis times [194–196]. Indeed, photolytic generation of $R_F\cdot$ from R_FI has been the method used to add $R_f\cdot$ to C_{60} and C_{70}, not for synthetic purpose, but to examine epr spectra of the resulting radical species [197–199]. A good comprehensive review of the early work on thermal and photochemically-induced free radical addition reactions to olefins can be found in Sosnovsky's book [60].

$$n\text{-}C_7F_{15}I \ + \ CH_2{=}CH\text{-}(CH_2)_5CH_3 \xrightarrow{h\nu} n\text{-}C_7F_{15}\text{-}CH_2CHI(CH_2)_5CH_3\ (87\%)$$

Scheme 39

In a recent related piece of work, Burton's group has discovered that one can obtain excellent yields of 1:1 adducts to electron deficient olefins such as ethyl acrylate by the use of low intensity 254 nm light [200].

$$\text{n-C}_3\text{F}_7\text{I} \quad + \quad \text{CH}_2\text{=CHCO}_2\text{Et} \quad \xrightarrow[\text{24 h}]{hv, \ 254 \ nm} \quad \text{n-C}_3\text{F}_7\text{-CH}_2\text{CHICO}_2\text{Et} \quad (100\%)$$

$$1 \ : \ 2$$

Scheme 40

Although the regiochemistry for such additions is usually such that the $R_F\cdot$ adds to the terminal, least-highly-substituted end of the olefin, unusual regio-chemistries can be observed for additions of perfluoroalkyl radicals, probably because of the intervention of polar effects [91, 201].

$$\text{CF}_3\text{I} \quad + \quad \text{CF}_2\text{=CH-CH}_3 \xrightarrow[\substack{hv, \ 4d, \\ 100\,^\circ}]{} \text{CF}_3\text{CH(CH}_3)\text{CF}_2\text{I} \quad (70\%)$$

$$\text{CF}_3\text{I} \quad + \quad \text{CF}_2\text{=CH-CF}_3 \xrightarrow{} \text{CF}_3\text{CH(CF}_3)\text{CF}_2\text{I} \quad (80\%)$$

Scheme 41

Good yields of addition to benzene and its derivatives have also been repor-ted under both thermal and photochemical conditions [202, 203].

$$\text{n-C}_7\text{F}_{15}\text{I} \quad + \quad \langle\bigcirc\rangle \quad \xrightarrow[\text{15 h}]{\Delta, \ 250\,^\circ} \quad \langle\bigcirc\rangle\text{-C}_7\text{F}_{15} \quad (62\%)$$

$$\text{CF}_3\text{I} \quad + \quad \langle\bigcirc\rangle \quad \xrightarrow[\text{100 h}]{hv, \ Hg} \quad \langle\bigcirc\rangle\text{-CF}_3 \quad (65\%)$$

$$1 \ : \ 5$$

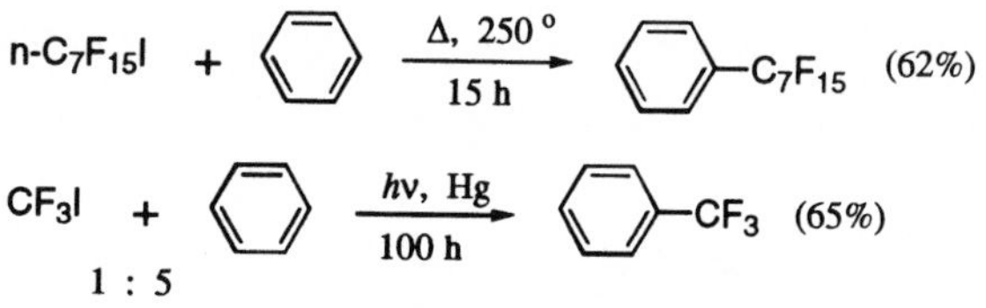

Scheme 42

6.3.2
Telomerization of Fluoroolefins

It has been recognized since the time of Haszeldine's first reports that the thermal and photochemically-induced additions of perfluoroalkyl radicals to olefins were prone to lead to telomeric products due to competition between the processes of chain-transfer and propagation. Indeed, extensive studies have demonstrated that the degree of telomerization is dependent upon a number of factors, including (a) the relative concentration of olefin and telogen (chain-transfer agent), (b) the relative steric effect for propagation vs chain-transfer, (c) reactivity factors regarding the propagating radical, (d) the reactivity of the chain transfer agent, particularly with respect to the strength of the bond which is broken in the chain-transfer step, (e) reaction temperature, and (f) reaction time.

For example, Haszeldine found that in the reaction of CF_3I with TFE, if the ratio of CF_3I:TFE is kept high, then formation of 1:1 adduct will be predo-minant [204].

$$CF_3I \quad + \quad CF_2=CF_2 \quad \xrightarrow{h\nu} \quad CF_3(CF_2CF_2)_nI$$

$$\text{ratio} = \begin{matrix} 10:1 \\ 5:1 \\ 1:1 \\ 1:10 \end{matrix} \quad \begin{matrix} 94\% \ (n=1), \ 4\%(n=2) \\ 81\% \ (n=1) \\ 16\% \ (n=1), \ 10\% \ (n=2), \ 5\% \ (n=3), \ 63\% \ (n>3) \\ n = 10\text{-}20 \end{matrix}$$

Scheme 43

Similarly thermally:

$$CF_3I \quad + \quad CF_2=CF_2 \quad \xrightarrow[8\ h]{\Delta, \ 200\,^\circ} \quad CF_3(CF_2CF_2)_nI$$

large excess of CF_3I 50% (n=1), 20% (n=2), 8% (n=3)

ratio = 1 : 2   9% (n=1), 3 % (n=1&2), 87% (n>3)

Scheme 44

The relative efficiencies of CF_3I, C_2F_5I, $n\text{-}C_3F_7I$ and $(CF_3)_2CFI$ as chain transfer agents were studied within the context of the telomerization reactions of $CH_2=CF_2$ and TFE and it was found under almost identical conditions and a 1:1 ratio of chain transfer agent to olefin that CF_3I was the poorest and $(CF_3)_2CFI$ the best at inducing 1:1 adduct formation, a result consistent with the relative C-I bond strengths for these iodides [205].

$$R_FI \quad + \quad CH_2=CF_2 \quad \xrightarrow[17\text{ - }45\ h]{185\text{ - }200\,^\circ} \quad R_F(CH_2CF_2)_nI$$

$$R_F = \begin{matrix} CF_3 \\ C_2F_5 \\ n\text{-}C_3F_7 \\ i\text{-}C_3F_7 \end{matrix} \quad \begin{matrix} 46\% \ (n=1), \ 33\%(n=2), \ 15\% \ (n=3), \ 7\% \ (n>3) \\ 92\% \ (n=1), \ 6\% \ (n=2), \ 2\% \ (n=3) \\ 70\% \ (n=1), \ 25\% \ (n=2), \ 5\% \ (n=3) \\ 90\% \ (n=1), \ 10\% \ (n=2) \end{matrix}$$

Scheme 45

Moreover, higher temperature was found to lead to greater formation of telomers at the expense of 1:1 adduct.

Other studies have looked at additions to perfluoropropylene where 84% 1:1 adduct and 16% higher telomers were observed when a 1:1 mixture of $n\text{-}C_3F_7I : C_3F_6$ were heated at 200 °C for 88 h, and where again $(CF_3)_2CFI$ was found to be a better chain-transfer agent than various $n\text{-}R_FIs$ [206, 207].

Recently a study of the use of various α, ω-diiodo telomers led to an interesting, selective 1:1 addition reaction with hexafluoropropene [208, 209].

$$IC_2F_4CH_2CF_2I \quad + \quad CF_2=CFCF_3 \quad \xrightarrow[65\ h]{230\ ^\circ C} \quad IC_2F_4CH_2CF_2CF_2CFICF_3 \ 1$$
$$1 \ : \ 0.12 \qquad\qquad\qquad\qquad\qquad\qquad\qquad 55\%$$

Scheme 46

In a thorough study of the effect of reactant ratio, initiator concentration, temperature and time of reaction, it was found that the optimal reaction conditions for 1:1 adduct formation in the reaction of perfluoro-*n*-hexyl iodide to vinyl acetate are [210]:

$$n\text{-}C_6F_{13}I \quad + \quad CH_2{=}CHOAc \quad \xrightarrow[80\,°,\ 60\text{ - }90\text{ min}]{AIBN\ (2.5\%)} \quad n\text{-}C_6F_{13}CH_2CHIOAc$$

$$1 : 1 \qquad\qquad\qquad\qquad\qquad\qquad \begin{array}{l} 80\% \text{ conversion} \\ 99\% \ 1:1 \text{ adduct} \end{array}$$

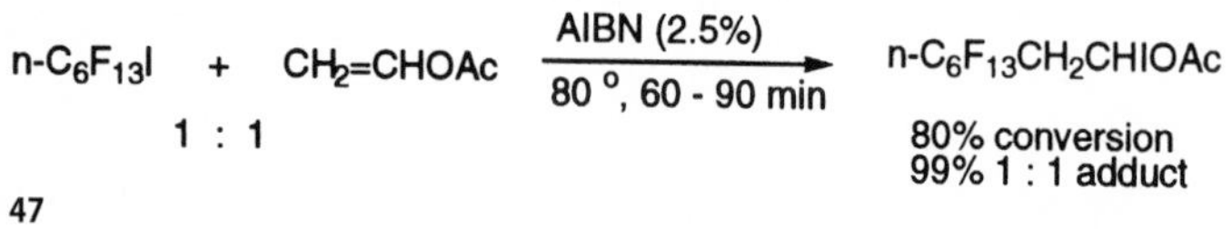

Scheme 47

What remains to be done in this area is to find conditions for control of chain transfer so that one can optimize production of telomers with various specific degrees of telomerization.

6.3.3
Polymerization of Fluoroolefins

In general, fluoropolymers possess the unique combination of high thermal stability, chemical inertness, unusual surface properties, low dielectric constants and dissipation factors, low water absorptivities, excellent weatherability and low flammabilities. Therefore there appears to be an ever-increasing market for fluoropolymers in spite of their relatively high cost [211, 212].

Ideal conditions for polymerization of a fluoroolefin are those where little, or better yet, *no* chain-transfer occurs. Thus perfluoroalkyl iodides are not used to initiate polymerizations. Instead, non-chain-transfer agents, particularly peroxide initiators, including perfluorodiacyl peroxides and ammonium persulfate, are used effectively for this purpose [67].

Indeed, free radical polymerization of fluoroolefins continues to be the only method which will produce high-molecular weight fluoropolymers. High molecular weight homopolymers of TFE, $CFCl{=}CF_2$, CH_2CF_2, and $CH_2{=}CHF$ are prepared by current commercial processes, but homopolymers of hexafluoropropylene or longer-chain fluoroolefins require extreme conditions and such polymerizations are not practiced commercially. Copolymerization of fluoroolefins has also led to a wide variety of useful fluoropolymers. Further discussion of the subject of fluoroolefin polymerization may be found elsewhere and is beyond the scope of this review [213–215].

6.3.4
Free Radical Initiator-Induced Additions

Shortly after Haszeldine's initial studies, Tarrant, Brace, and others began to use Kharasch's technique of diacyl peroxide initiation of such thermal additions [60, 216–227]. With such inducement, these addition processes could be run at lower temperature and usually with greater efficiency. Nevertheless, such radical-induced additions can also require relatively long reaction times to attain decent conversions, apparently because of remaining difficulties in the chain-transfer process.

$$CF_2Br_2 \quad + \quad CH_2=CHCH_3 \xrightarrow[85°,\,4\,h]{(BzO)_2} BrCF_2CH_2CHBrCH_3 \quad (67\%)$$

$$4.1 \,:\, 1$$

$$n\text{-}C_3F_7I \quad + \quad \text{[cyclohexene]} \xrightarrow[22\,h,\,50\%\text{ conversion}]{\Delta,\,50°,\,AIBN(3\%)} \text{[cyclohexane-}C_3F_7,\,I] \quad (90\%)\quad t:c=1:1$$

Scheme 48

6.3.5
Reductively-Initiated Additions

It was therefore a significant breakthrough when procedures involving electron-transfer initiation began to appear in the 1960s, and today reductive initiation constitutes the most commonly used method of accomplishing the addition of R_FI to olefins and alkynes [228]. Perhaps the first example of such a process was that of Kehoe and Burton in 1966 [229, 230].

$$n\text{-}C_3F_7I \quad + \quad CH_2=CHC_6H_{13} \xrightarrow[t\text{-BuOH, reflux, 24 h}]{CuCl,\,HOCH_2CH_2NH_2} n\text{-}C_3F_7CH_2CHIC_6H_{13}$$
$$(68\%)$$

$$CF_2Br_2 \quad + \quad CH_2=CHC_6H_{13} \xrightarrow{\quad\text{"}\quad} BrCF_2CH_2CHBrC_6H_{13}$$
$$(57\%)$$

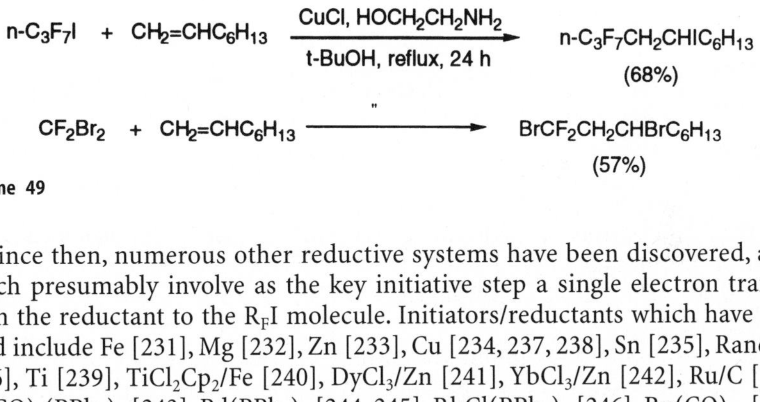

Scheme 49

Since then, numerous other reductive systems have been discovered, all of which presumably involve as the key initiative step a single electron transfer from the reductant to the R_FI molecule. Initiators/reductants which have been used include Fe [231], Mg [232], Zn [233], Cu [234, 237, 238], Sn [235], Raney Ni [236], Ti [239], $TiCl_2Cp_2/Fe$ [240], $DyCl_3/Zn$ [241], $YbCl_3/Zn$ [242], Ru/C [243], $Ni(CO)_2(PPh_3)_2$ [243], $Pd(PPh_3)_4$ [244, 245], $RhCl(PPh_3)_3$ [246], $Ru(CO)_{12}$ [247], $Fe_3(CO)_{12}$ [247], SmI_2 [248, 249], $PhSO_2Na$ [250], $Na_2S_2O_4$ [251, 252, 254], $(NH_4)_2S_2O_4$ [253], piperidine [255], Me_3Al [256], Bu_4NI [255], NaOEt [257].

A difficult process to execute successfully is addition of an electrophilic per-fluoroalkyl radical to an ethene which is itself substituted with a perfluoroalkyl group. However, Hu and Hu have reported a smooth conversion of this type using $NiCl_2 \cdot 6H_2O/Al/alkene/R_FI$ in a ratio of 0.2:2:1.5:1 in CH_3CN at RT [258].

$$ClC_8F_{16}I \quad + \quad CH_2=CHCF_2C_3F_7 \xrightarrow[CH_3CN]{NiCl_2 \cdot H_2O/Al} ClC_8F_{16}CH_2CH=CFC_3F_7$$
$$45\%$$

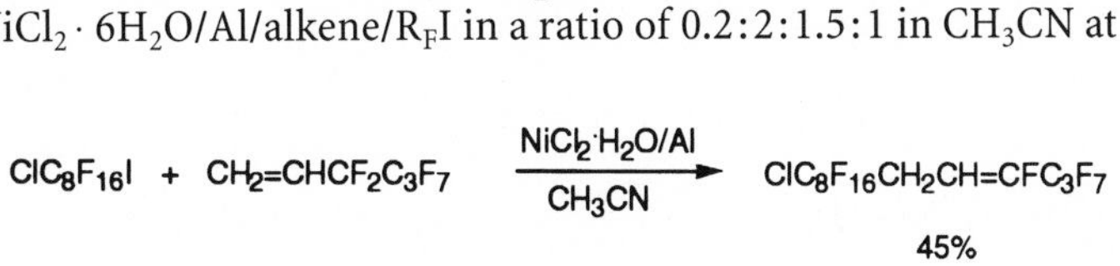

Scheme 50

CF$_2$Br$_2$ and HCF$_2$I have also been utilized in reductively-induced reactions to provide 1:1 adducts [229, 256, 258–260] A mechanistic study by Wu et al. provided strong evidence for the SET nature of the CF$_2$Br$_2$/Zn process for alkene addition [261].

$$CF_2Br_2 \xrightarrow[\substack{CuCl,\ HOCH_2CH_2NH_2 \\ \text{t-BuOH, reflux, 24 h}}]{CH_2=CHC_6H_{13}} BrCF_2CH_2CHBrC_6H_{13} \quad (68\%)$$

$$CHF_2I \ + \ CH_2=CHC_4H_9 \xrightarrow[\text{MeCN, H}_2\text{O, RT, 14 h}]{\text{Na}_2\text{S}_2\text{O}_4/\text{NaHCO}_3} HCF_2CH_2CHIC_4H_9 \quad (86\%)$$

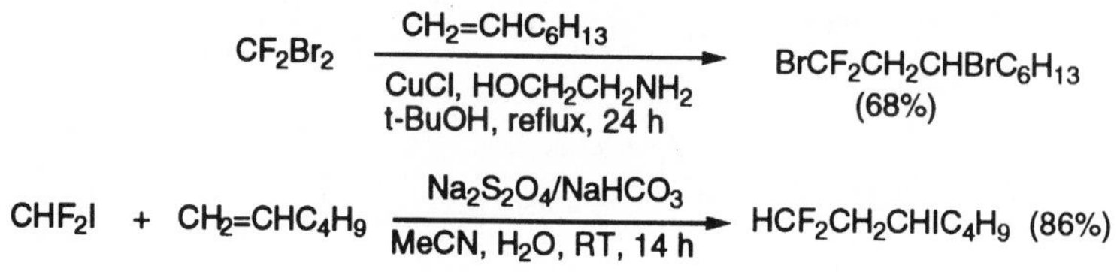

Scheme 51

Other examples indicate how additions of R$_F$I can be used to instigate more extensive chemical transformations such as an oxiranyl carbinyl radical ring-opening and cyclopropane ring formation [254, 262–264].

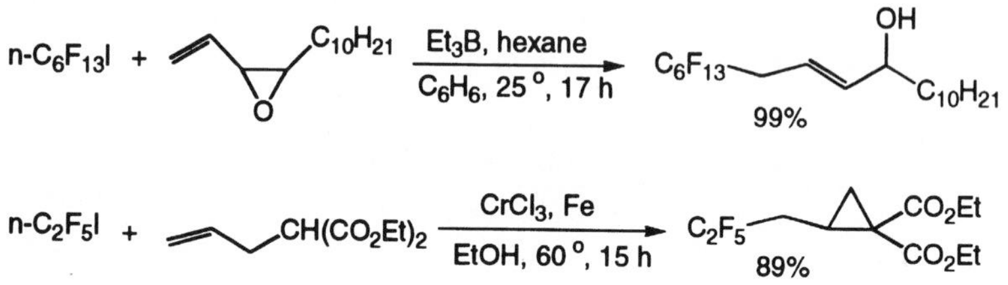

Scheme 52

Controlled electrolytic methodology, as discussed earlier, has also been used effectively for inducing perfluoroalkylation of olefins and alkynes [184, 265].

Examples of addition of perfluoroalkyl iodide to isonitriles have been reported [266, 267].

It has been found that reductive addition of R$_F$I to electron-deficient alkenes can lead to good yields of *hydro*perfluoroalkylation products [268, 269], and Hu

$$n\text{-}C_4F_9I + HC\!\equiv\!CC(Me)_2OH \xrightarrow{\text{e}^-,\ \text{DMF, LiCl}} \begin{array}{l} n\text{-}C_4F_9CH=Cl\text{-}CMe_2OH \quad (54\%) \\ + \\ n\text{-}C_4F_9CH=CH\text{-}CMe_2OH \quad (16\%) \end{array}$$

Scheme 53

$$n\text{-}C_4F_9I \ + \ n\text{-}C_4H_9N\!\!=\!\!C\!: \xrightarrow{\text{Cu}} n\text{-}C_4H_9\text{-N}=ClC_4F_9 \quad (90\%)$$

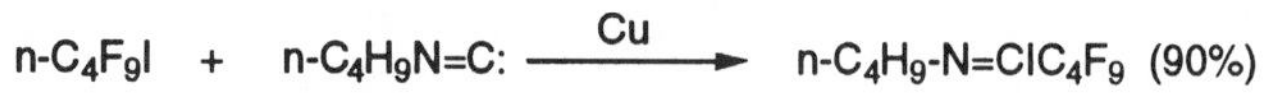

Scheme 54

and Chen have also found conditions under which CF_2Br_2 will productively undergo reductive addition to both electron-deficient and electron-rich olefins [270, 271].

$$\text{n-}C_6F_{13}I + CH_2{=}CHCO_2Et \xrightarrow[\text{Zn, 20}^\circ]{\text{bromo(pyridine) cobaloxime(III)}} \text{n-}C_6F_{13}CH_2CH_2CO_2Et \ (72\%)$$

$$CF_2Br_2 + CH_2{=}CHCO_2Et \xrightarrow[60^\circ\text{, 20 h}]{CrCl_3\text{, Fe}} BrCF_2CH_2CH_2CO_2Et \ (72\%)$$

Scheme 55

When one carries out additions of R_FI using a stoichiometric amount of aryl thiolates, selenates or tellurates, one obtains net *perfluoroalkyl thiolation* etc., with the tellurate being most reactive [272, 273].

$$\text{n-}C_8F_{17}I + CH_2{=}CHC_6H_{13} \xrightarrow[\text{RT, 2 h; X= S, Se, Te}]{PhX^-\text{, EtOH, }Et_2O} \text{n-}C_8F_{17}CH_2CH(XPh)C_6H_{13} \ (59-81\%)$$

Scheme 56

6.3.6
Additions of α,α-Difluoro Keto- or Ester Radicals

Recently Burton and Qiu have developed an excellent radical-based method of synthesis of α,α-difluoro-functionalized ketones by means of the addition of iododifluoroalkyl ketones to alkenes, presumably via the α,α-difluoroketo radical [274, 275].

There were also earlier reports of similar Cu-induced additions of methyl iododifluoroacetate which led to good synthetic procedures [276, 277].

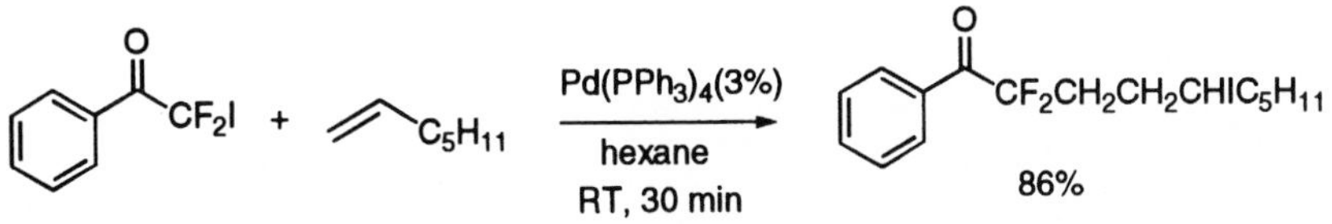

Scheme 57

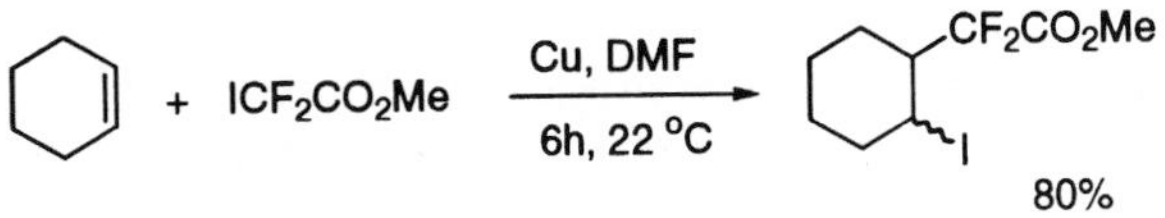

Scheme 58

6.3.7
SET-Induced Additions to Aromatic Systems

There are also a few examples of the application of the SET reductive methodology for perfluoroalkylation of *aromatics*, a reaction which although formally a substitution, mechanistically involves initial addition to the aryl substrate [278, 279].

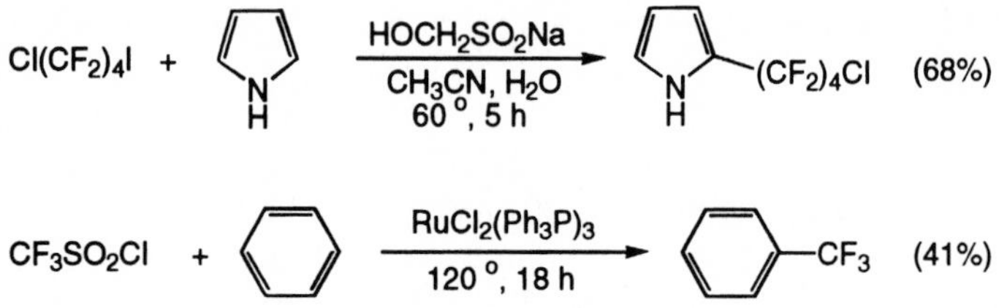

Scheme 59

The utilization of perfluorodiacyl peroxides for this purpose has been more widely developed. The rate of decomposition of perfluorodiacyl peroxides in the presence of electron-rich benzene derivatives is enhanced by a significant factor via a process of electron-transfer [66, 280]. As can be seen by the contrasting examples below [281], highly reactive arenes are capable of trapping the perfluoroalkyl carboxyl radical before it decarboxylates to R_F, a result which can diminish the synthetic utility of this process.

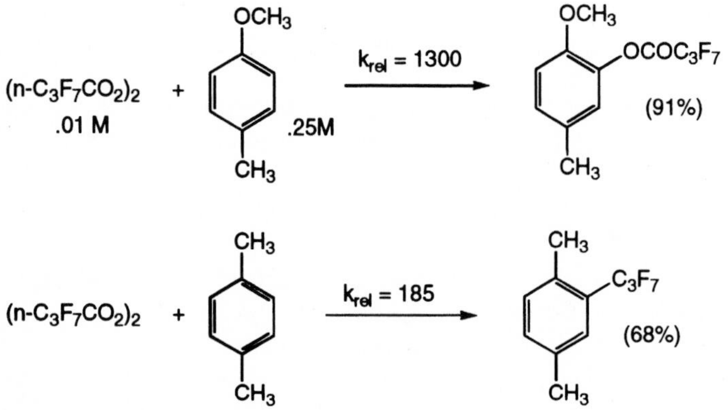

Scheme 60

Nevertheless, there are numerous productive examples of perfluoroalkylations of benzene derivatives, heteroaromatics, and uracil derivatives using diacyl peroxides, such as $(CF_3CO_2)_2$, $(n\text{-}R_FCO_2)_2$, and $(ClCF_2CO_2)_2$, as the source of the perfluoroalkyl group [282–286].

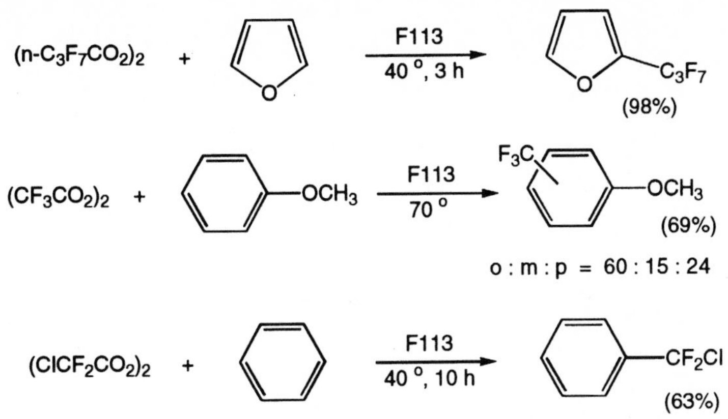

Scheme 61

6.3.8
Fenton-Type Initiation of Arene Addition

Fenton conditions, which result in eventual abstraction of I from R_FI by methyl radicals (see Sect. 4.1.5), can be used effectively to provide perfluoroalkylation of arenes [61].

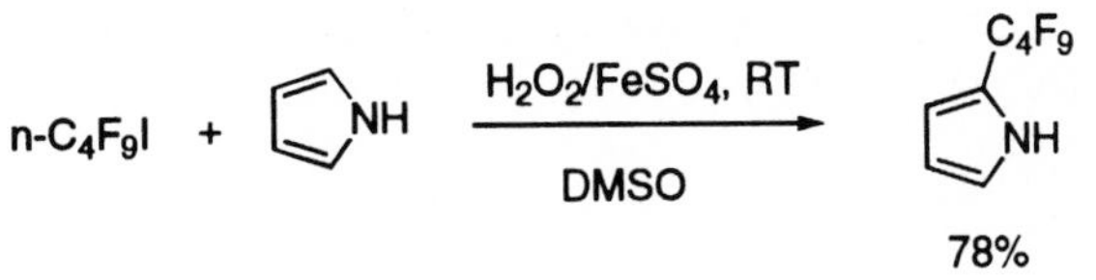

Scheme 62

6.3.9
Oxidatively-Induced Additions

Kolbe-type alkylations are, of course, oxidative in nature. There are a few other oxidative processes which lead to perfluoroalkylative addition, namely oxidation of carboxylic acids with xenon difluoride and oxidation of sodium perfluoroalkylsulfinates [63, 287].

6.3.10
Other Chain-Transfer Enhancing Methodologies

Although the synthetic usefulness of such reactions has not yet been widely recognized, it should be possible to carry out hydroperfluoroalkylation of olefins and alkynes very efficiently by the strategic use of an appropriate homolytic hydrogen-transfer agent. The rates of hydrogen transfer for many such agents

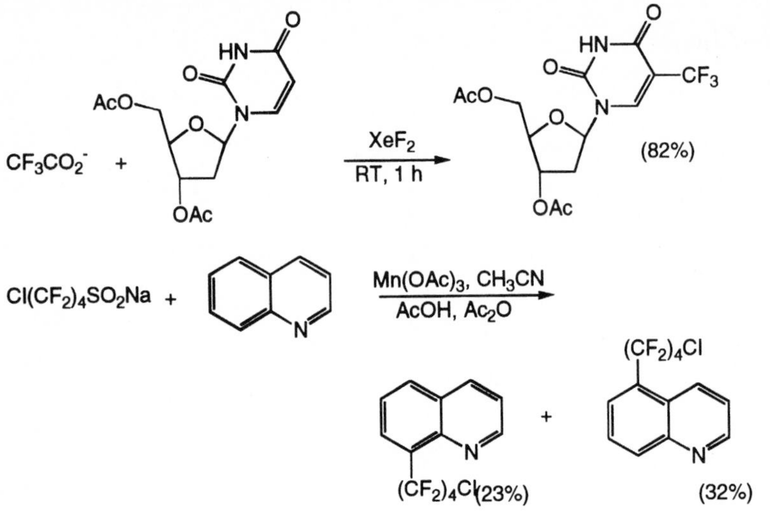

Scheme 63

have been recently reported [140, 141], and both Et_3SiH and n-Bu_3GeH appear to have kinetic properties which will allow addition to compete efficiently with reduction of $R_F\cdot$ to produce good yields of adduct.

Another reported way to facilitate the chain transfer process is addition of the propagating radicals to allylic stannanes [286]. Because of their efficiency, such processes as these will undoubtedly be used more frequently by synthetic chemists in the future.

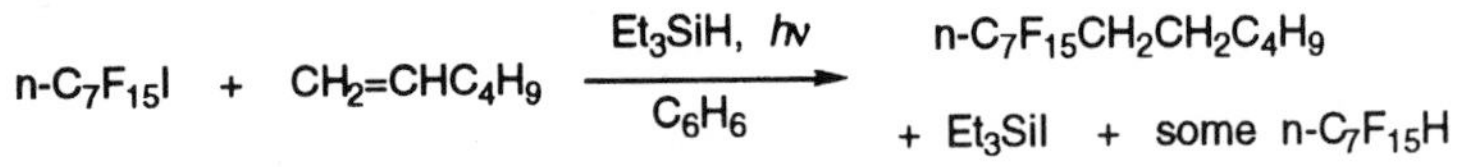

Scheme 64

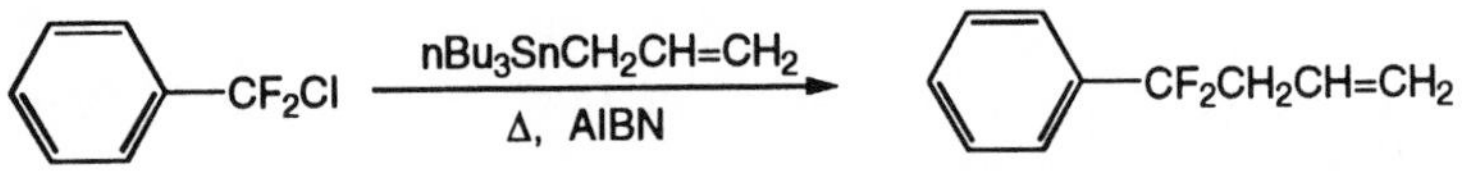

Scheme 65

6.4 Substitution Reactions

Perfluoroalkyl iodides are well-known for their ability to act as substrates in $S_{RN}1$ substitution reactions [288].

6.4.1
Substitution by Thiols and Thiolates

Whereas it has been demonstrated that both malonate ions and thiolate ions can catalyze the free radical chain addition reaction of perfluoroalkyl iodides to olefins [289, 290], under appropriate conditions one can obtain products deriving from substitution in such processes. Following early work carried out photolytically in liquid ammonia, recent reports have indicated that good yields of substitution products can be obtained in polar solvents at room temperature, without irradiation [291–296].

Perfluoroselenides and tellurides [297–299], and perfluoroalkyl sulfinates [300, 301] are also synthetically accessible via similar processes:

$$\text{n-C}_8\text{F}_{17}\text{I} + \text{PhS}^-\text{Na}^+ \xrightarrow{\text{DMF, 25}^\circ} \text{n-C}_8\text{F}_{17}\text{SPh (92\%)}$$

$$\text{n-C}_3\text{F}_7\text{I} + \text{CH}_3\text{S}^-\text{Na}^+ \xrightarrow{\text{DMF, 100}^\circ} \text{n-C}_3\text{F}_7\text{SCH}_3 \text{ (73\%)}$$

Scheme 66

$$\text{CH}_3\text{Se}^-\text{Na}^+ + \text{BrCF}_2\text{CF}_2\text{Br} \xrightarrow[\text{72 h}]{\text{NH}_3\text{, - 78}^\circ} \text{CH}_3\text{SeCF}_2\text{CF}_2\text{SeCH}_3 \text{ (70\%)}$$

$$\text{H(CF}_2)_8\text{I} + \text{Na}_2\text{S}_2\text{O}_4 \xrightarrow[\text{CH}_3\text{CN, 85}^\circ\text{, 7 h}]{\text{NaHCO}_3\text{, H}_2\text{O}} \text{H(CF}_2)_8\text{SO}_2\text{Na (94\%)}$$

Scheme 67

6.4.2
Substitution by Carbanions

In what has become a classic example of an $S_{RN}1$ reaction, the 2-propylnitronate anion undergoes efficient perfluoroalkylation in its reaction with perfluoroalkyl iodides [302]:

$$\text{n-C}_6\text{F}_{13}\text{I} + \text{Na}^+ \text{Me}_2\text{C=NO}_2^- \xrightarrow[\text{3 h}]{h\nu\text{, DMF}} \text{n-C}_6\text{F}_{13}\text{CMe}_2\text{NO}_2 \text{ (89\%)}$$

Scheme 68

Reactions with stabilized enolate species such as malonate ions also lead to alkylation, but because of the requisite basic conditions, the initial alkylation products are subsequently converted to secondary products [289].

$$C_2F_5I + CH_3COCH_2COCH_3 \xrightarrow{h\nu,\ NH_3} CH_3COCH{=}C(CF_3)NH_2$$

$$[\text{via } CH_3COCH(C_2F_5)COCH_3]$$

Scheme 69

In contrast, it has recently been found that *non-stabilized* enolates can also be perfluoroalkylated if the reaction is Et_3B-catalyzed, and if a chiral auxiliary is used, such reactions can lead to decent diastereomeric excess [303]:

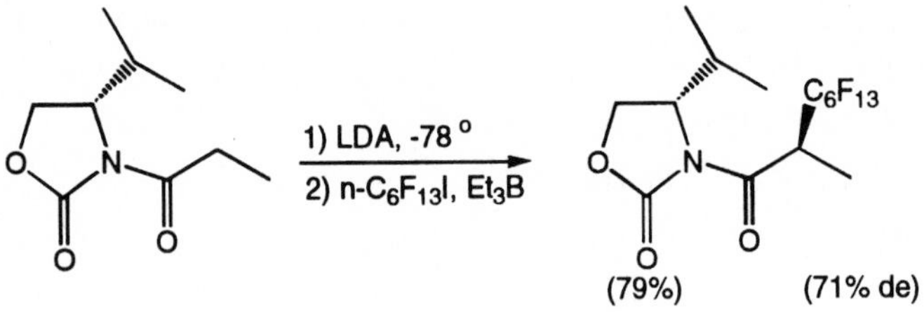

Scheme 70

Enamines are also observed to undergo reactions with perfluoroalkyl iodides which lead to overall α-perfluoroalkylation of ketones [304].

Lastly, electron transfer processes can also compete with nucleophilic acyl substitution in the reaction of perfluorodiacyl peroxides with Grignards [305]. Such a process can lead to a coupling of the Grignard with the perfluoroalkyl radical intermediate. (In contrast, benzyl lithium gives no perfluoroalkylation under the same conditions.)

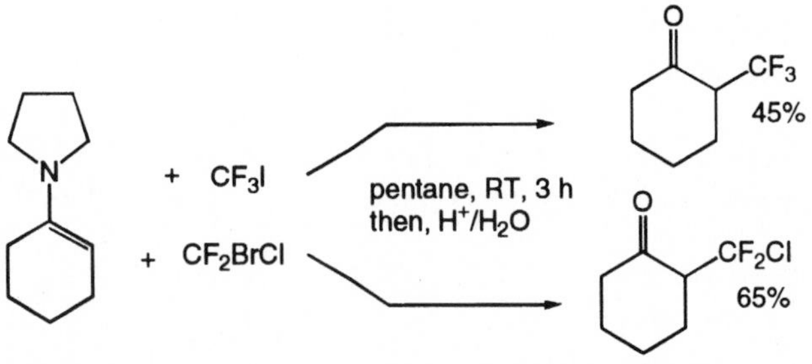

Scheme 71

$$(C_3F_7CO_2)_2 + PhCH_2MgBr \xrightarrow{Et_2O,\ F113} C_3F_7CH_2Ph\ (24\%) + C_3F_7COCH_2Ph\ (21\%)$$

Scheme 72

6.4.3
Hunsdiecker Reactions

Hunsdiecker reactions of salts of perfluoroalkanoic acids are known primarily as perhaps the best way of making perfluoroalkyl halides [58, 306, 307], but there have been other synthetic uses found for the perfluoroalkyl radicals which are formed by this decarboxylative process [308].

$$CF_3CO_2^-Ag^+ \xrightarrow{\Delta,\ I_2\ \text{(excess)}} CF_3I\ (91\%)$$

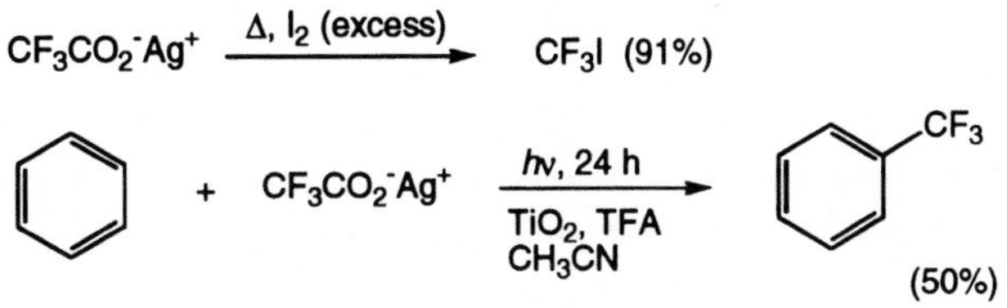

Scheme 73

6.4.4
Reductions Involving Perfluoroalkyl Radicals

Replacement of the iodine or bromine substituent of a perfluoroalkyl iodide or bromide with hydrogen is a process which is a side reaction in most of the reductively-catalyzed perfluoroalkylation processes described earlier. If one wishes to carry out such a reaction synthetically, it may be accomplished easily by use of any of a number of hydrogen atom transfer agents such as n-Bu$_3$SnH, n-Bu$_3$GeH, (TMS)$_3$SiH, or Et$_3$SiH.

Van der Puy et al. have examined the use of n-Bu$_3$SnH for synthetic purposes in the replacement of one or more halogens in polyhalofluorocarbons [309].

$$ClCF_2CFClCF_2Cl + n\text{-Bu}_3SnH \xrightarrow[110\ ^\circ C]{AIBN} HCF_2CFHCF_2H$$
$$80\%$$

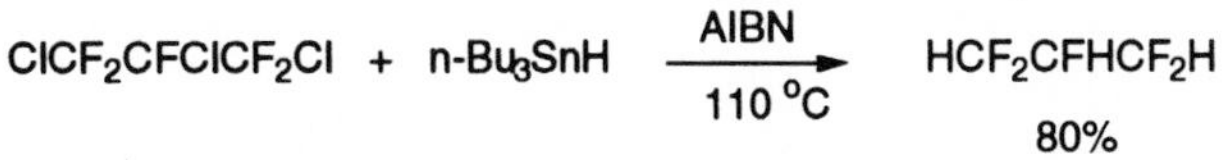

Scheme 74

The rate constants for each of these H-atom transfer agents have been determined, and they were presented and discussed earlier in Sect. 5.2 of this review [140, 141].

6.5
Defluorinations of Perfluorocarbons

Recently a number of groups, most significantly those of Crabtree and Richmond, have devised methods of carrying out reductive defluorination/ functionalization processes of saturated fluorocarbons. It is assumed that these processes involve successive single electron transfer processes [82]. Richmond and

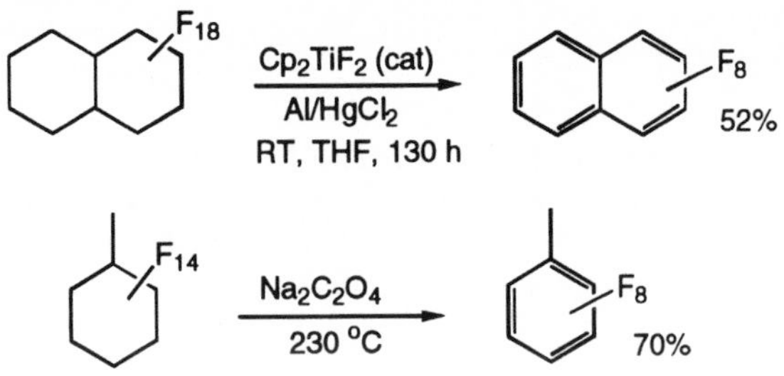

Scheme 75

then Crabtree have subsequently found ways to carry out such processes in the absence of nucleophiles to form perfluoroaromatics [83, 310].

6.6 Rearrangements of Fluorinated Radicals

Although radicals are not nearly so prone to rearrangement as are, for example, carbocations, there are a few such "rearrangements" which have become identified as *characteristic* of carbon radicals. These include radical cyclizations, particularly the 5-hexenyl radical cyclization, and radical C-C bond cleavages, particularly the cyclopropylcarbinyl to allyl carbinyl radical rearrangement. In hydrocarbon systems, as organic synthetic chemists have learned how to control rapid chain processes, such rearrangements have become important synthetic tools [176–179].

There are many fewer examples of perfluoro- or even partially-fluorinated radicals undergoing such reactions, although Brace utilized perfluoroalkyl radicals in his early studies of hydrocarbon radical cyclization reactions, for example [311, 312]:

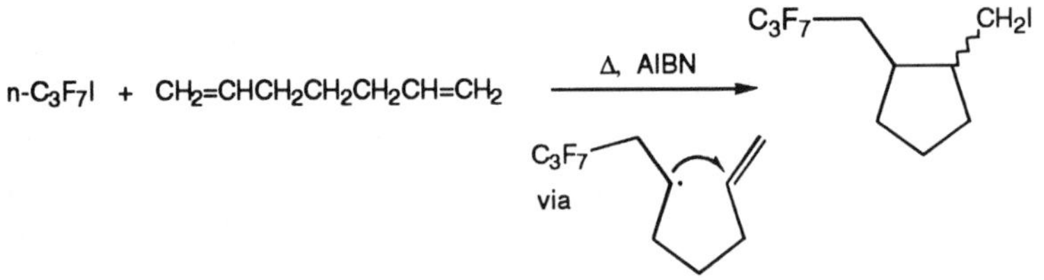

Scheme 76

6.6.1 *5-Hexenyl Radical Cyclizations*

The 5-exo cyclization reactions of hydrocarbon 5-hexenyl radicals comprise the most highly-studied of all cyclization processes, and such processes have been extensively and effectively exploited for synthetic purpose [176–179]. There have, however, been relatively few examples reported of comparable cyclizations of fluorinated 5-hexenyl systems. Piccardi et al. reported an early example of cyclization of a partially-fluorinated 5-hexenyl radical system [313].

The cyclization kinetics of a number of other partially-fluorinated systems as well as those for the cyclization of the parent perfluoro-5-hexenyl radical have been discussed earlier in Sect. 5.3.1 of this review [140, 162]. Other than these examples, the only remaining reports of cyclizations of radicals with fluorine proximate to the radical center involve some examples of α,α-difluoro-, β,β-difluoro- and α-trifluoromethyl-5-hexenyl radical systems.

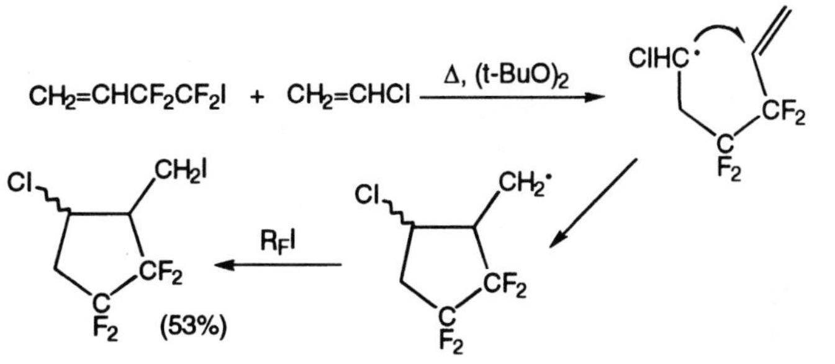

Scheme 77

6.6.1.1
α,α-Difluoro Radical Systems [128, 314–316]

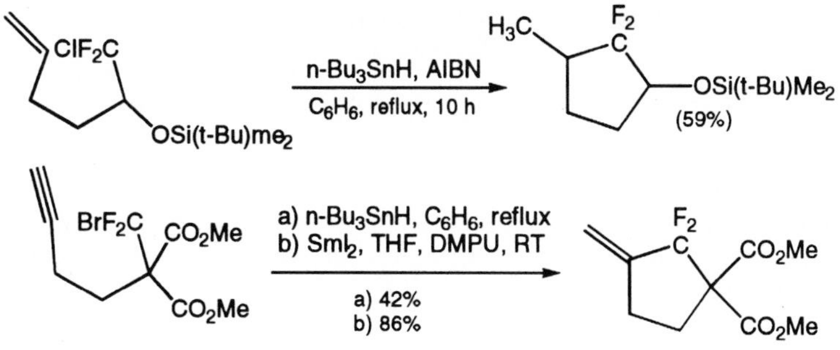

Scheme 78

6.6.1.2
β,β-Difluoro Radical Systems [317, 318]

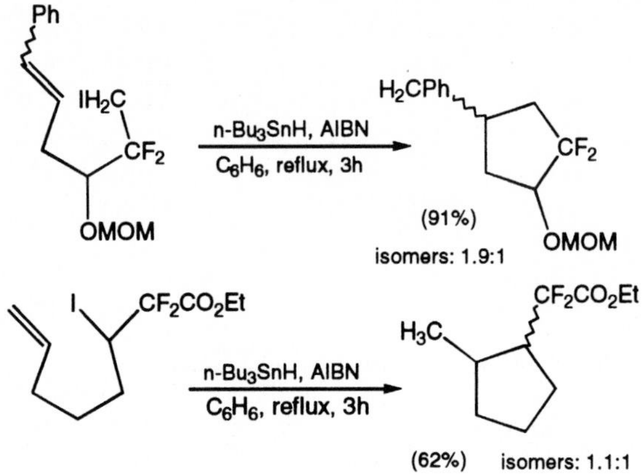

Scheme 79

6.6.1.3
α-Trifluoromethyl Radical Systems [318–320]

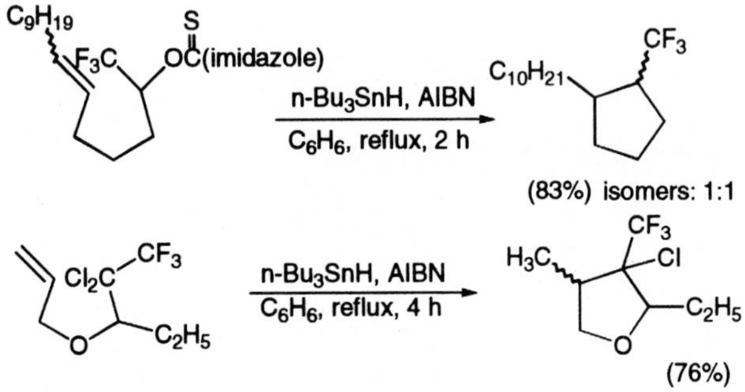

Scheme 80

It would appear that there is considerable potential in using fluorinated 5-hexenyl radical systems to synthesize specifically fluorine-substituted cyclopentyl ring systems.

6.6.2
6-Heptenyl Radical Cyclizations [317, 321, 322]

α,α-Difluoro- and *β,β*-difluoro-6-heptenyl radicals have been found to undergo *exo-trig* cyclization to form 6-membered rings in reasonable yields:

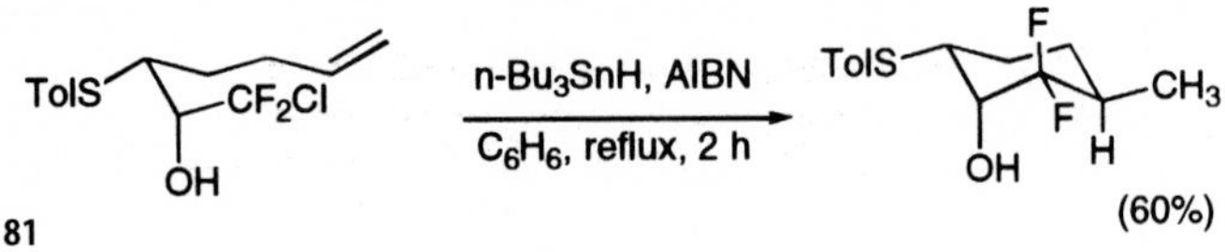

Scheme 81

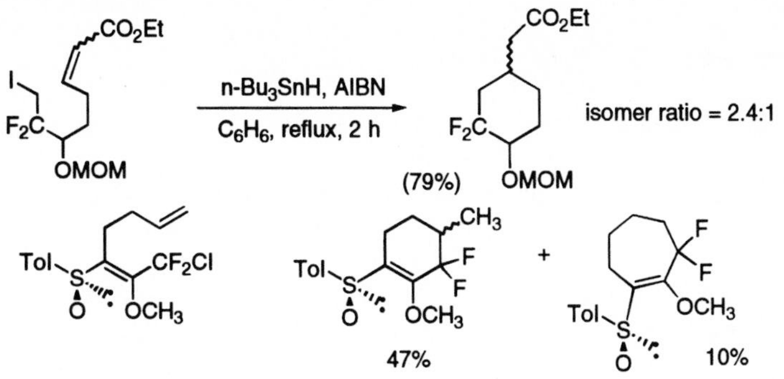

Scheme 81 (continued)

Because of the rate-enhancing effects of fluorine substitution on cyclization reactions of partially-fluorinated radical systems, it is likely that, unlike for pure hydrocarbon systems, it will be possible to utilize such processes to make six-, seven-, and even larger membered rings in radical systems which have appropriate fluorine substitution.

6.6.3
4-Pentenyl Radical Cyclizations

An interesting aspect of fluorinated 4-pentenyl radicals that distinguishes them from their hydrocarbon counterparts is their ability to cyclize to form four-membered rings. As mentioned in Sect. 5.3.2, Piccardi and his coworkers reported in 1971 that C_2F_5I underwent free radical addition to 3,3,4,4-tetrafluoro-1,5-hexadiene to form a four-membered ring product [173]. Subsequently they observed similar results in the addition of CCl_4 [323].

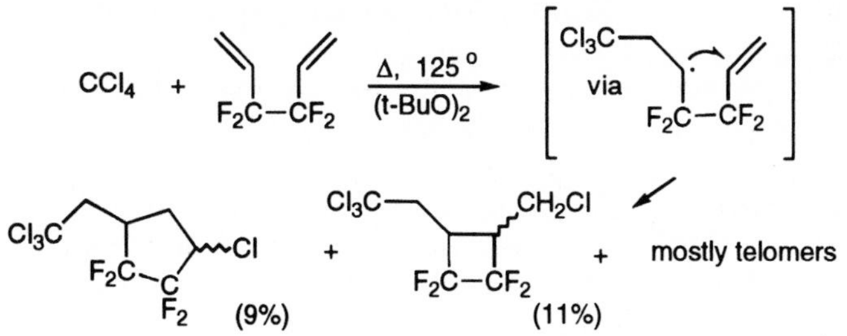

Scheme 82

There is much yet to be learned about the factors which determine whether particular fluorinated four-membered rings will be able to be formed via cyclization processes. Nevertheless there should be considerable synthetic utility to be found in this area.

Indeed, there will likely be considerable future research activity in the broad and potentially-fertile area of cyclizations of unsaturated fluorinated radicals.

6.6.4
Cyclopropylcarbinyl Radical Rearrangements

As discussed earlier in Sect. 5.3.3 of this review, because of the incremental strain imparted by fluorine substitutents to cyclopropane systems, the rate of ring-opening of the 2,2-difluorocyclopropylcarbinyl radical is substantially enhanced with respect to the already very fast analogous hydrocarbon system [174].

Although there are few examples, the facility of the ring-opening process, as well as its regiospecificity can be exploited for synthetic purpose [324, 325]:

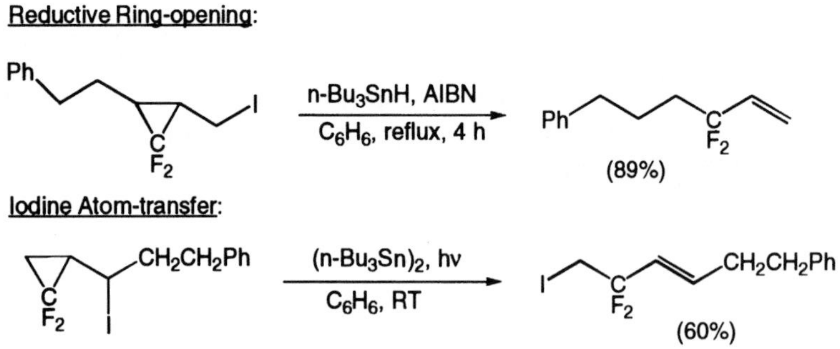

Scheme 83

7
Conclusions

It should be evident that radicals play a very important role within the realm of organofluorine chemistry. Fluorine substituents impart unique reactivity characteristics to free radical intermediates, and knowledge of how to generate and utilize such species is very important for those synthetic chemists who wish to incorporate fluorinated alkyl groups into organic substrates. It has been attempted in this review to provide a strategic overview of all aspects of organofluorine radical chemistry, with the hope that readers with an interest in the field will able to get their basic questions answered as well as be stimulated to dig deeper into specific aspects of the subject via the detailed references which have been provided.

8
References

1. Smart BE (1995) Properties of Fluorinated Compounds, Physical and Physicochemical Properties. In: Hudlicky M, Pavlath SE (eds) Chemistry of Organic Fluorine Compounds II, ACS Monograph 187. American Chemical Society, Washington, D. C, p 979
2. Lloyd RV (1988) The Structure of Alkyl Radicals. In: Alfassi ZB (ed) Chemical Kinetics of Small Organic Radicals, Vol. 1. CRC, Boca Raton, Florida, p 1

3. Fessenden RW, Schuler R H (1965) J Chem Phys 43:2704
4. Krusic PJ, Bingham RC (1976) J Am Chem Soc 98:230
5. Deardon DV, Hudgens JW, Johnson RD III, Tsai B P, Kafafi SA (1992) J Phys Chem 96: 585
6. Yamada C, Hirota E (1983) J Chem Phys 78:1703
7. Morokuma K, Pendersen L, Karplus M (1968) J Chem Phys 48:4801
8. Griller D, Ingold KU, Krusic PJ, Smart BE, Wonchoba ER (1982) J Phys Chem 86:1376
9. Bernardi F, Cherry W, Shaik S, Epiotis ND (1978) J Am Chem Soc 100:1352
10. Lloyd RV, Rogers MT (1973) J Am Chem Soc 95:1512
11. Hasegawa A, Wakabayashi T, Hayashi M, Symons MCR (1983) J Chem Soc Faraday Trans 1 79:941
12. Meakin P, Krusic PJ (1973) J Am Chem Soc 95:8185
13. Chen KS, Krusic PJ, Meakin P, Kochi JK (1974) J Phys Chem 78:2014
14. Griller D, Ingold KU, Krusic PJ, Fischer H (1978) J Am Chem Soc 100:6750
15. Edge DJ, Kochi JK (1972) J Am Chem Soc 94:6485
16. Chen Y, Rauk, A, Tschuikow-Roux E (1990) J Chem Phys 93:6620
17. Guerra M (1992) J Am Chem Soc 114:2077
18. Smart BE, Krusic PJ, Meakin P, Bingham RC (1974) J Am Chem Soc 96:7382
19. Korth HG, Trill H, Sustmann R (1981) J Am Chem Soc 103:4483
20. Lias SG, Bartmess JE, Liebman JF, Holmes JL Levin RD, Mallard WG (1988) J Phys Chem Ref Data 17:Suppl. 1
21. Kispert LD, Liu H, Pittman CU Jr (1973) J Am Chem Soc 95:1657
22. Yoshida M, Morishima A, Suzuki D, Iyoda M, Aoki K, Ikuta S (1996) Bull Chem Soc. Japan 69:2019
23. McMillen DF, Golden DM (1982) Ann Rev Phys Chem 33:493
24. Bordwell FG, Zhang XM (1993) Acc Chem Res 26:510
25. Pasto DJ (1988) J Am Chem Soc 110:8164
26. Lide DR (1995-1996) CRC Handbook of Chemistry and Physics, 76th edn. CRC Press, Boca Raton, p 9-63
27. Baldwin RR, Drewery GG, Walker RW (1984) J Chem Soc, Faraday Trans I 80:2827
28. Rodgers AS, Ford WG F (1973) Int J Chem Kinet 5:965
29. Seakins PW, Pilling MJ, Niiranen JT, Gutman D, Krasnoperov LN (1992) J Phys Chem 96:9847
30. Martell JM, Boyd RJ, Shi Z (1993) J Phys Chem 97:7208
31. Bartberger MD, Dolbier WR Jr, Lusztyk J, Ingold KU (1997) Tetrahedron (in Press)
32. Jiang XK, Li XY, Wang KY (1989) J Org Chem 54:5648
33. Dolbier WR Jr, Piedrahita CA, Al-Sader BH (1979) Tetrahedron Lett 20:2957
34. Dolbier WR Jr, Phanstiel OIV (1989) J Am Chem Soc 111:4907
35. Levy JB, Lehmann E J (1971) J Am Chem Soc 93:5790
36. Creary X, Sky AF, Mehrsheikh-Mohammadi ME (1988) Tetrahedron Lett 29:6839
37. Placzek DW, Rabinovitch BS (1965) J Phys Chem 69:2141
38. Dolbier WR Jr, McClinton MA (1995) J Fluorine Chem 70:249
39. Scherer KV Jr, Ono T, Yamanouchi K, Fernandez R, Henderson P, Goldwhite H (1985) J Am Chem Soc 107:718
40. Cherstkov VF, Avetisyan EA, Tumanskii BL, Sterlin SR, Bubnov NN, German LS (1990) Bull Acad Sci USSR, Div Chem Sci 39:2223
41. Avetisyan EA, Tumanskii BL, Cherstkov VF, Sterlin SR, German LS (1993) Russ Chem Bull 42:207
42. Avetisyan EA, Tumanskii BL, Cherstkov VF, Sterlin SR, German LS (1995) Russ Chem Bull 44:558
43. Tumanskii BL, Cherstkov VF, Delyagina NI, Sterlin SR, Bubnov NN, German LS (1995) Russ Chem Bull 44:962
44. Gross U, Rudinger S, Dimitrov A (1996) J Fluorine Chem 76:139
45. Allayarov SR (1994) Russ J Org Chem 30:1204
46. Sterlin SR, Cherstkov VF, Tumanskii BL, Avetisyan EA (1996) J Fluorine Chem 80:77

47. Pearson RG (1988) J Am Chem Soc 110:7684
48. Rodriquez CF, Sirois S, Hopkinson AC (1992) J Org Chem 57:4869
49. Danovich D, Apeloig Y, Shaik S (1993) J Chem Soc, Perkin Trans 2 1993:321
50. NIST Standard Reference Database 25. NIST Structures Properties Database and Estimation Program 1991. U.S. Department of Commerce, Gaithersburg
51. Buckley TJ, Johnson R III, Huie R, Zhang A, Kuo S, Klemm B (NIST) (unpublished data)
52. DePuy CH, Bierbaum VM, Damrauer R (1984) J Am Chem Soc 106:4051
53. Bartmess JE, Scott JA, McIver RT Jr (1979) J Am Chem Soc 101:6047
54. Christodoulides AA, McCorkle DL, Christophorou LG (1984) Electron Affinities of Atoms, Molecules, and Radicals. In: Christophorou LG (ed) Electron-Molecule Interactions and their Applications, Vol 2. Academic, Orlando, p 423
55. Buckley TJ, Johnson RD III (1997) Paper P56, 13th Winter Fluorine Symposium
56. Pearson RG (1989) J Org Chem 54:1423
57. Tarrant P (1984) J Fluorine Chem 25:69
58. Haszeldine RN (1986) J Fluorine Chem 33:307
59. Wakselman C (1994) In: Banks RE, Smart BE, Tatlow J C (eds) Organofluorine Chemistry, Principles and Commercial Applications. Plenum, New York, p 177
60. Sosnovsky G (1964) Free Radical Reactions in Preparative Organic Chemistry, Macmillan, New York
61. Baciocchi E, Muraglia E (1993) Tetrahedron Lett 34:3799
62. Huang WY (1992) J Fluorine Chem 58:1
63. Tanabe Y, Matsuo N, Ohno N (1988) J Org Chem 53:4582
64. Haszeldine RN (1951) J Chem Soc 1951:584
65. Barton DHR, Lacher B, Zard SZ (1986) Tetrahedron 42:2325
66. Yoshida M, Kamigata N (1990) J Fluorine Chem 49:1
67. Gumprecht WH, Dettre RH (1975) J Fluorine Chem 5:245
68. Zhao C, Zhou R, Pan H, Jin X, Qu Y, Wu C, Jiang X (1982) J Org Chem 47:2009
69. Serov SI, Zhuravlev MV, Sass VP, Lokolov SV (1980) J Org Chem, USSR 16:1360
70. Avila DV, Ingold KU, Lusztyk J, Dolbier WR Jr, Pan HQ (1996) J Org Chem 61:2027
71. Zhuravlev MV, Burmakov AI, Bloshchitsa FA, Sass VP, Sokolov S V (1983) J Org Chem, USSR 18:1597
72. Chambers WJ, Tullock CW, Coffman DD (1962) J Am Chem Soc 84:2337
73. Leventhal E, Simonds CR (1962) Can J Chem 40:930
74. Nakamura T, Yabe A (1995) Chem Lett 1995:533
75. Sangster JM, Thynne JC J (1969) J Phys Chem 73:2746
76. Charles SW, Whittle E (1960) Trans Farad Soc 56:794
77. Harris JF Jr (1965) J Org Chem 30:2182
78. Umemoto T, Ando A (1986) Bull Chem Soc Jpn 59:447
79. Naumann D, Wilkes B, Kischkewitz J (1985) J Fluorine Chem 30:73
80. Schwarz B, Cech D, Reefschlager J (1984) J Prakt Chem 326:985
81. Tortelli V, Tonelli C, Corvaja C (1993) J Fluorine Chem 60:165
82. Burdeniuc J, Chupka W, Crabtree RH (1995) J Am Chem Soc 117:10119
83. Kiplinger JL, Richmond, TG (1996) J Am Chem Soc 118:1805
84. Giese B (1983) Angew Chem Int Ed Engl 22:753
85. Tedder JM, Walton JC (1976) Acc Chem Res 9:183
86. Stefani AP, Herk L, Szwarc M (1961) J Am Chem Soc 83:4732
87. Leavitt F, Levy M, Szwarc M, Stannett V (1955) J Am Chem Soc 77:5493
88. Stefani A P (1971) Fluorinated Radicals. In: Tarrant P (ed.) Fluorine Chem Rev, Vol 5. Marcel Dekker, New York, p 115
89. Serov SI, Zhuravlev MV, Sass VP, Sokolov SV (1981) J Org Chem, USSR 17:48
90. Komendantov AM, Rondarev DS, Sass VP, Sokolov SV (1983) J Org Chem, USSR 19:1684
91. Haszeldine RN (1953) J Chem Soc 1953:3565 and references therein.
92. Tedder JM, Walton JC (1980) Tetrahedron 36:701
93. Tedder JM (1982) Angew Chem Int Ed Engl 21:401
94. Balague J, Ameduri B, Boutevin B, Caporiccio G (1995) J Fluorine Chem 73:237.

95. Abell PI (1976) In: Bamford CH, Tipper CFH (eds) Comprehensive Chemical Kinetics Vol. 18, Chapter 3. Amsterdam, p 111.
96. Dewar MJS, Olivella S (1978) J Am Chem Soc 100:5290
97. Zipse H, He J, Houk KN, Giese B (1991) J Am Chem Soc 113:4324
98. Wong MW, Pross A, Radom L (1993) J Am Chem Soc 115:11050
99. Wong MW, Radom L (1995) J Phys Chem 99:8582
100. Salem L (1968) J Am Chem Soc 90:543
101. Fukui K (1970) Fortschr Chem Forsch 15:1
102. Houk KN (19750 Acc Chem Res 8:361
103. Sustmann R, Trill H (1972) Angew Chem Int Ed Engl 11:838
104. Fischer H (1984) In: Landolt-Bornstein, New Series, Radical Reaction Rates in Liquids, Vol. 13, Parts a, b. Springer-Verlag, Berlin
105. Fischer H (1986) Substituent Effects on Absolute Rate Constants and Arrhenius Parameters for the Addition of Tert-Butyl Radicals to Alkenes. In: Viehe HG, Janousek Z, Merenyi R, (eds) Substituent Effects in Radical Chemistry. Reidel, Dordrect, p 123
106. Heberger K, Walbiner M, Fischer H (1992) Angew Chem Int Ed Engl 31:635
107. Zytowski T, Fischer H (1996) J Am Chem Soc 118:437
108. Riemenschneider K, Bartels H M, Dornow R, Drechsel-Grau E, Eichel W, Luthe H, Matter YM, Michaelis W, Boldt P (1987) J Org Chem 52:205
109. Giese B, He J, Mehl W (1988) Chem Ber 121:2063
110. Santi R, Bergamini F, Citterio A, Sebastiano R, Nicolini M (1992) J Org Chem 57:4250
111. Beranek I, Fischer H (1989) Polar Effects on Radical Addition Reactions: An Ambiphilic Radical. In: Minisci F (ed) Free Radicals in Synthesis and Biology. Kluwer, Amsterdam, p 303
112. Wong MW, Pross A, Radom L (1993) J Am Chem Soc 115:11050
113. Fueno T, Kamachi M Macromolecules 21:908
114. Avila DV, Ingold KU, Lusztyk J, Dolbier WR Jr, Pan HQ (1993) J Am Chem Soc 115:1577
115. Avila DV, Ingold KU, Lusztyk J, Dolbier WR Jr, Pan HQ, Muir M (1994) J Am Chem Soc 116:99
116. Citterio A, Arnoldi A, Minisci F (1979) J Org Chem 44:2674, data modified for temperature and other factors in Table III of ref 116.
117. Avila DV, Ingold KU, Lusztyk J, Dolbier WR Jr, Pan HQ (unpublished data)
118. Avila DV, Ingold KU, Lusztyk J, Dolbier WR Jr, Pan HQ (1996) Tetrahedron 52:12351
119. Johnston LJ, Scaiano JC, Ingold K U (1984) J Am Chem Soc 106:4877
120. From ref 101, part a, p 160.
121. Munger H, Fischer H (1985) Int J Chem Kinet 17:809
122. Jent F, Paul P, Roduner E, Heming M, Fischer H (1986) Int J Chem Kinet 18:1113
123. Wong MW, Pross A, Radom L (1994) J Am Chem Soc 116:11938
124. Madhaven V, Schuler RH, Fessenden RW (1978) J Am Chem Soc 100:888
125. Gilbert BC, Lindsay Smith JR, Milne EC, Whitwood AC, Taylor P (1993) J Am Chem Soc 116:11938
126. Jiang XK, Ji GZ, Xie JRY (1996) J Fluorine Chem 79:133
127. Takeuchi Y, Kawahara S, Suzuki T, Koizumi T, Shinoda H (1996) J Org Chem 61:301
128. Buttle LA, Motherwell WB (1994) Tetrahedron Lett 35:3995
129. Salikhov A, Fischer H (1993) Appl Magn Reson 5:445
130. Tedder JM (1982) Tetrahedron 38:313
131. Chatgilialoglu C, Guerrini A, Lucarini M (1992) J Org Chem 57:3405
132. Newcomb M (1993) Tetrahedron 49:1151
133. Kanabus-Kaminska JM, Hawari A, Griller D, Chatgilialoglu C (1987) J Am Chem Soc 109:5267
134. Clark KB, Griller D (1991) Organometallics 10:746
135. Burkey TJ, Majewski M, Griller D (1986) J Am Chem Soc 108:2218
136. Eksterowicz JE, Houk KN (1993) Tetrahedron Lett 34:427
137. Chen Y, Tschuikow-Roux E (1993) J Phys Chem 97:3742
138. Brace NO (1963) J Org Chem 28:3093

139. Andrieux CP, Gelis L, Medebielle M, Pinson J, Saveant J M (1990) J Am Chem Soc 112:3509
140. Rong XX, Pan HQ, Dolbier WR Jr, Smart BE (1994) J Am Chem Soc 116:4521
141. Dolbier WR Jr, Rong X X (1995) J Fluorine Chem 72:235
142. Dolbier WR Jr, Rong XX, Smart BE, Yang ZY (1996) J Org Chem 61:4824
143. Dolbier WR Jr, Rong XX (1994) Tetrahedron Lett 35:6225
144. Kim SS, Kim SY, Ryou SS, Lee CS, Yoo KH (1993) J Org Chem 58:192
145. Chatgilialoglu C, Ingold KU, Lusztyk J, Nazran A S, Scaiano JC (1983) Organometallics 2:1332
146. Rong XX (1995) (Ph. D. Thesis, University of Florida)
147. Malatesta V, Scaiano JC (1982) J Org Chem 47:1455
148. Griller D, Howard JA, Marriott PR, Scaiano JC (1981) J Am Chem Soc 103:619
149. Dolbier WR Jr, Li AR (unpublished results)
150. Chatgilialoglu C, Ingold KU, Scaiano JC (1981) J Am Chem Soc 103:7739
151. Chatgilialoglu C (1992) Acc Chem Res 25:188
152. Baldwin JE (1976) J Chem Soc, Chem Commun 1976:734
153. Beckwith AL J, Easton CJ, Serelis A (1980) J Chem Soc, Chem Commun 1980:482
154. Beckwith ALJ, Schiesser CH (1985) Tetrahedron Lett 26:373
155. Beckwith ALJ, Schiesser C H (1985) Tetrahedron 41:3925
156. Spellmeyer DC, Houk KN (1987) J Org Chem 52:959
157. Beckwith ALJ, Zimmermann JJ (1991) J Org Chem 56:5791
158. Beckwith ALJ, Easton CJ, Lawrence T, Serelis AK (1983) Aust J Chem 36:545
159. Wilt JW, Lusztyk J, Peeran M, Ingold KU (1988) J Am Chem Soc 110:281
160. Newcomb M, Horner JH, Filipkowski MA, Ha C, Park SU (1995) J Am Chem Soc 117:3674, and references therein
161. Dolbier WR Jr, Rong XX (1996) Tetrahedron Lett 37:5321
162. Dolbier WR Jr, Rong XX, Bartberger MD, Koroniak H, Smart BE, Yang ZY (1997) J Am Chem Soc (submitted)
163. Low HC, Tedder JM, Walton JC (1976) J Chem Soc, Faraday Trans 72:1707
164. Beckwith ALJ (1981) Tetrahedron 37:3073
165. Smart BE (1986) In: Liebman JF, Greenberg A (eds) Molecular Structure and Energetics, Vol.3: Studies of Organic Molecules. VCH, New York
166. Chiu NS, Burrow PD, Jordan KD (1979) Chem Phys Lett 68:121
167. Dolbier WR Jr, Medinger KS (1982) Tetrahedron 38:2415
168. Pasto DJ, Krasnansky R, Zercher C (1987) J Org Chem 52:3062
169. Fischer H (1977) In: Fischer H, Hellwege K H (eds) Magnetic Properties of Free Radicals, Landolt-Bornstein, New Series Vol 9, Part b, Chapt 3. Springer-Verlag, Berlin
170. Bartberger MD, Dolbier WR Jr (unpublished work)
171. Park SU, Varick TR, Newcomb M (1990) Tetrahedron Lett 31:2975
172. Jung ME, Trifunovich ID, Lensen N (1992) Tetrahedron Lett 33:6719
173. Piccardi P, Modena M, Cavalli L (1971) J Chem Soc (C) 1971:3959
174. Dolbier WR Jr, Al-Sader BH, Sellers SF, Koroniak H (1981) J Am Chem Soc 103:2138
175. Dolbier WR Jr, (1981) Acc Chem Res 14:195
176. Giese B (1986) Radicals in Organic Synthesis: Formation of Carbon-Carbon Bonds. Pergamon, Oxford
177. Jasperse CP, Curran DP, Fevig TL (1991) Chem Rev 91:1237
178. Curran DP (1988) Synthesis 1988:417 & 489.
179. Ramaiah M (1987) Tetrahedron 43:3541
180. Welch JT, Eswarakrishnan S (1991) Fluorine in Bioorganic Chemistry. John Wiley and Sons, New York
181. Filler R, Kobayashi Y, Yagupolskii LM (eds) (1993) Organofluorine Compounds in Medicinal Chemistry and Biomedical Applications. Elsevier, Amsterdam
182. Golitz P, de Meijer A (1977) Angew Chem Int Ed Engl 16:854
183. Uneyama K, Nanbu H (1988) J Org Chem 53:4598
184. Dapremont C, Calas P, Commeyras A, Amatore C (1992) J Fluorine Chem 56:249

185. Medebielle M, Pinson J, Saveant JM (1992) Tetrahedron Lett 33:7351
186. Muller N (1986) J Org Chem 51:263
187. Renaud RN, Chanpagne PJ (1975) Can J Chem 53:529
188. Brookes CJ, Coe PL, Owen DM, Pedler AE, Tatlow J C (1974) J Chem Soc, Chem Commun 1974:323
189. Emeleus HJ, Haszeldine RN (1949) J Chem Soc 1949:2948
190. Haszeldine RN (1949) J Chem Soc 1949:2856
191. Haszeldine RN (1953) J Chem Soc 1953:1199
192. Jeanneaux F, Le Blanc M, Cambon A, Guion J (1974) J Fluorine Chem 4:261
193. Cirkva V, Paleta O, Ameduri B, Boutevin B (1995) J Fluorine Chem 75:87
194. Tiers GVD (1962) J Org Chem 27:2261
195. Moore LD (1964) J Chem Eng Data 9:251
196. Tarrant P, Stump EC Jr (1964) J Org Chem 29:1198
197. Fagan PJ, Krusic PJ, McEwen CN, Lazar J, Parker DH, herron N, Wasserman E (1993) Science 262:404
198. Morton JR, Preston KF (1994) J Phys Chem 98:4993
199. Borghi R, Lunazzi L, Placucci G, Krusic PJ, Dixon DA, Matsuzawa N, Ata M (1996) J Am Chem Soc 118:7608
200. Qiu ZM, Burton DJ (1995) J Org Chem 60:3465
201. Haszeldine RN, Steele BR (1955) J Chem Soc 1955:3005
202. Tiers GVD (1960) J Am Chem Soc 82:5513
203. Cowell AB, Tamborski C (1981) J Fluorine Chem 17:345.
204. Haszeldine RN (1953) J Chem Soc 1953:3761
205. Chambers RD, Hutchinson J, Mobbs RH, Musgrave WKR (1964) Tetrahedron 20:497
206. Hauptchein M, Braid M, Lawler FE (1957) J Am Chem Soc 79:2549
207. Balague J, Ameduri B, Boutevin B, Caporiccio G (1995) J Fluorine Chem 74:49
208. Manseri A, Ameduri B, Boutevin B, Chambers RD, Caporiccio G, Wright AP (1995) J Fluorine Chem 74:59
209. Manseri A, Ameduri B, Boutevin B, Chambers RD, Caporiccio G, Wright AP (1996) J Fluorine Chem 78:145
210. Napoli M, Fraccaro C, Conte L, Gambaretto GP, Legnaro E (1992) J Fluorine Chem 57:219
211. Banks RE, Tatlow JC (1986) In: Banks RE, Sharp DWA, Tatlow JC (eds) Fluorine, The First Hundred Years. Elsevier, Lausanne, p 272
212. Chambers RD (1973) Fluorine in Organic Chemistry. John Wiley and Sons, New York, p 176
213. Wall LE (ed) (1972) Fluoropolymers. Wiley-Interscience, New York
214. Ref. 140, Chapters 15–22.
215. Schmiegel WW in Ref 1, p 1101
216. Tarrant P, Lovelace AM (1954) J Am Chem Soc 76:3466
217. Tarrant P, Lovelace AM (1955) J Am Chem Soc 77:2783
218. Brace NO (1964) U. S. Patent 3,145,222
219. Brace NO (1962) J Org Chem 27:3093
220. Brace NO (1973) J Org Chem 38:3167
221. Napoli M, Scipioni A, Conte L, Legnaro E, Krotz LN (1994) J Fluorine Chem 66:249
222. Cirkva V, Ameduri B, Boutevin B, Kvicala J, Paleta O (1995) J Fluorine Chem 74:97
223. Baum K, Bedford CD, Hunadi R J (1982) J Org Chem 47:2251
224. Brace NO (1982) J Fluorine Chem 20:313
225. Brace NO (1963) J Org Chem 28:2093
226. Brace NO (1962) J Org Chem 27:3033
227. Brace NO (1962) J Org Chem 27:4491
228. Wakselman C (1992) J Fluorine Chem 59:367
229. Burton DJ, Kehoe LJ (1966) Tetrahedron Lett 7:5163
230. Burton DJ, Kehoe LJ (1970) Tetrahedron 35:1339
231. Chen QY, He YB, Yang ZY (1986) J Fluorine Chem 34:255
232. Chen QY, Qiu ZM, Yang ZY (1987) J Fluorine Chem 36:149

233. Abou-Ghazaleh B, Laurent Ph, Blancon H, Commeyras A (1994) J Fluorine Chem 68:21
234. Kotora M, Hajek M, Amenduri B, Boutevin B (1994) J Fluorine Chem 68:49
235. Kuroboshi M, Ishihara T (1988) J Fluorine Chem 39:299
236. Chen QY, Yang ZY (1986) J Chem Soc, Chem Commun 1986:498
237. Chen QY, Yang ZY (1985) J Fluorine Chem 28:399
238. Yang ZY, Nguyen BV, Burton DJ (1992) Synlett 1992:141
239. Davis CR, Burton DJ, Yang ZY (1995) J Fluorine Chem 70:135
240. Hu CM, Qiu YL (1992) J Chem Soc, Perkin Trans 1 1992:1569
241. Ding Y, Zhao G, Huang WY (1992) Tetrahedron Lett 33:8119
242. Ding Y, Zhao G, Huang WY (1993) Tetrahedron Lett 34:1321
243. von Werner, K.J. Fluorine Chem. 1985, 28, 229–233.
244. Ishihara T, Kurobashi M, Okada Y (1986) Chem Lett 1986:1895
245. Chen QY, Yang ZY, Zhao CX, Qiu ZM (1988) J Chem Soc, Perkin Trans 1 1988:563
246. Chen QY, Yang ZY (1988) J Fluorine Chem 39:217
247. Fuchikami T, Ojima I (1984) Tetrahedron Lett 25:303
248. Lu X, Ma S, Zhu J (1988) Tetrahedron Lett 29:5129
249. Ma S, Lu X (1990) Tetrahedron 46:357
250. Feiring, AEJ. Org. Chem. 1985, 50, 3269–3274.
251. Li X.; Provencher L; Singh SM. Tetrahedron Lett. 1994, 35, 9141–9144.
252. Huang WY, Hu LQ, Ge WZ (1989) J Fluorine Chem 43:305
253. Hu CM, Qiu YL, Qing FL (1991) J Fluorine Chem 51:295
254. Rong G, Keese R (1990) Tetrahedron Lett 31:5615
255. Brace NO (1979) J Org Chem 44:212
256. Maruoka K Sano H Fukutani Y Yamamoto H. Chem Lett 1985, 1689
257. Zhu S, Qing C, Zhou C, Zhang J, Xu B (1996) J Fluorine Chem 79:77
258 Hu QS, Hu CM (1996) J Fluorine Chem 76:117
259. Cao P, Duan JX, Chen QY (1994) J Chem Soc, Chem Commun 1994:737
260. Wu FH, Huang BN, Lu L, Huang WY (1996) J Fluorine Chem 80:91
261. Wu SH, Liu WZ, Jiang XK (1994) J Org Chem 59:854
262. Ichinose Y, Oshima K, Utimoto K (1988) Chem Lett 1988:1437
263. Takeyama Y, Ichinose Y, Oshima K, Utimoto K (1989) Tetrahedron Lett 30:3159
264. Hu CM, Chen J (1993) Tetrahedron Lett 34:5957
265. Medebielle M, Oturan MA, Pinson J, Saveant JM (1996) J Org Chem 61:1331
266. Tordeux M, Wakselman C (1981) Tetrahedron 37:315
267. Yu HB, Huang WY (1996) Tetrahedron Lett 37:7999
268. Hu CM, Qiu YL (1991) Tetrahedron Lett 32:4001
269. Hu CM, Qiu YL (1992) J Org Chem 57:3339
270. Hu CM, Chen J (1993) J Chem Soc, Chem Commun 1993:72
271. Chen J, Hu CM (1994) J Chem Soc, Perkin Trans 1 1994:1111
272. Uneyama K, Kitagawa K (1991) Tetrahedron Lett 32:3385
273. Uneyama K, Kanai M (1991) Tetrahedron Lett 32:7425
274. Qiu ZM, Burton DJ (1994) Tetrahedron Lett 35:1813
275. Qiu ZM, Burton DJ (1995) J Org Chem 60:5570
276. Kitagawa O, Miura A, Kobayashi Y, Taguchi T (1990) Chem Lett 1990:1011
277. Yang ZY, Burton DJ (1989) J Fluorine Chem 45:435
278. Huang BN, Liu JT (1993) J Fluorine Chem 64:37
279. Chen QY, Qiu ZM (1988) J Fluorine Chem 39:289
280. Kamigata N, Fukushima T, Yoshida M (1990) Chem Lett 1990:649
281. Zhao CX, El-Taliawi GM, Walling C (1983) J Org Chem 48:4908
282. Sawada H, Yoshida M, Hauii H, Aoshima K, Kobayashi M (1986) Bull Chem Soc Jpn 59:215
283. Sawada H, Nakayama M, Yoshida M, Yoshida T, Kamigata N (1989) J Fluorine Chem 46:423
284. Yoshida M, Moriya K, Sawada H, Kobayashi M (1985) Chem Lett 1985:755
285. Nishida M, Fujii S, Kimoto H, Hayakawa Y, Sawada H, Cohen L A (1993) J Fluorine Chem 63:43

286. Yoshida M, Morinaga Y, Ueda M, Kamigata N, Iyoda M (1992) Chem Lett 1992:287
287. Huang WY, Liu JT, Li J (1995) J Fluorine Chem 71:51
288. Rossi RA, Pierini AB, Palacios SM (1990) Advances in Free Radical Chemistry Vol. 1. JAI Press, New York, p 193
289. Chen QY, Qiu ZM (1986) J Fluorine Chem 31:301
290. Feiring AE (1984) J Fluorine Chem 24:191
291. Yagupolskii LM, Matyushecheva GI, Pavlenko NV, Boiko VN (1982) J Org Chem USSR (Engl. Transl.) 18:10
292. Popov VI, Boiko VN, Yagupolskii LM (1982) J Fluorine Chem 21:365
293. Haley B, Haszeldine RN, Hewitson B, Tipping AE (1976) J Chem Soc 1976:525
294. Boiko VN, Shchupak GM (1994) J Fluorine Chem 69:207
295. Chen QY, Chen MJ (1991) J Fluorine Chem 51:21
296. Koshechko VG, Kiprianova LA, Fileleeva CI (1992) Tetrahedron Lett 33:6677
297. Bhasin KK, Cross RJ, Rycroft DS, Sharp DWA (1979) J Fluorine Chem 14:171
298. Voloshchuk VG, Boiko VN, Yagupolskii LM (1977) J Org Chem USSR (Engl Transl) 13:1866
299. Kondratenko NV, Popov VI, Kolomeitsev AN, Sadekov ID, Minkin VJ, Yagupolskii LM (1979) J Org Chem USSR (Engl Transl) 15:1394
300. Huang W, Huang B, Wang W (1985) Huaxue Xueboa 43:663; Chem Abs (1986) 104:206683z
301. Tordeux M, Langlois B, Wakselman C (1989) J Org Chem 54:2452
302. Feiring AE (1983) J Org Chem 48:347
303. Iseki K, Takaashi M, Asada D, Nagai T, Kobayashi Y (1995) J Fluorine Chem 74:269
304. Cantacuzene D, Wakselman C, Dorme R (1977) J Chem Soc, Perkin Trans 1 1977:1365
305. Yoshida M, Shimokoshi K, Kobayashi M (1987) Chem Lett 1987:433
306. Henne AL, Finnegan WG (1950) J Am Chem Soc 72:3806
307. Paskovich D, Gaspar P, Hammond GS (1967) J Org Chem 32:833
308. Lai C, Mallouk TE (1993) J Chem Soc, Chem Commun 1993:1359
309. Van der Puy M, Belter RK, Borowski RJ, Ellis LAS, Persichini PJ III, Poss AJ, Rygas TP, Tung HS (1995) J Fluorine Chem 71:59
310. Burdeniuc J, Crabtree RH (1996) Science 271:340
311. Brace NO (1964) J Am Chem Soc 86:523
312. Brace NO (1966) J Org Chem 31:2879
313. Piccardi P, Massardo P, Modena M, Santoro E (1974) J Chem Soc , Perkin Trans 1 1974:1848
314. Barth F, O-Yang C (1991) Tetrahedron Lett 32:5873
315. Arnone A, Bravo P, Cavicchio G, Frigerio M, Viani F (1992) Tetrahedron 48:8523
316. Itoh T, Ohara H, Emoto S (1995) Tetrahedron Lett 36:3531
317. Morikawa T, Kodama Y, Uchida J, Takano M, Washio Y, Taguchi T (1992) Tetrahedron 41:8915
318. Morikawa T, Uejima M, Kobayashi Y, Taguchi T (1993) J Fluorine Chem 65:79
319. Morikawa T, Uejima M, Kobayashi Y (1989) Chem Lett 1989:623
320. Watanabe Y, Yokozawa T, Takata T, Endo T (1988) J Fluorine Chem 39:431
321. Arnone A, Bravo P, Frigerio M, Viani F, Cavicchio G, Crucianelli M (1994) J Org Chem 59:459
322. Arnone A, Bravo P, Viani F, Zanda M, Cavicchio G, Crucianelli M (1996) J Fluorine Chem 76:169
323. Piccardi P, Massardo P, Modena M, Santoro E (1973) J Chem Soc , Perkin Trans 1 1973:982
324. Morikawa T, Uejima M, Kobayashi Y (1988) Chem Lett 1988:1407
325. Morikawa T, Uejima M, Yoda K, Taguchi T. (1990) Chem Lett 1990:467

Telomerisation Reactions of Fluorinated Alkenes

Bruno Améduri · Bernard Boutevin

ESA 5076 (CNRS) Ecole Nationale Supérieure de Chimie de Montpellier 8, Rue Ecole Normale F-34296 Montpellier Cedex 5 (France). *E-mail: ameduri (or boutevin)@cit.enscm.fr*

The synthesis of fluorinated oligomers by radical telomerisation of fluoroalkenes is discussed. After a brief presentation of the various parameters which direct this reaction and the kinetic laws, this review outlines the traditional telogens efficient in telomerisation and also considers other transfer agents which undergo less known cleavages. Telomerisation involving main fluorinated alkenes (vinylidene fluoride, tetrafluoroethylene, chlorotrifluoroethylene, trifluoroethylene and hexafluoropropene) and less used agents is then reviewed, considering either the method of initiation including new activation systems, or specific cleavage of the telogens used. Some examples of cotelomerisation are given constituting interesting models of copolymerisation. General concepts of reactivity regarding the orientation of addition of free radicals to unsymmetrical fluorinated alkenes and also comparison of the reactivity of various fluorinated alkenes with the telogens are discussed and several examples given. Beside the traditional process of telomerisation, the living telomerisation methods are also described. In these methods, the living character is mentioned and well chosen transfer agents with specific cleavable bonds described. The living behaviour of the iodine transfer polymerisation or stepwise cotelomerisation occurs in the presence of perfluoroalkyl iodides or α,ω-diiodoperfluoroalkanes as the telogens and shows high interest in the synthesis of structurally well defined telomers. Finally, the application of fluoropolymers synthesised from these telomers is presented showing the high added value of these materials.

Keywords: Telomerisation, fluoroalkene, telogen, free radical, livingness.

Topics in Current Chemistry, Vol. 192

List of Symbols and Abbreviations

AIBN	azobisisobutyronitrile
BEPH	butyl ethyl peroxyhexanoate
BDE	bond dissociation energy
Bu	n-butyl
*t*Bu	*tertio*-butyl
CR	counter radical
C_T	transfer constant
CTFE	chlorotrifluoroethylene
d	day
DBP	dibenzoyl peroxide
DIPC	diisopropyl peroxydicarbonate
$\overline{DP_n}$	average degree of polymerisation in number
DTBP	di*tertio*-butyl peroxide
F113	1,1,2-trifluorotrichloroethane
HFPO	hexafluoropropylene oxide
iPr	isopropyl
IR_FI	α, ω-diiodoperfluoroalkane
ITP	iodine transfer polymerisation`
M	monomer
[M]	concentration of monomer
$\overline{M_n}$	average molecular weight in number
$\overline{M_\omega}$	average molecular weight in weight
NMR	nuclear magnetic resonance
PCTFE	poly(chlorotrifluoroethylene)
PHFP	poly(hexafluoropropene)
PTFE	poly(tetrafluoroethylene)
PVDF	poly(vinylidene fluoride)
PVF	poly(vinyl fluoride)
$R^\cdot$	alkyl radical
R	alkyl group
R_F	perfluorinated group
$R_{F,Cl}$	chlorofluorinated group
R_FBr	perfluoroalkyl bromide
R_FI	perfluoroalkyl iodide
RT	room temperature
sc	supercritical
Tempo	2,2,6,6-tetramethyl piperidinyl-1-oxyl
TFE	tetrafluoroethylene
Tg	glass transition temperature
THF	tetrahydrofuran
UV	ultra violet
VDF	vinylidene fluoride
VF	vinyl fluoride
Δ	heating

1 Introduction

Fluorinated polymers are considered high value-added materials, due to their outstanding properties which open up various applications [1–9]. Such polymers exhibit high thermostability and chemical inertness, low refractive index and coefficient of friction, good water and oil repellence, low surface energy and valuable electrical properties. In addition, they are non-sticky and resistant to UV, ageing, and to concentrated mineral acids and alkalies.

The great value of the unique characteristics of fluorinated polymers in the development of modern industries has ensured an increasing technological interest since the discovery of the first fluoropolymer, poly(chlorotrifluoroethylene) in 1934. Hence, their fields of applications are numerous: paints and coatings [10] (for metals [11], wood and leather [12], stone and optical fibers [13, 14]), textile finishings [15], novel elastomers [5, 6, 8], high performance resins, membranes [16, 17], functional materials (for photoresists and optical fibers), biomaterials [18], and thermostable polymers for aerospace.

But the processing of these products is still difficult because several of them are not usually soluble, and others are not meltable or exhibit very high melting points. In addition, their prices are still high and even if their properties at high temperatures are still preserved, this is not so at low temperatures because their glass transition temperature is too high or because of their high crystallinity compared to those of silicones for example.

It was thus worth finding a model of the synthesis of fluoropolymers in order to predict their degree of polymerisation, their structure, and the mechanism of the reaction.

One of the most interesting strategies is that of telomerisation. Such a reaction, introduced for the first time by Hanford in 1942 [19], in contrast to polymerisation, usually leads to low molecular weight polymers, called telomers, or even to monoadducts with well-defined end-groups. Such products *1* are obtained from the reaction between a telogen or a transfer agent (X-Y), and one or more (n) molecules of a polymerisable compound M (called a taxogen or monomer) having ethylenic unsaturation, under radical polymerisation conditions, as follows:

$$X{-}Y + nM \xrightarrow{\text{free radicals}} X{-}(M)_n{-}Y$$

$\underline{1}$

Telogen X-Y is easily cleaved by free radicals (formed according to the conditions of initiation) leading to an $X^{\cdot}$ radical which can react further with the monomer. After the propagation of monomer, the final step consists of the transfer of the telogen to the growing telomeric chain. Telomers *1* are intermediate products between organic compounds (e.g. n=1) and macromolecular species (n=100). Hence, in certain cases, end-groups exhibit a chemical importance which can provide useful opportunities for further functionalisations.

The scope of telomerisation was first outlined by Freidlina [20] in 1966, then improved upon by Starks in 1974 [21], and further developed by Gordon and

Loftus [22]. Recently, we have reviewed such a reaction [23, 24] in which the mechanism and kinetics of radical and redox telomerisations are described.

All monomers involved in radical polymerisation can be utilised in telomerisation. In this chapter, only monomers containing a fluorine atom or a short perfluorinated group linked to an unsaturated carbon atom will be taken into account.

Below, the initiation step and mechanisms involved in telomerisation are described, followed by the various processes used, the telogens, the monomers and their respective reactivities. The following part deals with the pseudoliving radical telomerisation of fluorinated monomers and finally different applications of fluorinated telomers are given.

2
Initiation and Mechanisms

A telomerisation reaction is the result of four steps: initiation, propagation, termination and transfer. These four steps have been described in previous reviews [21–24] summarised below.

2.1
Methods of Initiation

Telomerisation can be initiated from various processes: photochemical (in the presence of UV, X or γ rays), in the presence of radical initiator or redox catalysts, or thermally. For each case, different mechanisms have been proposed. Below are explained the initiation processes and some examples are listed in Table 1.

2.1.1
Photochemical Initiation

An interesting example was proposed by Haszeldine et al. [25] in the photochemical induced addition of alcohol RH to hexafluoropropene (HFP) knowing that no propagation of HFP occurs:

Initiation

$$C_3F_6 \xrightarrow{UV} (C_3F_6)^*$$
$$(C_3F_6)^* + RH \longrightarrow R^{\cdot} + C_3F_6H^{\cdot}$$
$$R^{\cdot} + C_3F_6 \longrightarrow RC_3F_6^{\cdot}$$

Transfer

$$C_3F_6H^{\cdot} + RH \longrightarrow CF_3CFHCHF_2$$
$$RC_3F_6^{\cdot} + RH \longrightarrow R(C_3F_6)H + R^{\cdot}$$

$R(C_3F_6)H$ is composed of two isomers: RCF_2CFHCF_3 (major) and $F_3CCFRCHF_2$ (minor).

Tedder and Walton [26] suggested almost a similar mechanism.

Table 1. Different methods of cleavage in the telogens

Bond Cleaved	Method	References
	$Cl_3CH + xF_2C{=}CH_2 \xrightarrow{rad} Cl_3C{-}(C_2F_2H_2)_xH$	(85)
C-H	$THF + CF_2{=}CFCF_3 \xrightarrow[peroxide]{UV\ or}$  $-CF_2CFHCF_3$	(86)
	$CH_3{-}OH + nF_2C{=}CFX \xrightarrow{peroxide\ or\ UV} H{-}(CFX{-}CF_2)_n{-}CH_2OH$; $X{=}Cl,F,CF_3$	(25, 87, 88)
C-F	$Cl_3CF + F_2C{=}CFCl \xrightarrow{AlCl_3} Cl_3CCF_2CF_2Cl$	(89)
	$HCCl_3 + nF_2C{=}CFCl \xrightarrow{FeCl_3} HCCl_2{-}(CF_2CFCl)_n{-}Cl$	(90)
C-Cl	$CCl_4 + F_2C{=}CFH \xrightarrow{CuCl} Cl_3CCF_2CFHCl + Cl_3CCFHCF_2Cl$	(91)
	$RCCl_3 + nF_2C{=}CH_2 \xrightarrow[or\ rad]{redox} RCCl_2(CH_2CF_2)_nCl$ (R: Cl, CH_2OH, CO_2CH_3)	(92)
C-Br	$CFBr_3 + nF_2C{=}CFH \xrightarrow{UV} Br_2CFCFHCF_2Br$(maj) + Br_2CFCF_2CFHBr(min)	(93)
	$BrCF_2CFClBr + (m + n-1)F_2C{=}CFCl \xrightarrow{UV} Br(CF_2CFCl)_n(CFClCF_2)_m{-}Br$	(94)
C-Br	$BrCF_2Br + xF_2C{=}CFCF_3 \xrightarrow{thermal} BrCF_2(CF_2{-}CF(CF_3))_x{-}Br$	(95)

Table 1 (continued)

Bond Cleaved	Method	References
C-I	$i\text{-}C_3F_7I + nF_2C{=}CH_2 \xrightarrow{\text{thermal}} i\text{-}C_3F_7(C_2F_2H_2)_nI \quad n = 1-4$	(96, 97)
	$C_4F_9I + xHFC{=}CF_2 \xrightarrow{\text{thermal}} C_4F_9(C_2F_3H)_xI \quad x=1-3$	(98)
	$C_6F_{13}I + xF_2C{=}CFCF_3 \xrightarrow{\text{thermal}} C_6F_{13}(C_3F_6)_xI \quad x=1-3$	(99)
S-H	$CF_3SH + F_2C{=}CFH \xrightarrow{\text{UV}} CF_3SCFHCF_2H\ (98\%) + CF_3SCF_2CFH_2\ (2\%)$	(100)
	$HOC_2H_4SH + xF_2C{=}CH_2 \xrightarrow{\text{rad}} HOC_2H_4S(C_2F_2H_2)_xH \quad x=1-10$	(101)
S-S	$RS{-}SR + nM \xrightarrow{\text{UV}} RS(M)_nSR$ $R: CH_3$ or $CF_3 \quad M: C_2F_4, C_3F_6, C_2F_3Cl, C_2F_2H_2$	(84, 102, 103)
S-Cl	$FSO_2Cl + xF_2C{=}CH_2 \xrightarrow{\text{peroxide}} FSO_2(C_2F_2H_2)_xCl \quad x=1-3$	(104)
P-H	$PH_3 + xF_2C{=}CH_2 \xrightarrow{\text{UV}} H_{3-x}P(C_2H_2F_2)_xH \quad x=1, 2$	(105)
	$H\,P(=O)(OEt)_2 + xCF_2{=}CF_2 \xrightarrow{\gamma\text{ rays}} (EtO)_2{-}P(=O){-}(CF_2{-}CF_2)_x{-}H$	(106)

Table 1 (continued)

Bond Cleaved	Method	References
P-Cl	$PCl_5 + nF_2C{=}CFCl \xrightarrow{\text{peroxide}} Cl_4P\text{-}(C_2F_3Cl)_n{-}Cl$	(107)
	$POCl_3 + nF_2C{=}CFCl \xrightarrow{\text{rad}} Cl_2P(O)(C_2F_3Cl)_n{-}Cl$	(108)
Si-H	$Cl\,Si(CH_3)(Cl){-}H + nH_2C{=}CHC_6F_5 \xrightarrow{H_2PtCl_6} Cl{-}Si(CH_3)(Cl){-}CH_2CH_2C_6F_5$	(109)
	$Cl_3SiH + xCF_3CH{=}CH_2 \xrightarrow{\text{UV}} Cl_3Si(CH_2CH(CF_3))_x{-}H \quad x=1,2$	(110)
Br-H	$H{-}Br + F_2C{=}CFCl \xrightarrow{\text{UV}} BrCFClCF_2H$	(111)
I-F	$IF_5 \xrightarrow{I_2} [I{-}F] \xrightarrow{xF_2C=CFCl} I(C_2F_3Cl)_xF$	(112, 113)
I-Cl	$I{-}Cl + F_2C{=}CFCF_3 \xrightarrow[\Delta \text{ or UV}]{\text{thermal}} CF_3CFICF_2Cl\ (92\%) + CF_3CFClCF_2I\ (8\%)$	(114)
	$I{-}Cl + F_2C{=}CFCl \xrightarrow[\text{or redox or rad}]{} ICFClCF_2Cl\ (95-100\%) + ICF_2CFCl_2(0-5\%)$	(115)
I-Br	$I{-}Br + nF_2C{=}CFCl \xrightarrow{\text{thermal}} I(C_2F_3Cl)_nBr$	(116)
I-I	$I_2 + nF_2C{=}CF_2 \xrightarrow{\Delta} I(C_2F_4)_nI$	(117)
Cl-Cl	$SO_2Cl_2 + nF_2C{=}CFCl \xrightarrow{\text{rad}} Cl(C_2F_3Cl)_n{-}Cl$	(118)
Br-Br	$Br_2 + nF_2C{=}CFCl \xrightarrow{\text{redox}} Br(C_2F_3Cl)_n{-}Br$	(119)

2.1.2
Radical telomerisation

The general process of a telomerisation involving the radical initiator A_2 has been suggested as follows (A_2 can be an azo or peroxide or perester initiator which undergoes a homolytic cleavage under heating and generates two radicals [27]).

Initiation

$$A_2 \xrightarrow[\Delta]{k_i} 2\,A^{\cdot}$$

$$A^{\cdot} + XY \longrightarrow X^{\cdot} + AY$$
$$X^{\cdot} + F_2C{=}CRR' \longrightarrow X(C_2F_2RR')^{\cdot}$$

Propagation

$$XC_2F_2RR'^{\cdot} + F_2C{=}CRR' \xrightarrow{k_{p1}} X(C_2F_2RR')_2^{\cdot}$$

$$X(C_2F_2RR)_n^{\cdot} + F_2C{=}CRR' \xrightarrow{k_{pn}} X(C_2F_2RR')_{n+1}^{\cdot}$$

Termination

$$X(C_2F_2RR')_n^{\cdot} + X(C_2F_2RR')_p^{\cdot} \xrightarrow{k_{T_e}} X(C_2F_2RR')_{n+p}X$$

Transfer

$$X(C_2F_2RR')_n^{\cdot} + XY \xrightarrow{k_{tr}} X(C_2F_2RR')_nY + X^{\cdot}$$

XY, $F_2C{=}CRR'$, k_i, k_{p1}, k_{Te} and k_{tr} represent the telogen, the monomer, the rate-constant of initiation, the rate-constants of propagation, of termination and of transfer, respectively.

The kinetic law, regarding the reverse of average degree of telomerisation, $\overline{DP_n}$, depends upon the transfer constant of the telogen (C_T), the telogen [XY] and monomer [M] molar concentrations, as follows [21–24]:

i) For high $\overline{DP_n}$: $\dfrac{1}{\overline{DP_n}} = C_T \dfrac{[XY]}{[M]}$ with $C_T = \dfrac{k_{tr}}{kp}$

According to this above equation, transfer constants of various telogens were determined [28].

Concerning the nature of the initiator no prediction can be made on the efficiency of such a reactant in the telomerisation of fluoroalkenes. Non fluorinated initiators led to unsuccessful results whereas investigations on the synthesis of highly fluorinated peresters or peroxides were performed by Rice and Sandberg [29] and Sawada [30], respectively.

ii) For low $\overline{DP_n}$

If one considers F_n as the molar fraction in telomer having n as the degree of polymerisation, David and Gosselain [31] have proposed the following equation (where R and n represent the (Telogen)/(Monomer) molar ratio and the degree of telomerisation, respectively). $F_n = \dfrac{C_T^n R}{\prod_{j=1}^{j=n} (1 + C_T^j . R)}$

Actually C_T^n varies with n, increasing from C_T^1 to C_T^5 and then becomes constant since the contribution of the telogen on the reactivity of the n order-radical is negligible.

2.1.3
Redox Catalysis

In this case, the reaction requires the use of a catalyst MeL_x (salt of transition metal or metallic complex) which participates in the initiation with an increase of the degree of oxidation of the metal as follows (being usually a halogen):

Initiation

$$Me^{n+}L_x + XY \xrightarrow{k_i} Me^{(n+1)}L_xY + X^{\cdot}$$
$$X^{\cdot} + F_2C{=}CRR' \longrightarrow X(C_2F_2RR')^{\cdot}$$

Propagation

$$X(C_2F_2RR')^{\cdot} + F_2C{=}CRR' \xrightarrow{k_{p1}} X(C_2F_2RR')_2^{\cdot}$$
$$X(C_2F_2RR')_n^{\cdot} + F_2C{=}CRR' \xrightarrow{k_{pn}} X(C_2F_2RR')_{n+1}^{\cdot}$$

Transfer

$$X(C_2F_2RR')_{n+1}^{\cdot} + Me^{(n+1)+}L_xY \xrightarrow{k_{tr}} X(C_2F_2RR')_{n+1}Y + Me^{n+}L_x$$

Termination

$$X(C_2F_2RR')_n^{\cdot} + X(C_2F_2RR')_p^{\cdot} \xrightarrow{k_{Te}} X(C_2F_2RR')_{n+p}X$$

Table 2. Transfer constants of catalysts and telogen in the redox telomerisation of chlorotrifluoroethylene with CCl_4 [32]

Catalyst	C_{Me}	C_{CCl4}
$FeCl_3$/benzoin	75	0.02
CuCl	700	0.02
$RuCl_2(PPh_3)_3$	800	0.02

The kinetic law applied to redox telomerisation depends on both the transfer constants of the catalyst and of the telogen but the former one is much greater as shown is Table 2 [32]:

$$\frac{1}{\overline{DP}_n} = C_{Me}\frac{[MeL_x]}{[M]} + C_{XY}\frac{[XY]}{[M]}$$

As noted in both radical and redox mechanisms of telomerisation of fluoroalkenes, termination reactions always occur by recombination and not by disproportionation [29].

In addition, it has been shown that, for the same telogen and fluoroolefin, a radical telomerisation leads to higher molecular weights than those obtained from redox catalysis [21–24].

2.1.4
Thermal Initiation

In certain conditions, sufficient energy supplied by temperature causes fission of the X-Y bond of the telogen.

The mechanism is rather similar to that of the radical telomerisation except for the initiation step in which the radicals are produced from the telogen on heating as follows:

$$X{-}Y \xrightarrow[\Delta]{} X^{\cdot} + Y^{\cdot}$$

$$X^{\cdot} + CF_2{=}CRR' \longrightarrow XC_2F_2RR'^{\cdot}$$

2.2
Cotelomerisation

In cotelomerisation, the problem is slightly more complex and the kinetics have been little investigated. However, the equation proposed by Tsuchida [33] relating the instantaneous $\overline{DP}_n$ *versus* the kinetic constants has been confirmed [34]:

$$(\overline{DP}_n)_i = (r_1[M_1]^2 + 2[M_1][M_2] + r_2\,[M_2]^2)/(r_1C_{T1}[M_1][T] + r_2C_{T2}[M_2][T])$$

where $[M_i]$, $[T]$, r_i and C_{Ti} represent the concentrations of the monomer M_i and of the telogen, the reactivity-rate of the monomer M_i and the transfer constant of the telogen to the monomer.

Furthermore, for known r_i, the conversion-rates α_1 and α_2 of both monomers *vs* time could be determined.

The composition of the cotelomer i.e., the repartition of the monomers in the cotelomer can be predicted from known kinetic constants r_i and C_{Ti} and from $[M_i]_o$ according to the statistical theory of the cotelomerisation. Moreover, the probability to obtain cotelomers of well-defined composition and then the functionality of the cotelomer were calculated [21–24].

2.3
Specific Initiations for Fluoroalkenes

In contrast to classic initiating systems, other less known processes have been used successfully in telomerisation of fluoroalkenes: these systems involve hypofluorites, hypobromites, or concern ionic (anionic and cationic) or unusual telomerisations depicted below.

2.3.1
From Hypohalites

It has been known that highly fluorinated alkyl-hypochlorites and hypofluorites R_FOX (with X=Cl or F) are of synthetic value in the preparation of many fluorine containing products [35–39]. Actually, between –150 and +50 °C, such compounds generate $R_FO^{\cdot}$ radicals able to initiate a telomerisation. However, from higher temperatures (above 70 °C) certain reactions may be explosive.

CF_3OX (X=F,Cl) derivatives are the most extensively studied products with respect to their preparation and chemistry. For example, was the most studied perfluoroalkyl hypofluorite, CF_3OF, synthesised for the first time by Kellogg and Cady more than 45 years ago [40]. It offers interesting results with vinylidene fluoride [41, 42], chlorotrifluoroethylene (CTFE) [43–45, 48], hexafluoropropene [46], 1,1-dichlorodifluoroethylene [41, 47], *trans*-1,2-dichlorodifluoroethylene [48], perfluorobut-2-ene [49] according to the general following reaction:

$$CF_3OF + n\,F_2C{=}CRR' \longrightarrow R_1(C_2F_2RR')_nF + CF_3O(C_2F_2RR')_{2n}OCF_3$$
$$(R_1{=}CF_3O \text{ major}\,;\ R_1{=}F \text{ minor}).$$

Johri and DesMarteau [41] have previously claimed that such reactions underwent a free radical mechanism. Actually, the radical mechanism was demonstrated by Di Loreto and Czarnowski [50], the polyfluoroalkyl fluoroxy compound being usually broken homolytically producing $R_FO^{\cdot}$ and $^{\cdot}X$ radicals in the initiation step, as presented in Scheme 1.

Interestingly, Marraccini et. al. [43, 51] reacted CF_3OF, $C_nF_{2n+1}OF$ or $R_{F,Cl}CF_2OF$ to CTFE yielding the ten first telomers whereas they obtained the three first adducts only from the telomerisation of 1,2-dichlorodifluoroethylene with CF_3OF [48]. However, higher molecular weight-telomers were produced by modifying the pressures of both gaseous reactants [47].

Initiation

$$R_FOF \xrightarrow[\Delta]{} R_FO^{\cdot} + {}^{\cdot}F$$

$$R_FO^{\cdot} + F_2C{=}CRR' \longrightarrow R_FO(C_2F_2RR')^{\cdot}$$

Propagation

$$R_FO(C_2F_2RR')^{\cdot} + F_2C{=}CRR' \longrightarrow R_FO(C_2F_2RR')_2^{\cdot}$$

$$R_FO(C_2F_2RR')_i^{\cdot} + F_2C{=}CRR' \longrightarrow R_FO(C_2F_2RR')_{i+1}^{\cdot}$$

Transfer

$$R_FO(C_2F_2RR')^{\cdot} + R_FOF \longrightarrow R_FO(C_2F_2RR')F + R_FO^{\cdot}$$

$$R_FO(C_2F_2RR')_{i+1}^{\cdot} + R_FOF \longrightarrow R_FO(C_2F_2RR')_{i+1}F + R_FO^{\cdot}$$

Termination

$$R_FO(C_2F_2RR')_p^{\cdot} + R_FO(C_2F_2RR')_q^{\cdot} \longrightarrow R_FO(C_2F_2RR')_{p+q}OR_F$$

Scheme 1. Mechanism of telomerisation of fluoroalkene $F_2C{=}CRR'$ with R_FOF (in order to generalise the formulae, the expression C_2F_2RR' has been prefered to CF_2CRR' or $CRR'CF_2$ since the nature of R and R′ (which can be either electron-withdrawing or electron-donating groups) influences the sense of addition of $R_FO^{\cdot}$ to a preferential site of the fluoroolefins).

DesMarteau et al. synthesised new perfluoroalkyl hypofluorites [44] and hypobromites [52]. The first ones were produced from perfluorinated acid fluorides or halogenated ketones as depicted in the following schemes [44]:

$$R_{F,Cl}C(O)F + XF \xrightarrow{CsF} R_{F,Cl}CF_2OX \quad (X=F, Cl)$$

where $R_{F,Cl}$ represents $ClCF_2$, HCF_2, $ClCF_2CFCl$, $BrCF_2CFBr$, FSO_2CF_2 or $FSO_2CF(CF_3)$

$$R_{F,Cl}C(O)R'_{F,Cl} + XF \longrightarrow R_{F,Cl}R'_{F,Cl}CFOX$$

with $R_{F,Cl}=YCF_2$ (Y=H or Cl) and $R'_{F,Cl}=CF_3$ or $ClCF_2$.

All these above polyfluoroalkyl hypohalites successfully reacted to $CF_2=CH_2$ and to $CF_2=CFCl$ [44], in equimolar ratios, from −145 °C to room temperature for 18–24 h, with better yields in the presence of CTFE. They mainly led to monoadducts composed of both normal and reverse isomers. The authors observed that branched hypohalites were more reactive than linear ones and that for the same $R_{F,Cl}$ group, $R_{F,Cl}OF$ gave higher yields than those obtained from $R_{F,Cl}OCl$.

In addition, the same team described the synthesis of perfluoroalkyl hypobromites R_FOBr, achieved by reacting $BrOSO_2F$ to R_FONa (where R_F represents $(CF_3)_3C$ or $C_2F_5C(CF_3)_2$ groups) [52]. These compounds reacted with vinylidene fluoride, tetrafluoroethylene, 1,1-dichlorodifluoroethylene and *cis*-1,2-difluoroethylene in the −93 °C- room temperature range for 8–12 h. Monoadducts were produced mainly, the yields ranging from 20 to 37% based on $BrOSO_2F$ used. $(CF_3)_3COBr$ seems more reactive than $C_2F_5C(CF_3)_2OBr$ whatever the fluoroalkene, but less reactive than hypochlorites or fluoroxy compounds mentioned above. These novel hypobromites readily decompose above −20 °C by an assumed free radical β-elimination [52].

All these polyfluorinated ethers produced are stable, inert, and colourless liquids at room temperature.

2.3.2
Ionic Initiations

Many investigations concerning the ionic or the ring opening polymerisations of numerous monomers have been detailed in many books. However, to our knowledge very few studies about cationic or anionic polymerisations of fluoroalkenes have been investigated, except the well known anionic polymerisation of perfluorobutadiene as studied by Narita [53].

In the examples described below, it can be considered that the anionic telomerisation of hexafluoropropylene oxide (HFPO) and of tetrafluoroethylene, or the cationic telomerisation of fluorinated oxiranes or oxetanes were successful.

2.3.2.1
Anionic Telomerisation

2.3.2.1.1 Perfluorinated Oxiranes
The oldest investigation was conducted by Warnell [54] in 1967 who telomerised HFPO or tetrafluoroethylene epoxide by ring opening with two ω-iodofluoro-carbon ether acid fluorides producing the corresponding adducts, as follows:

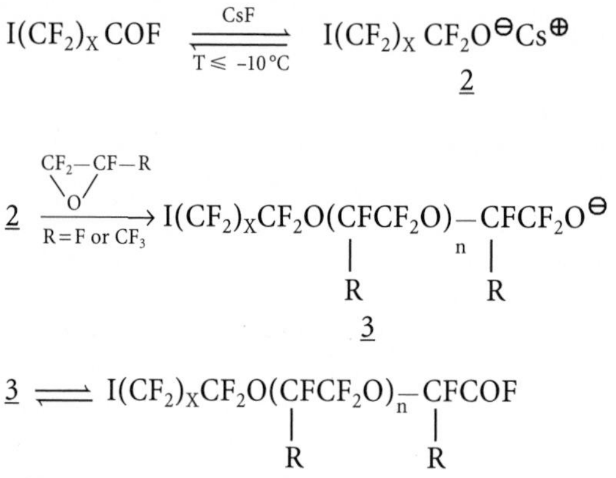

$$I(CF_2)_X COF \underset{T \leqslant -10\,°C}{\overset{CsF}{\rightleftharpoons}} I(CF_2)_X CF_2O^{\ominus}Cs^{\oplus}$$
$$\underline{2}$$

$$\underline{2} \xrightarrow[R=F \text{ or } CF_3]{\overset{CF_2-CF-R}{\diagdown O \diagup}} I(CF_2)_XCF_2O(CFCF_2O)_n-CFCF_2O^{\ominus}$$

with R groups:

$$\underline{3}$$

$$\underline{3} \rightleftharpoons I(CF_2)_XCF_2O(CFCF_2O)_n-CFCOF$$

$$R=CF_3 \ n=1-4 \text{ and } R=F \ n=0-2$$

Initiation by ionising radiation or by various catalysts [55] such as mono-valent metal fluorides can be achieved.

Several years later, Ito et al. [56] used such a concept to cotelomerise successfully the 4-bromo and 4-chloroheptafluoro-1,2-epoxybutane with HFPO. In the same way, Kvicala et al. [57] prepared the first three telomers $\underline{4}$ by telomerising HFPO. Similar telomerisations were also achieved from polyfluoroketones [58].

$$R_{F,Cl}\!\left[\!\!\begin{array}{c}CF_2OCF\\ |\\ CF_3\end{array}\!\!\right]_n\!\!CF_2O$$

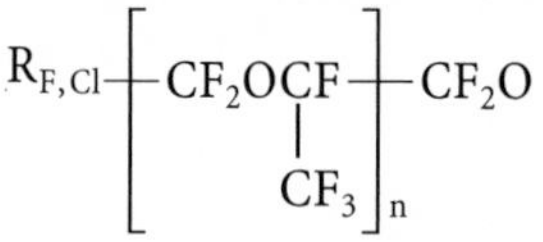

The extrapolation of this work has led to KRYTOX products commercially available by the Dupont Company [59–61].

Other triblock copolymers were obtained by a similar process from a fluo-rinated oxetane instead of HFPO [62] leading to DEMNUM commercialised by Daikin.

2.3.2.1.2 Tetrafluoroethylene

Using similar catalysts as above, Fielding [63] suggested the synthesis of novel branched alkenes from TFE as follows:

$$CF_2=CF_2 + CsF \longrightarrow C_nF_{2n}$$

The most abundant oligomer is the pentamer. These perfluorinated olefins are not reactive since the double bond is internal in the chain. However, the ICI Company [64] functionalises these alkenes by adding phenol onto the perfluorinated olefins:

$$C_{10}F_{20} + RC_6H_4OH \xrightarrow[R'NH_2]{-HF} C_{10}F_{19}OC_6H_4R$$

The telomerisation of tetrafluoroethylene was photochemically initiated in the presence of oxygen, leading to telechelic fluorinated polyethers 5 [65]:

$$F_2C{=}CF_2 \xrightarrow[h\nu]{O_2} F(O{=})C(CF_2O)_X-(C_2F_4O)_y-CF_2C({=}O)F$$

5

with $0.6-1.5 \times 10^3 < \overline{M_n} < 4 \times 10^4$

Such compounds have been commercialised by the Montefluos-Ausimont Company, called FOMBLIN [66]. They are composed of two groups: Fomblin Y, produced by a fluorination of 5, are non functional polyethers $CF_3(CF_2O)_x(C_2F_4O)_yC_2F_5$ whereas Fomblin Z are obtained by chemical change of 5 leading to the following telechelics:

$$\textcircled{G}-(CF_2O)_X(C_2F_4O)_y CF_2-\textcircled{G} \text{ with } \textcircled{G}: OH, CO_2R \text{ or } NH_2$$

Similarly, HFP [67] or perfluorobutadiene [68] can also be polymerised.

2.3.2.2
Cationic Telomerisation

To our knowledge, no work dealing with the cationic telomerisation of fluorinated alkenes has been described in the literature in contrast to successful works achieved from fluorinated epoxides [69, 70] or oxetanes [71].

2.3.3
Other Initiations

Beyond well established methods of initiation, other efficient systems have been scarcely used for initiating telomerisation reactions. Pulsed carbon dioxide laser was used to induce telomerisation of tetrafluoroethylene (TFE) with perfluoroalkyl iodides and led to telomeric distributions [72]. It is an exothermic radical chain reaction, producing $R_F(C_2F_4)_nI$ with low n. Dimer proportion dominated in the case of IR multiphoton dissociation. However, at low pressures, n was increased. Similarly, the telomerisation of TFE with BrC_2F_4Br was successful from a KrF laser or a xenon-lamp and molecular weight distributions depend upon the light intensity and the light source [73]. Krespan and Petrov [74] have shown that the initiation by strong fluoroxidisers such as XeF_2, COF_3, AgF_2, MnF_3, CeF_4 or a mixture of them allows telomerisation of various fluoroalkenes. In addition, the same team [75] has recently performed the telomerisation of TFE with perfluorobutyl iodide initiated by fluorographite intercalated with SbF_5 at 80 °C leading to a telomeric distribution showing the five first adducts. Furthermore, a high electrical field during the vacuum-deposition process has yielded the formation of ultrathin ferroelectrical vinylidene fluoride telomeric films [76].

Very interesting surveys were investigated by Paleta's group on the use of aluminium trichloride which favours the cleavage of a C-F bond [77, 78] as follows:

$$Cl_3CF + F_2C{=}CFCl \xrightarrow{AlCl3} Cl_3CCF_2CF_2Cl + FCCl_2CF_2CFCl_2$$

To our knowledge no biochemical initiation was efficient whereas it was successful for the addition of perfluoroalkyl iodides to allyl acetate [79].

Less relevant to the intiating system but important to process of telomerisation, is the use of supercritical CO_2, which is particularly efficient because it allows a better solubility of fluoroalkenes in organic media [80], leading to controlled telomerisation with narrower molecular weight distributions.

3 Cleavable Telogens

This section deals with the various telogens or transfer agents involved in telomerisation of fluoroalkenes. As mentioned above, a telogen must fulfil three targets:

- i It requires a weak X-Y linkage able to be cleaved by heating, by γ, UV radiation, X-ray or microwave or has to react with another species (very often a radical produced by initiator) to generate a telogen-radical;
- ii This telogen radical must be able to initiate the telomerisation (i.e., such a radical produced has to react with the fluoroolefin).
- iii The telogen must be efficient enough to transfer to the growing telomeric chain.

More details about the reactivity of telogens are given is Sect. 5.2, mostly investigated by Tedder and Walton [81–83].

This is mainly linked to the transfer reaction of the telogen to the growing chain. Theoretical rules have been mentioned in Sect. 2.1.2. The transfer constant is an intrinsic value of the telogen associated with a monomer and varies between 10^{-4} and 35. The presence of an electron withdrawing group adjacent to the cleavable bond allows a higher efficiency of the telogen and offers a high selectivity. This is the case of CF_3SSCF_3 with leads to more interesting results than CH_3SSCH_3 [84].

3.1 Classical Cleavage

Several examples of telomerisation of fluoroalkanes with various telogens are reported in Table 1. All the telogens employed in telomerisation of hydrogenated olefins [21–24] are also efficient for fluoroalkenes and the system of initiator also works well with these telogens. Consequently, a large variety of cleavages of X-Y bonds is possible; these bonds can be located at the chain end or in the centre of the molecule (e.g., disulphides) [84]. Hence, they offer a wide range of opportunities of products.

The most known cleavable bonds are C-H, C-X (X=Cl, Br, I), X-X, S-H, S-S, P-H and Si-H (Table 1).

However, less classic bonds of telogens have shown to be also efficient in certain conditions and are described below.

3.2 More Specific Cleavages

Hereafter are described three characteristic examples.

3.2.1 *Cleavage of C-F Bond*

Bond dissociation energy of the C-F bond (486 KJ · mol^{-1}) is the highest and consequently shows how strong such a bond is. However, interesting surveys were performed by Paleta's group. Cl_3CF was used as the telogen in the telomerisation of CTFE [89] and 1,2-dichlorodifluoroethylene [77] in the presence of $AlCl_3$. Such a catalyst is efficient for the cleavage of the C-F bond [76, 120], but not selective when the telogen contains C-F and C-Cl bonds (see Sect. 2.2.3).

3.2.2 *Cleavage of Si-H Bond*

To our knowledge, the telomerisation of vinyl acetate with dimethylchlorosilane, initiated by AIBN, is the only example [121] described in the literature which allows the synthesis of real telomers from a Si-H containing telogen. Nevertheless, very low molecular weight TFE telomers were obtained from activated trichlorosilane [122] or methyldichlorosilane [123] or even surprisingly dimethyl silane [124]. But, it is well-known that the hydrosilylation reactions lead mainly to monoadducts, in the presence of hexachloroplatinic acid, even from fluorinated monomers: VDF [110], TFE, HFP, CTFE or VF_3 [125], $C_6F_5CH{=}CH_2$ [109] or $C_6F_{13}CH{=}CH_2$ [126].

However, peroxides, UV or γ radiation also allow these hydrosilylations and, interestingly, monoadduct and diadduct were produced from 1,1,1-trifluoropropene [110] as follows:

$$Cl_3SiH + CF_3{-}CH{=}CH_2 \longrightarrow Cl_3Si(CH_2{-}\underset{\displaystyle CF_3}{\underset{|}{C}H})_n{-}H \quad n=1,2$$

3.2.3 *Cleavage of O-F Bond*

As mentioned in Sect. 2.3.1, hypohalites are efficient transfer agents for the telomerisation of various fluoromonomers. In contrast to both examples above, such a reaction is performed readily and does not require any catalyst or initiator.

Most investigations have shown that various hypofluorites have led to mono-adducts but in certain cases telomers were obtained. Although expected $R_FO(M)_nF$ were synthesised, by-products such as $R_FO(M)_nOR_F$ or $F(M)_nF$ were noted.

4
Fluoroalkenes Used in (Co)Telomerisation

In this section, two main parts are considered. The first one deals with the telomerisation of fluoroolefins the most used e.g. vinylidene fluoride, tetrafluoroethylene and chlorotrifluoroethylene for which considerable work has been performed, trifluoroethylene and hexafluoropropene. The second part summarises investigations performed on less used fluoromonomers e.g., vinyl fluoride, 1,1,1-trifluoropropene, 1,1-difluorodichloroethylene and so on.

4.1
Classic or Most-Known Fluoroolefins

4.1.1
Vinylidene Fluoride

The telomerisation of vinylidene fluoride (1,1-difluoroethylene) has been investigated by many authors. Almost all kinds of transfer agents have been used, as mentioned below, requiring various ways of initiation: thermal, photochemical or from systems involving redox catalysts or radical initiators (Tables 3 and 4) and also hypofluorites as mentioned in Sect. 2.3.1.

4.1.1.1
Thermal Initiation

In contrast to photochemical or radical initiation few transfer agents have been used in thermal telomerisation of VDF (Table 3). Only two series have to be considered: those which exhibit C-Br or a C-I cleavable bond and hypofluorites (see Sect. 4.1.1.5) whereas no work involving a telogen with a C-Cl bond was described in the literature.

Hauptschein et al. [127] have shown that VDF reacts with CF_2Br_2 and $CF_3CFBrCF_2Br$ at 190 °C and 220 °C, respectively yielding telomeric distributions $R_F(CH_2CF_2)_nBr$ n = 1 – 8 ($CF(CF_3)Br$ group being more reactive than CF_2Br in the case of the second telogen).

Concerning iodinated transfer agents almost all perfluoroalkyl iodides and, α, ω-diiodoperfluoroalkanes were successfully utilised in thermal telomerisation of VDF.

One of the pioneers of such work was Hauptschein et al. [127, 128] who used CF_3I, C_2F_5I, nC_3F_7I, i-C_3F_7I, $ClCF_2CFClI$ and $ClCF_2CFICF_3$ at 185 – 220 °C range of temperature leading to high telogen-conversions.

Later, Apsey et al. [96] and Balagué et al. [97] used i-C_3F_7I and linear $C_nF_{2n+1}I$ (n=4,6,8) telogens, leading to telomeric distributions, with a better reactivity of the branched transfer agent. Recently, we have shown that, while the monoad-

Table 3. Thermal and radical telomerisations of vinylidene fluoride (DTBP, DBP and sc-mean-ditertiarybutyl peroxide, dibenzoyl peroxide and supercritical, respectively)

Telogen	Method of initiation	Structure of telomers	References
CF_2Br_2 $CF_3CFBrCF_2Br$	190 °C 220 °C	$R_F(C_2H_2F_2)_nBr$ n = 1–8	127 127
$ClCF_2CFClI$	181 °C/26h	$ClCF_2CFCl(VDF)_nI$ n=1–6	127
$C_nF_{2n+1}I$ n=1–8	180–220 °C	$R_F(C_2H_2F_2)_nI$ n=1–7	97, 127, 128
iC_3F_7I	185–220 °C	id n=1–5	96, 97, 129
$(CF_3)_3CI$	Thermal	n=1, 2	132
$IC_nF_{2n}I$ n=2,4,6	180–200 °C	$I(VDF)_pC_nF_{2n}(VDF)_qI$ p, q=1, 2, 3	130
$Cl_3C\text{-}H$	DTBP	$Cl_3C(C_2H_2F_2)_nH$ n > 3	85, 133
$HOCH_3$	DTBP, DBP or AIBN	$HOCH_2(C_2H_2F_2)_nH$ n > 10	134, 135
$RCCl_3$	peroxide	$RCCl_2(C_2H_2F_2)_nCl$	85, 92, 135
KBr	peroxide	variable n	133, 136–139, 141, 142
$BrCF_2CFClBr$	DTBP	$BrCF_2CFCl(C_2H_2F_2)_nBr$	133, 140
RI	peroxide	variable n	131, 140, 143, 144
C_4F_9I	AIBN, $scCO_2$	n=1–9	79
HOC_2H_4SH	AIBN, DBP (or DTBP)	n=1, 2 (or 1–6)	101, 145
FSO_2Cl	DBP	$FSO_2(C_2H_2F_2)_nCl$ (n=1–3)	104, 146
Cl_3CSO_2Br	DBP	$Cl_3C(C_2H_2F_2)_nBr$ (n=1, 2)	147
$(EtO)_2P(O)H$	DTBP or perester	$(EtO)_2P(O)(VDF)_nH$ (n=1–5)	148, 149

duct exhibits the only structure $R_FCH_2CF_2I$, the diadduct is composed of two isomers [97] (see Sect. 5.2).

An original telogen $(CF_3)_3CI$ was successfully used by a chinese team [132] as follows:

$$(CF_3)_3CI + nH_2C{=}CF_2 \longrightarrow (F_3C)_3C(CH_2CF_2)_nI \ (n=1,2)$$

In conclusion on the thermal initiation, two main series of transfer agents have been used having specific cleavable bonds (C-X with X=Br and I). They lead to telomeric distributions, in good yields, with more or less high $\overline{DP_n}$ according to the nature of the telogen (and especially the electrophilicity of the radical generated), the experimental conditions (initial pressures and (VDF)/(Telogen) molar ratio, temperature). However, if our recent investigations have shown that the monoadduct is composed of $R_FCH_2CF_2X$ as the sole isomer, the diadduct already consists of two isomers.

Table 4. Photochemical, redox and special telomerisation of vinylidene fluoride

Telogen	Method of initiation	Structure of telomers	References
$CFBr_3$	UV	$Br_2CF(C_2H_2F_2)_nBr$ (n=1, 2)	93
Cl_3CBr, CF_2Br_2	UV	low n	152, 153
CF_2Br_2, CF_3CFBr_2, CF_3CBr_3	γ ray	$R_F(C_2H_2F_2)Br$ variable n	154
CF_3I	UV/28 d/RT	n=1	150
CF_3I	UV/140 °C	$CF_3CH_2CF_2I$ (major) $CF_3CF_2CH_2I$ (minor)	151
CF_2HI	UV	$HCF_2CH_2CF_2I$	131
H_2S	X ray	$H_{2-x}S(CH_2CF_2H)_x$ x=1, 2	100
CF_3SH	X ray	$CF_3SCH_2CF_2H$	155
RSSR(R=CH_3 or CF_3)	UV	$RS(C_2H_2F_2)_nSR$ n=1–6	84,102,103
PH_3	UV	$H_{3-x}P(CH_2CF_2H)_x$ x=1, 2	105
ICl	UV	ICH_2CF_2Cl	156
$RCCl_3$ R=Cl, CO_2CH_3, CH_2OH	$FeCl_3$/Ni (or benzoin) or $CuCl_2$	n=1–4	92
C_4F_9I	$FeCl_3$/Ni	$C_4F_9CH_2CF_2I$	97
R_FOX (X=F or Br)	- 145 °C to RT	$R_FOCH_2CF_3$ mainly	41, 42, 44, 52
HI (gas)	–	CH_3CF_2I	157
R_3SiH	H_2PtCl_6 or UV	$R_3SiCH_2CF_2H$	110

4.1.1.2 Radical Initiation

Various transfer agents, easily cleavable by radical initiation have been used successfully (Table 3). In contrast to numerous works involving telogens which exhibit a weak C-H bond, and especially for those which have a cleavable C-halogen bond, no investigations have been performed on transfer agents with a C-F bond. Two main telogens having a cleavable C-H group have been utilised in the radical telomerisation of VDF: chloroform and methanol. For the first one, investigation was already mentioned in 1967, by Toyoda et al. [85] requiring di t-butyl peroxide as the initiator. Beside the formation of several by-products, the obtained $Cl_3C(VDF)_n$-H telomers were used as surfactants after a chemical change of the trichloromethyl end-group into carboxyl extremity. We have recently performed similar telomerisation [133]: even with a large excess of chloroform, rather high molecular weight-telomers were produced with yet poor yields.

The radical telomerisation of VDF with methanol was formerly investigated by Oku et al. [134] in 1986 (yield=40%) in order to synthesise the original $H_2C{=}CHCO_2CH_2(VDF)_nH$ macromonomers as follows:

$$HOCH_3 + n\, H_2C{=}CF_2 \xrightarrow{(tBuO)_2} HOCH_2(C_2H_2F_2)_n H$$

$$\xrightarrow{H_2C=CHCOCl} H_2C{=}CHCO_2CH_2(VDF)_n H$$

Recently, similar telomerisation was extensively investigated in our laboratory [135]. Whatever the nature of the initiator (azo, peroxides, peresters, percarbonates) or even the way of initiation (photochemical, thermal or in the presence of redox catalysts) rather high molecular weight-telomers were obtained in medium yields.

However, even if only CCl_4,$Cl_3CCO_2CH_3$ and Cl_3CCH_2OH [92] were used as telogens involving a cleavage of the C-Cl bond [85, 92, 133], all kinds of brominated telogens were successfully used in the presence of peroxides: CCl_3Br [133], $ClCF_2Br$ [136], $BrCF_2Br$ [133, 137], CF_3CF_2Br [138], CF_3CFBr_2 [138], $BrCF_2CF_2Br$ [138, 139], $BrCF_2CFClBr$ [133, 140] for which CFClBr is quite reactive contrarily to $BrCF_2$ end group, $CHBr_3$ [141], CBr_4 [141] or ω-bromoperfluoropolyethers $CF_3(OC_2F_4)_n(OCF_2)_pBr$ yielding original block copolymers [142].

Yet, in contrast to work using iodinated telogens in thermally initiated reactions, only five of them were attempted using peroxides: $ClCH_2I$, CH_2I_2 [143], CH_3I [144] and $ClCF_2CFClI$ [140] and, interestingly, including the use of C_4F_9I in the presence of AIBN in supercritical carbon dioxide as the solvent [80].

Quite a few sulphurated transfer agents have been studied. In our laboratory, 2-hydroxyethyl mercaptan reacted with VDF leading mainly to the first two telomers in the presence of dibenzoyl peroxide or to the six first telomers with ditertiarybutyl peroxide, used with an excess of VDF [101, 145] ; whereas Grigor'ev [104] or Tiers [146] used FSO_2Cl producing $FSO_2(CH_2CF_2)_nCl$ ($n=1-3$). Zhu et al. [147] tried Cl_3CSO_2Br initiated by dibenzoyl peroxide and obtained the first two adducts in low yields. Neither disulphides nor thiocarbamates were tested in the radical telomerisation of VDF.

Among the phosphorylated telogens, diethyl phosphate behaves effectively for the telomerisation of VDF initiated by di t-butyl peroxide [148, 149] or perester [149].

4.1.1.3
Photochemical Initiation

Photoinduced telomerisation of VDF has been investigated by many authors (Table 4). The first work started in 1954 by Haszeldine and Steele [150] who used CF_3I as the transfer agent (using an exposure with a wave length higher than 300 nm) led to the monoadduct after 28 days of irradiation at room temperature. Cape et al. [151] studied the same reaction at 140 °C for 12 h and determined the Arrhenius parameters for the addition of $CF_3^{\cdot}$ radicals to both sites of VDF. That of the isomer $CF_3CH_2CF_2I$ was about three times higher than that of $CF_3CF_2CH_2I$ one.

Tedder and Walton also used brominated methane derived-telogens: $CFBr_3$ [93] even in excess, led to expected normal and reverse monoadducts, to five fluorobrominated by-products, (mainly formed by recombination of radicals) and to a diadduct. As described above, the kinetics of reaction was investigated, as for CCl_3Br [152] and CF_2Br_2 [153].

More recently, various chain length-telomers were produced by γ-ray-induced telomerisation of VDF with CF_2Br_2, CF_3CFBr_2 or CF_3CBr_3 [154] by monitoring the initial reactants molar ratio.

Telogens containing sulphur atom(s) were also utilised and, as for the above halogenated transfer agents, led mainly to monoadducts. For instance, Harris and Stacey [100] chose a large excess of H_2S leading to $HSCH_2CF_2H$ in 69% yield whereas in the presence of a lower amount of such a transfer agent, the monoadduct and diadduct were produced in 62 and 19% respectively. The former product is a fluorinated mercaptan, able to further react with other fluoro monomers [100]. These authors [155] also used CF_3SH as the telogen with X-ray-initiation yielding the monoadduct selectively.

Under UV irradiation, CH_3SSCH_3 [84] and CF_3SSCF_3 [102] were also investigated. The hydrogenated transfer agent led to the monoadduct, but by-products were also formed, whereas the fluorinated one allowed a total conversion of VDF and yielded a telomeric distribution $CF_3S(C_2H_2F_2)_nSCF_3$ n=1–6 with reverse and normal adducts [84, 102, 103]. Interestingly, the diadduct has the structure $CF_3SCH_2CF_2CF_2CH_2SCF_3$ and shows that the telomers are formed by recombination of primary radicals.

In addition, PH_3 has successfully reacted with VDF under photochemical initiation leading to mono and diadduct from an equimolar starting materials ratio [105].

VDF and I-Cl were photolysed producing ICH_2CF_2Cl as the sole product in fair yield [156]. Such a monoadduct was characterised from ^1H and ^{19}F NMR spectra of CH_3CF_2Cl, obtained after addition of $SnBu_3H$.

In conclusion, photochemical telomerisation of VDF with iodoalkanes involving the cleavage of the C-I bond was kinetically investigated in a very interesting way and mainly produced the monoadduct. Addition onto the CH_2 side of VDF was favoured but a non negligible formation of various by-products was also observed. From brominated telogens, the presence of several by-products, besides the expected telomers, was noted. However, the literature does not mention any work involving telogen with a C-Cl cleavable bond. But, sulphurated and phosphorylated telogens were successfully utilised.

4.1.1.4
Redox Telomerisation

Few investigations were performed on the redox telomerisation of VDF (Table 4). Several works conducted in our laboratory [92] have dealt with the use of CCl_4 in the presence of $FeCl_3$/Ni; $FeCl_3$/benzoin mixtures or $CuCl_2$ as the catalysts with two fold excess of VDF about CCl_4. The first four telomers were produced and the diadduct was composed of two isomers: $Cl_3C(C_2H_2F_2)_2Cl$ and $ClCF_2CH_2CCl_2CH_2CF_2Cl$. In similar conditions, $CCl_3CO_2CH_3$ led to the first three adducts (mono 60%, di 32% and triadduct 8%) whereas Cl_3CCH_2OH, catalysed by $FeCl_3$/benzoin in the same conditions, yielded 81% of monoadduct and 19% of diadduct. Usually, the yield of these reactions were poor to medium.

From C_4F_9I and $FeCl_3$/Ni system, only $C_4F_9CH_2CF_2I$ was obtained in 55% yield [97]. The structure of this monoadduct is consistent with that produced in a thermal way.

4.1.1.5
Other Methods of Initiation

Hypohalites and silanes have also been efficient for the telomerisation of VDF.

Concerning fluorinated hypohalites used as telogens (also mentioned in Sect. 2.3.1), DesMarteau et al. have performed the most interesting research. First, Johri and DesMarteau [41] achieved the addition of trifluoromethylhypofluorite to $CH_2{=}CF_2$ leading to $CF_3OCH_2CF_3$, $CF_3OCF_2CH_2F$ and CF_3CH_2F in 97.5, 2 and 0.5%, respectively. Similar investigation conducted by Sekiya and Ueda [42] produced the first isomer selectively.

Secondly, linear CF_3CFZCF_2OX (Z=Cl, Br and X=F,Cl) and branched $R_{F,Cl}R'_{F,Cl}CFOX$ ($R_{F,Cl}=R'_{F,Cl}=CF_2Y$ with Y=H,Cl or $R_{F,Cl}=CF_2Y$; $R'_{F,Cl}=CF_3$ polyhalogenoalkyl hypochlorite and hypofluorite were reacted with VDF in equimolar ratio from −145 °C to room temperature for 18–24 h giving the monoadduct in 20–70% yield with a major amount of normal $R_{F,Cl}OCH_2CF_3$ isomer [44]. The authors noted that for the same $R_{F,Cl}$ group, $R_{F,Cl}OF$ led to better yields than those obtained from $R_{F,Cl}OCl$, and that $CF_3CFBrCF_2OF$ was more reactive than $CF_3CFClCF_2OF$ and finally that branched hypohalites reacted more easily than linear ones.

Thirdly, the same group [52] synthesised original branched perfluorohypobromites which reacted with VDF from −93 °C to room temperature for 8–12 h as follows:

$$R_FOBr + H_2C{=}CF_2 \longrightarrow R_F\,OCF_2CH_2Br + \text{by products } (R_FOH') \quad (20\text{–}31\%)$$

$$R_F = (CF_3)_3C \text{ or } C_2F_5C(CF_3)_2$$

Surprisingly, both hypobromites produced monoadducts composed exclusively of the reverse $R_FCF_2CH_2Br$ isomer.

As for silanes, hydrosilylation mainly led to monoadducts [110].

4.1.1.6
Conclusion

Many investigations have been made on the telomerisation of VDF with a wide range of transfer agents from various ways of initiation: thermal, photochemical (mainly UV and few works on γ-rays or X-rays), or from redox catalysts or radical initiators.

Thermal initiation has the advantage to lead to the formation of well-defined telomers with either selective production of monoadduct or higher $\overline{DP_n}$ than those obtained from photochemical initiation. It appears as an easy and attractive way since, except the autoclave, no special equipment or solvent or expensive reactants (e.g., initiators) is required. In addition, this route provides "clean" reactions without the formation of any by-products.

Works from photoinduced telomerisation have led to most impressive investigations, with interesting kinetics performed by Tedder and Walton [93, 151–153]. VDF is used for the obtaining of lower telomeric distributions mainly for monoadducts, but yielded several by-products from brominated telogens.

Radical initiation is an intermediate way involving many kinds of telogens but sometimes it occurs with the formation of by-products coming from the presence of initiator end-group.

In the redox type telomerisation, few surveys were developed. Most of them involve CCl_3R type but also C_4F_9I producing monoadduct mainly, in the presence of ferric catalysts.

Concerning the structure of the telomers, defects of chaining have been observed since already a tail to tail addition occurred for the diadduct. The opportunity of monitoring new processes or the use of several ligands should be worth investigating and might afford regioselective telomers.

Because most industrial production of VDF is generated from F141b (Cl_2CFCH_3) or F142b ($ClCF_2CH_3$) for which the process is not ready to stop, it can be imagined that (co)telomerisation and (co)polymerisation of such an olefin still keep a prosperous future.

4.1.2
Tetrafluoroethylene

Tetrafluoroethylene is obviously the most used monomer in telomerisation because of its good ability to polymerise (by comparison to other monomers), its low price and the properties of the obtained perfluorinated chains. Furthermore, such an olefin is symmetrical and does not lead to problems of regioselectivity and defects in chaining. However, it is the most difficult to handle owing to its hazardous behaviour towards oxygen and pressure.

Whatever the ways of initiation, successful results have been obtained involving all kinds of telogens. A summary of the literature is listed in Tables 5, 6, 7, 8 and 9 corresponding to photochemical, thermal, radical, redox and miscellaneous initiations, respectively.

It is noteworthy that on an industrial scale, thermal telomerisation appears the most widely used, thanks to the thermal stability of the fluorinated telomers

Table 5. Photochemical telomerisation of tetrafluoroethylene

Telogen	Method of initiation	$\overline{DPn}$	Ref.
$R_1R_2C(OH)$-H (MeOH, EtOH, iPrOH)	UV	n=1, 2	87
CH_3COCH_3	γ ray	n=5, 6	158
$Cl_2C=CCl_2$	γ ray	n=5–15	159
X_3CS-SCX_3 X=H,F	UV	n=1–5	102, 103
Cl_3Si-H	UV	n ≥ 1	122
$(CH_3)_3Si$-H	UV	n ≥ 1	124
$(CH_3)_3Si$-H	γ ray	n=1(66%), 2(34%)	125
$ClCF_2CFClI$	γ ray	n < 3	160
CF_3I	UV	n=1–10	161, 162
R_FI	UV (185–220 °C)	1.3	129
CF_3I, C_2F_5I, C_3F_7I	γ ray	2–3	163

Table 6. Thermal telomerisation of TFE

Telogen	Catalyst	T (°C)	$\overline{DPn}$	Ref.
$C_6H_5SSC_6H_5$	I_2	175	n=1, 2	164
ICF_2COF	–	220	n≤5	54
PCl_5	–	178	n>4	107, 165
$POCl_3$	–	150	n=1–10	108
SF_5Cl	–	150	n>2	166
CCl_4	PCl_5	180	2–5	107
$BrCF_2Cl$	Ni	360	1.4	167
SF_5Br	–	90	n=1–16	168
$ClCF_2CF_2I$	–	180	n=2–5	169
$ClCF_2CFICF_3$	–	170	n≤6	170
CF_3CCl_2I	–	142	1.3	171
C_3F_7I	–	190–240	n=1–3	129
$C_nF_{2n+1}I$	–	>300	n=1–8	172
$I(C_2F_4)_nI$ n=0, 1	–	230	n=2–4	117
IC_2F_4I/F_5SSF_5	–	150	n=4–9	173

produced. Moreover, it is well admitted that the poor reactivity brought about by the redox catalysts has led to low molecular weight telomers just like the thermal initiation at high temperature. On the other hand, thermal initiation at low temperatures and chemical induced radical telomerisation have offered telomers with high $\overline{DP_n}$ and wider polydispersities.

4.1.3
Chlorotrifluoroethylene

4.1.3.1
Introduction

Chlorotrifluoroethylene (CTFE) has been one of the most used monomers after tetrafluoroethylene and vinylidene fluoride even if this fluoroalkene is endangered because of the recession of its precursor $ClCF_2CFCl_2$. All kinds of telogens have been successfully utilised.

The first and most important part reviews the telogens having cleavable C-X bond, except Paleta's work [89] in which C-F linkage is cleaved in the presence of $AlCl_3$ as the catalyst. The results concerning the cleavage of C-Cl, C-Br and C-I bonds are reviewed in Tables 10, 11, 12 and 13 respectively. Then, the second part of this literature approach describes other minor kinds of telogens, such as hypohalites, disulphides, alcohols, boranes, SO_2Cl_2 (Table 14).

Table 7. Radical telomerisation of TFE (DBP, DIPC, BEPH, DTBP and F 113 represent dibenzoyl peroxide, diisopropyl peroxydicarbonate, butyl ethyl peroxyhexanoate, ditertiarybutyl peroxide and $ClCF_2CFCl_2$, respectively)

Telogen	Initiator	Solvent	T (°C)	$\overline{DP}n$	Ref.
Alkane	DBP	F 113	125	1–3	174
Methylcyclohexane	DTBP	–	150	32	175
CH_3OH, C_2H_5OH,	DBP	–	100	n = 3	176
$(CH_3)_2CHOH$	DIPC	F 113	45	–	177
CH_3OH	DTBP	–	125–150	$n \leq 4$	88, 178
CH_3OH	H_2O_2	–	70	n=3–5	179
$(CH_3)_2CHOH$	BEPH	–	80	1.2	180
$CH_3OC_2H_4OH$	DTBP	–	115–130	n=1, 2	181
CH_3CO_2H	DBP	–	100	$n \leq 3$	74
Cl_3CBr, CH_3OH	AIBN	–	94–100	n=1–5	182
CH_3Cl	DBP	–	125	> 10	183
CF_3CCl_3	DTBP	–	150	20	184
CBr_4, $CHBr_3$	DTBP	–	150	n=1	185
Cl_3CSO_2Br	DBP	–	90–150	n=1–12	147
$ClSO_2F$	DTBP	–	135	n=1–10	146
$(EtO)_2P(O)H$	DBP	–	130	$n \leq 4$	106, 148
R_F-H	DTBP	–	150	260–500	175
CH_3CF_2I	acyl peroxide	–	–	n=1–7	157
C_2F_5I	CH_3CO_3H	F113	78–125/P	variable n	186
C_2F_5I	DIPC or AIBN or DBP	–	40–60	1–3	187
C_2F_5I	$(Cl_2C{=}CClCO_2)_2$	–	70–75	1.9	188
C_2F_5I	R_FCO_3H	–	60–150	3	189
C_2F_5I	$(C_mF_{2m+1}COO)_2$ m=3, 7, 8, 11	–	25–55	1.8–2.5	190
C_2F_5I,iC_3F_7I	$(RC_6H_{10}OCO_2)_2$	–	60–90	n=1–6	191
iC_3F_7I, R_FI	DTBP	–	130	n=1–10	192
R_FI	AIBN or DTBP	–	90–116	2.2	193
IC_2F_4I	DBP	–	–	1–3	194
R_F/R_FI_2	ammonium persulfate	–	80	high n	195

Table 8. Redox telomerisation of TFE

Telogen	Catalyst	T (°C)	$\overline{DPn}$	References
CCl_4	Cu, $CuCl_2$ or $Fe(CO)_5$	100	3[$Fe(CO)_5$]	196
CCl_4	$FeCl_3$/benzoin	140	1.7–5.0	197
C_2F_5I	Cu	80–100	1–5	198
$C_nF_{2n+1}I$ (n=2, 4, 6)	IF_5/SbF_5	60	1.6–4.0	199
CF_3I	$CuCl/H_2NC_2H_4OH$	200	1–10	200
R_FI	$ZrCl_4/H_2NC_2H_4OH$	150	1–6	201
C_2F_5I	Metal Salt/amine $AlCl_3$	140	3	202

Table 9. Telomerisation of TFE (miscellaneous)

Telogen	Catalyst or method of initiation	T (°C)	$\overline{DPn}$	References
CCl_4, CCl_3H	Et_4NF/pinene	150	n=10–24	203
iPrOH	–	–	n≤4	204
I_2	$As(C_2F_5)_2$	–	n=1–6	205
BrC_2F_4Br	KrF laser or Xe lamp or low P Hg lamp	RT	variable according to light source	73
CF_3I	pulsed CO_2 laser	RT	–	72
C_4F_9I	$MetF_n$ (Met=Xe, Co, Ag, Mn, Ce)	RT to 80	n=1–9	74
C_4F_9I	SbF_5/fluorographite	80	n=1–5	75
$FSO_2C_2F_4I$	–	–	–	206
$FSO_2C_2F_4OC_2F_4I$	–	–	–	207
$(CF_3)_3COBr$	–	–85	n=1	52
$C_2F_5C(CF_3)_2OBr$	–	–93	n=1	52

4.1.3.2
Telomerisation of CTFE with Telogens Containing C-Cl Bonds

Redox telomerisation of CTFE with CCl_4 has been extensively studied either in the presence of copper salts (whatever their degree of oxidation) or from iron salts (mainly ferric). In this latter case, a reducing agent is required: organic reducers such as benzoin were traditionally employed in our laboratory whereas later, the Occidental Chemical Company preferred using nickel metal.

Table 10. Telomerisation of CTFE with telogens containing cleavable C-Cl bond

Telogen	Catalyst	Solvent	T (°C)	$\overline{DPn}$	Ref.
CCl_4	CuCl or $CuCl_2$	CH_3CN	130	n=1	208–211
CCl_4	$FeCl_3$/benzoin	CH_3CN	110	n=1–20	32, 212
CCl_4	$FeCl_3$/benzoin	MeOH	110	n=1–20	213
CCl_4	$FeCl_3$/Ni	CH_3CN	110	n=1–10	214
CF_3CCl_3	$FeCl_3$/benzoin or CuCl	CH_3CN	130–150	n=1–5	215
CCl_4	$AlCl_3$		50	n=1	216
$RCCl_3$	$FeCl_3$/benzoin or $CuCl_2$	CH_3CN	110–150	n=1–10	217
$CFCl_2CF_2Cl$	$FeCl_3$/Ni_2	CH_3CN	150–200	n=1–5	218
RO_2C-CCl_3	$FeCl_3$/benzoin or $CuCl_2$	CH_3CN	110–150	n=1–10	219
Cl-(CFCl-CF_2)$_n$-CCl_3	$FeCl_3$/benzoin or $CuCl_2$	CH_3CN	110–150	n=1–3	209

Table 11. Telomerisation of CTFE with telogens having C-Br bonds (DBP, means dibenzoyl peroxide)

Telogen	Catalyst	Solvent	T (°C)	$\overline{DPn}$	Ref.
HBr	UV	–	RT	1	111, 222
Br_2	–	CF_2Cl-$CFCl_2$	RT	1	223, 224
$BrCF_2$-CFClBr	thermal $CuBr_2$/Cu	MeCN	120	1–10	119
$BrCF_2$-CFClBr	UV	–	RT	n=1	94, 225
CF_3-$CClBr_2$	$FeBr_3$/Ni	CH_3CN	110	n=1–10 52% n=1	226
CF_2Br_2, CF_2ClBr	DBP	–	100	6–50	227
CCl_3Br	$AlCl_3$	–	RT	n=1	228
CCl_3Br	UV	–	RT	n=1	111
CCl_3Br	UV	–	RT	n=1, 2	225, 229
CCl_3Br	$FeCl_3$/benzoin $CuCl_2$, DBP	CH_3CN	110–130	n=1	230
CCl_2Br_2	UV	–	RT	n=1	225

But, other telogens have also been efficient such as $RCCl_3$ (where R represents CCl_3, $CHCl_2$, CH_2Cl, CH_3 groups), $CCl_3CO_2CH_3$, $ClCF_2CFCl_2$ or CF_3CCl_3. Usually, the presence of copper salts favours the formation of the monoadduct, in contrast to iron salts which lead to polydispersed telomeric distributions.

Interestingly, when the telogen was totally consumed in the telomerisation of CTFE with CCl_4, it was observed that the telomers produced, containing CCl_3 end-group, are able to reinitiate telomerisation yielding products with the following formulae: $Cl(CFClCF_2)_nCCl_2(CF_2CFCl)_pCl$ [209].

Table 12. Telomerisation of CTFE with telogens bearing -CFClI and -CCl_2I end groups

Telogen	Catalyst	Conditions		$\overline{DPn}$	Ref.
$CF_2Cl\ CCl_2I$	none	18 h	140 °C	10–15	114
CF_3CCl_2I	none	17 h	145°C	n=4–6	233, 234
CF_3CFClI	t-butylperoxipyvalate	16 h	70°C	n=1–10	235
CF_3CFClI	none	18 h	180°C	n=1–3	235
$ClCF_2CFClI$	none	5 h	180°C	n = 1–20	273
$ClCF_2CFClI$	UV		RT	1–11	237
$FCl_2CCFClI$	DBP		140–150 °C	n = 1	238
ICF_2CFClI	sun light or UV or γ ray		RT	n = 2–10	239

Table 13. Telomerisation of CTFE with telogens exhibiting a cleavable C-I bond

Telogen	Catalyst	Conditions	$\overline{DPn}$	Ref.
CHI_3	$(C_2F_5CO)_2O_2$	12 h 50–55 °C	n=13	234
CF_3I	UV	RT	n = 1 (85%)	240, 241
CF_3I	t-Butylperoxipyvalate	8 h 100 °C	n=1–4	242
C_2F_5I	–	–	n=1–6	242, 243
$C_nF_{2n+1}I$	UV	RT		244
iC_3F_7I or C_2F_5I	$TiCl_4$ or $ZnCl_2$ or $YrCl_3$ with H_2N-C_2H_4OH	150 °C	n=1–5 n ≅ 2 (yield=25%)	201
$C_nF_{2n+1}I$ (n=4, 6)	CuCl/$CuCl_2$	150 °C	n=1–6	235

In addition, the chlorofluorinated $Cl(CFClCF_2)_n$ chain activates the CCl_3 end group. Thus, the telomers produced can be used as original telogens for various telomerisations of non fluorinated monomers leading e.g. to form diblock cotelomer 6 [220, 221]:

$$Cl(CFClCF_2)_nCCl_2(CH_2\underset{\displaystyle CO_2R}{\underset{|}{C}H})_mCl$$

6

Main results are listed in Table 10.

4.1.3.3
Telomerisation of CTFE with Telogens Containing C-Br Bonds

Brominated telogens have also been widely used in the literature and Table 11 lists various brominated telogens with first of all HBr and Br_2. In that latter

Table 14. Other telogens usesd in the telomerisation of CTFE

Telogen	Catalyst	Conditions		$\overline{DPn}$	Ref.
SO_2Cl_2	DBP	CCl_4	95 °C	n=2–5	245, 246
SO_2ClF	DTBP			n=1–10	146
$CF_3S\text{-}SCF_3$	UV	–		n=1–3	102, 103
$HSiCl_2CH_3$	H_2PtCl_6	160 °C		n>1	122
$H\text{-}PO(OEt)_2$	DTBP	6 h	130 °C	n=1 (mainly)	148
RR'CH-OH	UV	–		$n \leq 3$	247
CH_3OH	UV	–		n=1–20	248
iPrOH	γ	–		n=1	249
Boranes	DTBP	4 h	135 °C	variable	250
CF_3OF	–	–78 °C		$F(C_2F_3Cl)_nF$ (7%) $F_3CO(C_2F_3Cl)_nF$ (43%) n=1–4	43
C_2F_5OF	–	–72 °C		n=1–6	45

case, the $BrCFClCF_2Br$ adduct is easily obtained and has led to compounds containing an active CFClBr end group, offering a broader telomeric distribution.

The most studied telogen is certainly CCl_3Br initiated either by UV, peroxide or by redox systems. However, we have demonstrated that in this last case, the redox catalyst, especially ferric chloride, induces a disproportionation which leads to a mixture of new telogens as follows [230]:

$$BrCCl_3 \rightleftharpoons Br_2CCl_2 + CCl_4$$

In addition to these polybrominated telogens, CF_3CClBr_2 has also been shown to be efficient. It was also observed that the higher the number of bromine atoms in the telogen, the lower the $\overline{DP}_n$ and the polydispersity.

However, all these products are very interesting because they allow easy C-Br chemical interconversions in order to obtain functional compounds (e.g. RCO_2H, R-OH etc.) [231, 232].

4.1.3.4
Telomerisation of CTFE with Telogens Containing CFClI and CCl_2I End Groups

In order to increase the reactivity of the telogens having C-I groups (as it is seen below in Sect. 4.1.3.5), the authors have introduced one or two chlorine atoms into the iodinated extremity.

First, it is noteworthy describing the synthesis of these new products which are obtained by addition of IX (X being Cl or F) to chlorofluoroalkenes as follows:

$$ICl + Cl_2C{=}CF_2 \longrightarrow ICCl_2CF_2Cl + \text{traces of } Cl_3CCF_2I \ [114]$$
$$IF + Cl_2C{=}CF_2 \longrightarrow ICCl_2CF_3 + \text{traces of } Cl_2CFCF_2I \ [114, 236]$$
$$IF + ClCF{=}CF_2 \longrightarrow ICFClCF_3 + \text{traces of } ICF_2CFCl_2 \ [112, 113, 236]$$

Thus, versatile telogens have been produced with an increased activation of the cleavable bond. It is interesting to note that, already in 1955, Haszeldine [237] observed a kind of living character of the telomerisation of CTFE according to the following reaction:

$$ClCF_2CFClI + F_2C{=}CFCl \longrightarrow Cl(CF_2CFCl)_n\,I$$

The synthesis of the telogen was recently optimised [115].

Non exhaustive surveys dealing with the telomerisation of CTFE with these above telogens are summarised in Table 12.

4.1.3.5
Telomerisation of CTFE with Perfluoroalkyl Iodides

In contrast to the telomerisation of tetrafluoroethylene with perfluoroalkyl iodides (R_FI), the addition of these telogens to this olefin has not been quite investigated. This may be explained by the fact that the $R_F(CF_2CFCl)_nI$ telomers produced are more efficient telogens than the starting R_FI transfer agents as it can be seen in the previous section. Table 13 sums up the results of the literature.

When CF_3I is used as the telogen, the problem arises regarding the sense of addition of the $^{\cdot}CF_3$ radical to CTFE [240, 241] leading to CF_3CF_2CFClI (mainly, 90%) and $CF_3CFClCF_2I$.

Interestingly, Gumbrech and Dettre [234] used a mixture of $(C_2F_5CO_2)_2$ and CHI_3 (as a iodine donor) for the obtaining of $C_2F_5(C_2F_3Cl)_nI$ telomers. However, rather high $\overline{DP_n}$ (ca. 13) was observed.

More recent work [235] has indicated that redox catalysis provides a better selectively in polydispersity and $\overline{DP_n}$. However, some transfers from the catalyst occurred and led to by-products.

It can be concluded from all these results that the reaction between R_FI and CTFE is very complex and difficult to perform successfully.

4.1.3.6
Telomerisation of CTFE with Other Telogens

Concerning hypohalites, CTFE seems very efficient in contrast to other fluoroalkenes which have led to monoadduct mainly. The reaction is as follows:

$$R_FOX + CF_2{=}CFCl \xrightarrow{-75\,^\circ C} R_FO(C_2F_3Cl)_n\,X + X\,(C_2F_3Cl)_m\,X$$
$$+ R_FO(C_2F_3Cl)_pOR_F \quad X = F, Cl \text{ or Br and } R_F = CF_3, C_2F_3$$

Various other telogens have also been successfully attempted undergoing *S*-Cl, *S*-S, P-H, C-H cleavages or from boranes (Table 14).

Table 15. Telomerisation of trifluoroethylene with different telogens (DBP and DTBP, mean dibenzoyl peroxide and ditertiarybutyl peroxide, respectively)

Telogens	Initiation	Structures of Telomers	References
MeOH, Et_2O, MeCHO, THF	γ rays	$ROCHR'(C_2F_3H)_nH$ n=1–10	251
H_2S	X rays	HCF_2CFHSH (85%) H_2CFCF_2SH (15%)	100
CF_3SH	hυ	$CF_3SCFHCF_2H$ (98%) $CF_3SCF_2CFH_2$ (2%)	155
Cl_3SiH	γ rays	$Cl_3SiCHFCF_2H$ (100%)	125
$(CH_3)_3SiH$	γ rays	$(CH_3)_3SiCF_2CH_2F$ (50%) $(CH_3)_3SiCFHCF_2H$ (50%)	125
CF_3I, $(CF_3)_2CFI$	hυ	R_FCHFCF_2I (96%) R_FCF_2CFHI (4%)	252–255
	Thermal (190 °C)	R_FCHFCF_2I (85%) R_FCF_2CHFI (15%)	98, 252–255
CCl_4	$CuCl_2$ (80 °C)	CCl_3CHFCF_2Cl $CHFClCF_2CCl_3$	91
ICl	UV	$ClCF_2CFHI$	256
ICl	Thermal	$ClCF_2CFHI$ (95%) $ClCFHCF_2I$ (5%)	257
$ClCF_2CFClC_2F_4I$	DBP (3h; 110 °C)	$ClCF_2CFClC_2F_4(CFHCF_2)_nI$ n=1, 2, 3	259
$Cl_2CFCFClI$	DBP	$Cl_2CFCFCl(C_2F_3H)_nI$	260
CF_2Br_2	DBP	$BrCF_2CF_2CFHBr$ (Min) $BrCF_2CFHCF_2Br$ (Maj)	258
$CFBr_3$	hυ	$CFBr_2CFHCF_2Br$ (Maj) $CFBr_2CF_2CFHBr$ (Min)	93
CBr_4, $CHBr_3$	DTBP	CBr_3CFHCF_2Br (Maj) CBr_3CF_2CFHBr (Min)	141

4.1.4
Trifluoroethylene

Unlike vinylidene fluoride, tetrafluoroethylene and chlorotrifluoroethylene which have been extensively used in polymerisation and telomerisation, trifluoroethylene has been investigated in telomerisation by few authors. Non exhaustive results are listed in Table 15. Such an olefin exhibits an unsymmetric aspect which is particularly interesting.

Recently, Powell and Chambers [251] have prepared telomers by radiochemical initiation in the presence of γ-rays. Such a reaction was successful with transfer agents activated by oxygen atom, leading to variable $\overline{DP_n}$ (2.75–5.75) according to experimental conditions.

Sloan et al. [93, 141] noted the formation of reverse adducts when trifluoroethylene was telomerised either photochemically with fluorotribromomethane [93] or in the presence of radical initiator with CBr_4 or CBr_3H [141]. Haszeldine et al. [252–255] investigated the telomerisation of such a monomer with different perfluoroalkyl iodides (CF_3I, i-C_3F_7I) and they showed that the thermal initiation led to a higher amount of reverse adduct in contrast to the photochemical induced reaction. Recently, we have shown that a real telomerisation occurs when the reaction was initiated thermally since the first five adducts were formed.

Bissel [256] investigated the redox telomerisation of trifluoroethylene with iodine monochloride and observed the selective formation of $ClCF_2CFHI$ as the sole product. When thermal or photochemical initiations were used, such an isomer was produced in major amount (95–97%) but the formation of the reverse $ClCFHCF_2I$ was also observed [257]. Kotora and Hajek [91] showed that the redox telomerisation from carbon tetrachloride also led to two isomers. Cosca [258] and Anhudinov et al. [259] used brominated or polyhalogenated telogens, respectively whereas Harris and Stacey [100, 155] and Haran and Sharp [103] investigated such a reaction from sulphur containing transfer agents. In all these above four cases, reverse isomers were also produced.

According to this literature survey, most investigations have been performed photochemically and in most cases, the monoadduct is composed of two isomers, the ratio of which depends upon the electrophilicity of the telogen radical. In the case of fluoroalkyl iodides, such an olefin exhibits the same reactivity towards these telogens, whatever their structure. In addition, trifluoroethylene seems less reactive than vinylidene fluoride but more reactive than hexafluoro-propene.

4.1.5
Hexafluoropropene

4.1.5.1
Introduction

Among most of fluorinated monomers (and especially in contrast to those mentioned above), hexafluoropropene (HFP) is one of those which is the most difficult to telomerise. Actually, HFP is regarded as one of the most electrophilic of fluorinated olefins and may be considered to be polarised as follows [155, 261]:

$$\overset{\delta+}{F_2C}=\overset{\delta-}{CF}-CF_3$$

Even if several articles and patents deal with the cotelomerisation or copolymerisation of HFP with other fluorinated olefins (e.g., with vinylidene fluoride leading to well-known Viton elastomers), its homotelomerisation is not so easy. This may come from the fact that hexafluoropropene is a monomer which

Table 16. Telomerisation of hexafluoropropene with different telogens

Telogens	Initiation	Structure of telomers	References
$(CF_3)_2CFI$	YCl_3, $NH_2C_2H_4OH$	$(CF_3)_2CF(C_3F_6)_nNHC_2H_4I$ n = 1–3	264
RSH	UV	$RSCF_2CFHCF_3/RSCF(CF_3)CF_2H$	155
(R = CH_3, CF_3CH_2, CF_3)		(variable ratio)	
CF_3SSCF_3	UV	$CF_3SCF_2CF(CF_3)SCF_3$	102
$HSiCl_3$	γ rays	$Cl_3SiCF_2CFHCF_3$ (35%)	125
		$Cl_3SiCF(CF_3)CF_2H$ (65%)	
PH_3	UV	$PH_2CF_2CF(CF_3)H$ (66%)	105
		$PH_2CF(CF_3)CF_2H$ (34%)	
$(EtO)_2P(O)H$	Peroxides	$(EtO)_2P(O)CF_2CFHCF_3$ (80%)	148
		$(EtO)_2P(O)CF(CF_3)CF_2H$ (20%)	
ROH	UV (96 h)	$HOCR_1R_2CF_2CFHCF_3$	25, 267
R = Me, Et, iPr, Bu	Peroxides (150 °C)	$HCF_2CF(CF_3)CR_1R_2OH$	25, 265
$(CF_3)_2CFH$, CF_3CH_2	Thermal (280 °C)		25, 267
$R_{F,Cl}H$	275 °C(3–6 d)	$R_{F,Cl}C_3F_6H$	267
CF_3I	UV	$\lambda < 3000$ Å, n = 1	268
		$\lambda > 3000$ Å, n = 1–2	
C_2F_5I	$C_nX_{2n+1}CO_3H$	$C_2F_5(C_3F_6)_pI$	189
	X = H, F, Cl	variable p	
	γ rays or		86, 266
	Peroxides		
ICI	Thermal (98 °C)	CF_3CFICF_2Cl (92%)	114
		$CF_3CFClCF_2I$ (8%)	
R_FI	Thermal (200 °C)	$R_F(CF_2CFCF_3)_nI$ n = 1–10	95, 268, 271,
R_F = CF_3, C_3F_7			272
$CF_2ClCFCl$			
CF_3CFClI	Th(200–230 °C)	$CF_3CFCl(C_3F_6)I$	235
$C_nF_{2n+1}I$ (n = 4, 6, 8)	Thermal	$C_nF_{2n+1}(C_3F_6)_pI$ p = 1–3	99
i-C_3F_7I	Thermal (250 °C)	$(CF_3)_2CF(C_3F_6)I$ poor yield	99
$C_4F_9(C_3F_6)I$	250 °C/48 h	$C_4F_9(C_3F_6)_2I$, 8%	99
$I(C_2F_4)_nI$	Thermal (210 °C)	$I(C_3F_6)_m(C_2F_4)_n(C_3F_6)_pI$	117, 275
n = 2–4		m = 0–2, n = 2–4, p = 1–2	
CF_3Br, CF_2Br_2,	Thermal (250 °C)	$R_F(C_3F_6)_nBr$ n = 1–6	95
$CF_3CF_2CF_2Br$	R_0 = 1–5		
H_2CO	Th(130–150 °C)	$HOCH_2CF(CF_3)CO_2H$, 41%	276
CF_3OF	20–75 °C	$CF_3O(C_3F_6)F$	46

homopolymerises only with great difficulty in the presence of free radicals [237] as shown by various studies (actually the reactivity ratio of HFP in copolymerisation is close to zero [262]). In fact, only one investigation has successfully led to high molecular weight PHFP when the pressures used were greater than 10^8 Pa [263].

Non exhaustive results illustrating telomerisations of HFP are listed in Table 16.

4.1.5.2
Redox Telomerisation of HFP

The only result obtained by redox catalysis was performed by Jaeger [264] in the presence of yttrium salts, without achieving the synthesis of the expected telomers but yielding a secondary fluorinated amine containing HFP base units. However, the microstructure of HFP chaining was not detailed.

4.1.5.3
Photochemical Telomerisation of HFP

In contrast, the addition of various telogens containing heteroatoms was successfully initiated photochemically. The addition of various mercaptans (CF_3SH,CF_3CH_2SH and CH_3SH) to HFP led to relative amounts of normal and reverse isomers of 45:55; 70:30 and 91:9 [155]. These results can be correlated to relative electrophilicity of RS˙ radicals, the decreasing order being the following : $CF_3S^{\cdot} > CF_3CH_2S^{\cdot} > CH_3S^{\cdot}$. Dear and Gilbert [102] only obtained monoadduct $CF_3SCF_2CF(CF_3)SCF_3$ (in contrast to the six first VDF telomers) just like Burch et al. [105] and Haszeldine et al. [148] who synthesised two phosphorylated isomers from PH_3 and $HP(O)(OEt)_2$, respectively. Actually, the electrophilic $PH_2^{\cdot}$ or $(EtO)_2P(O)^{\cdot}$ radicals mainly react to the CF_2 side of HFP, such an observation being stressed in the case of the bulkier latter radical.

A series of various alcohols were also investigated in the photochemical induced telomerisation of HFP at room temperature and led mainly to monoadducts composed of both isomers. This can be explained according to the mechanism proposed by Haszeldine et al. [25] presented in Sect. 2.1. Such a monoadduct is now a commercially available fluoroalcohol produced by several companies (e.g. Hoechst).

No reaction occurred from hexafluoroisopropanol whereas trifluoroethanol was efficient with HFP [25]. Nevertheless, up to 45% of reverse adduct $HOCR_1R_2[CF(CF_3)CF_2]H$ was produced.

Interestingly, for a hydrogenated series of telogens, Lin et al. [265] suggested the similar following decreasing order: $CH_3OH > C_2H_5OH > (CH_3)_2CHOH$. Furthermore, initiation by peroxides led to the same results.

In the same way, γ rays (or peroxides) allowed the addition of one HFP unit to THF [86] or one or two units to cyclic ethers [266] as achieved by Chambers et al.:

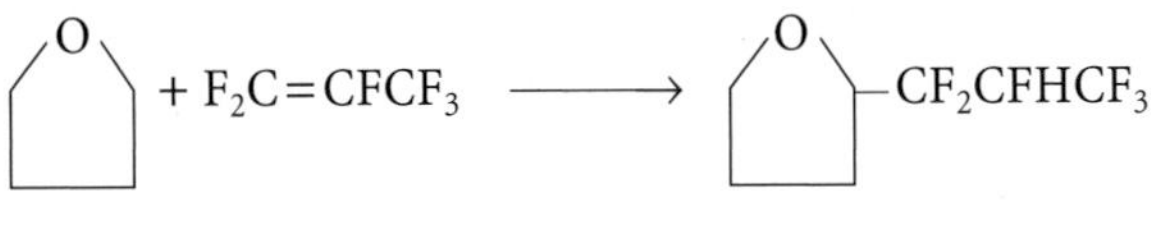

Such a procedure was successfully applied to poly(THF) leading to novel fluorinated telechelic diols precursor of α,ω-diacrylates useful for coating optical fibers [13]:

$$PTHF + F_2C{=}CFCF_3 \longrightarrow HO-(C_4H_8O)_x-(C_4H_7O)_y-H \quad \text{with } x = y = 4.5 \; | \; C_3F_6H$$

The hydrosilylation reaction of trichlorosilane with HFP initiated by γ-rays led to both expected isomers with a rather high amount of reverse adduct [125].

Haszeldine et al. [267] also performed the photochemical addition of various R_1R_2CHCl (with $R_1,R_2{=}CH_3$, F; C_2H_5, H; CH_3, CH_3; n-C_3H_7, H and C_2H_5, CH_3) telogens to HFP at about 40 °C. They observed, beside the formation of monoadducts mainly produced from the cleavage of C-H bond, few amounts of compounds obtained by the cleavage of C-Cl bond and cyclic or rearranged products.

Concerning transfer agents involving a C-I bond, for trifluoroiodomethane, the wave length λ plays an important role on the telomeric distribution since at λ lower than 3000 Å only $CF_3C_3F_6I$ was formed; whereas at higher wave lengths, the two first telomers were produced [268]. However, the photochemically induced addition of perfluoro-t-butyl iodide to HFP, even at 163 °C was unsuccessful [269].

4.1.5.4
Radical Initiation

Beside the telomerisation of HFP initiated by peroxides as presented above, interesting investigations were performed by Rudolph and Massonne [189] who used a fluorinated carboxylic peracid leading to the three first telomers. On the contrary, with hydrogenated peroxides (*t*-butyl peroxypivalate), perester (dibenzoyl peroxide) or azo (AIBN) initiators, no reaction occurred [270].

An interesting comparison between the photochemical and radical initiations was proposed by Low et al. [271] who studied the competitive addition of CF_3I to HFP and ethylene at different temperatures (up to 170 and 189°C from di-t-butylperoxide and UV induced reactions, respectively). These authors noted that in all cases $CF_3C_2H_4I$, $CF_3(C_3F_6)I$ and $CF_3(C_3F_6)C_2H_4I$ were formed. The two last products are composed of two isomers, the normal one ($CF_3CF_2CFICF_3$) being in higher amount than the reverse one ($(CF_3)_2CFCF_2I$). They also observed that the higher the temperature, the higher the yields of these three products and that below 140 °C, the photochemical initiation was more efficient. Above such a temperature the radical way led to better results. In addition, from UV induced reaction, $CF_3(C_3F_6)I$ was always produced in the highest amounts whereas the contrary was noted in the other process.

4.1.5.5
Thermal Initiation

In most studies concerning the telomerisation of HFP the reactions were carried out thermally.

At rather low temperatures ($-50\,^{\circ}$C), Dos Santos Afonso and Schumacher [46] succeeded in adding trifluoromethyl hypofluorite to HFP according to the following scheme:

$$CF_3OF + CF_2 = CFCF_3 \xrightarrow{20-75\,^{\circ}C} CF_3OCF_2CF_2CF_3 + CF_3OCF(CF_3)_2$$
$$\phantom{CF_3OF + CF_2 = CFCF_3 \xrightarrow{20-75\,^{\circ}C}} \underset{68\%}{} \underset{32\%}{}$$

But, except for ICl which led to both normal CF_3CFICF_2Cl and reverse $CF_3CFClCF_2I$ isomers at $98\,^{\circ}$C [114], most other transfer agents required much higher temperatures, especially higher than $200\,^{\circ}$C.

Concerning telogens involving a C-H cleavable bond, two series can be considered: first, chlorinated and fluorinated methanes or ethanes R-H [266] were shown to react with HFP from $275\,^{\circ}$C for $3-6$ days in $14-97\%$ HFP conversion. Such reaction mainly led to the monoadduct usually composed of two isomers (normal RCF_2CFHCF_3 and reverse $RCF(CF_3)CF_2H$) except for chloroform and fluorine containing alkanes which yielded the normal adduct selectively. However, the reaction between HFP and chlorinated propanes or butanes gave by-products (e. g., cyclic or thermally rearranged derivatives or compounds produced from hydrogen abstraction or from the C-Cl cleavage) [266]. These telogens gave better yields from photochemical reactions. Second, a series of alcohols was shown to be efficient from $280\,^{\circ}$C [25] thanks to the adjacent oxygen electron-withdrawing atom.

Allyl chloride or polychlorinated ethanes with a cleavable C-Cl bond were also investigated thermally ($250-290\,^{\circ}$C) by Haszeldine et al. [267]. Beside normal and reverse monoadducts produced in poor yields, cyclic or thermal rearranged compounds were obtained.

Perfluoroalkyl bromide and CF_2Br_2 led successfully to telomeric distributions from $250\,^{\circ}$C [95].

But the most interesting results come from perfluoroalkyl iodides or α, ω-diiodoperfluoroalkanes for which the terminal CF_2-I group has a lower bond dissociation energy than that of CF_2-Br. The pioneers of this investigation were Hauptschein and colleagues in 1958 [95] using CF_3I, C_3F_7I and $ClCF_2CFClI$ as the transfer agents. The originality of such a thermal telomerisation lies on the fact that the mechanism of this reaction does not proceed by propagation. The authors suggested that the mechanism is composed of a succession of addition-steps according to the following scheme:

$$nC_3F_7I + F_2C=CFCF_3 \longrightarrow nC_3F_7CF_2CFICF_3$$
$$nC_3F_7CF_2CFICF_3 + F_2C=CFCF_3 \longrightarrow nC_3F_7[CF_2CF(CF_3)]_2I$$

and so on.

Kirschenbaum et al. [272] proposed that each step implies a complex of addition such as:

$$CF_3CF_2\overset{.}{C}F_2\text{-----}\overset{.}{I}$$

$$\overset{.}{C}F_2\text{------}\overset{.}{C}FCF_3$$

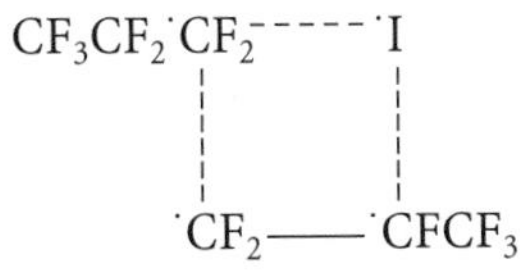

As for the telomerisation involving other fluorinated transfer agents, R_FCFClI telogens are more reactive than R_FCF_2I as shown by Hauptschein et al.

[95] or by Amiry et al. [235]. Tortelli and Tonelli [117] and Baum and Malik [275] have confirmed the step-wise mechanism suggested by Hauptschein et al. [273]. However, we did not confirm the work of Haupstchein who showed that the thermal addition of i-C_3F_7I to HFP was as successful as that of linear transfer agents [95]. In our case, such a telogen reacted with HFP in a low conversion, up to 36% at 250 °C [99]. In addition, in order to confirm such a result, the telomerisation of HFP with the steric hindered $C_4F_9(C_3F_6)I$, carried out at 250 °C for 48 h, led to 8% of $C_4F_9(C_3F_6)_2I$. Apparently, this was not in good agreement with Hauptschein's results [95] who obtained better yields.

But the nature of the autoclave has a great influence on the formation of by-products. A Russian team [274] has shown that for a vessel made of nickel, the addition of CF_3I to HFP led mainly to $(CF_3)_3CI$ whereas for similar reaction in Hastelloy reactor, no by-product was observed.

Interestingly, Tortelli and Tonelli [117] and Baum and Malik [275] succeeded in telomerising HFP with α,ω-diiodoperfluoroalkanes leading to normal $I(C_3F_6)_x(CF_2)_nI$ and "false" $I(C_3F_6)_y(CF_2)_n(C_3F_6)_zI$ adducts (where x, y or z = 1 or 2).

Finally, a particular case deals with the condensation of paraformaldehyde with HFP in the presence of chlorosulfonic acid at 130–150 °C leading to α-fluoro α-trifluoromethylhydracrylic acid *7* in 41% yield as follows [276]:

$$F_2C{=}CFCF_3 + H_2CO + H_2O \longrightarrow [HOCF_2CF(CF_3)CH_2OH] \longrightarrow HO_2CCF(CF_3)CH_2OH$$

<u>7</u>

4.1.5.6 Conclusion

Except the redox catalysis for which only one investigation was performed, several initiations are possible for reacting HFP with various transfer agents. However, usually such a reaction is rather selective since low molecular weight-telomers are produced. The thermal initiation appears as the most efficient, especially for R_FX (X = Br or I) and IR_FI. If the telomeric distribution is limited to the two or three first telomers, the monoadduct is composed of two isomers, the amount of which depends upon the temperature and the electrophilic character of the telogenic radical. On the other hand, linear R_FI are much more reactive than branched ones, linked probably to the high electrophilicity of the telogen radical and to the steric hindrance of the CF_3 side group.

4.2 Less Used Monomers

Beside these above fluoroolefins, numerous other fluorinated alkenes have been involved in telomerisation. Most of them are unsymmetrical and thus lead to adducts composed of at least two isomers. Among them, vinyl fluoride and 1,1,1-trifluoropropene have been attempted with various telogens.

Table 17. Telomerisation of vinyl fluoride with various transfer agents ((a) both monoadduct isomers with a high amount of normal adduct)

Telogen	Initiation	Structure of telomers (%)	Ref.
Cl_3C-Br	UV/150°C	{Cl_3CCH_2CHFBr (95) {$Cl_3CCHFCH_2Br$ (5)	277
$BrCF_2Br$	UV	{$BrCF_2CH_2CHFBr$ (96) {$BrCF_2CHFCH_2Br$ (4)	278
Br_2CF-Br	UV/100–200 °C	a)	93
F_3C-I	UV/14 d/RT	{F_3CCH_2CFHI (84) {$F_3C(CH_2CFH)_2I$ (7)	279
F_3C-I	UV/variable T	a)	151
F_3C-I	$C_4F_9IF_2$ cat/120 °C	$F_3C(C_2H_3F)_nI$	280
nC_pF_{2p+1}-I/iC_3F_7I p = 2, 3, 4, 7, 8	UV/variable T	$R_F(C_2H_3F)I$ a)	281
$(F_3C)_3C$-I	UV/72–163 °C	a) slightly increase of reverse adduct with temperature	268, 282
F_3CO-F	- 50 °C	{$F_3COCH_2CHF_2$ (87–90) {$F_3COCHFCH_2F$ (13–10)	41, 42

4.2.1
Vinyl Fluoride

Vinyl fluoride is an interesting monomer, precursor of PVF or Tedlar (produced by the Dupont Company), known for its good resistance to UV radiation. But in telomerisation, the most intensive work was achieved by Tedder and Walton who used several telogens exhibiting cleavable C-Br or C-I bonds, under UV at various temperatures (Table 17). Their surveys were mostly devoted to the obtaining of monoadduct and to their kinetics (e.g., determination of relative rate constants of formation of normal and reverse isomers and of Arrhenius parameters).

However, telomers were obtained by Rondestvedt [280] who added $C_4F_9IF_2$ (prepared by reacting C_4F_9I with ClF_3) as the catalyst at 120 °C. Although original investigation from trifluoromethyl hypofluorite led mainly to the monoadduct, to our knowledge no work was performed from alcohol, mercaptan, disulfide, and phosphorous or silicon containing telogens.

4.2.2
1,1,1-Trifluoropropene

Such a fluoroalkene, prepared for the first time by Haszeldine [283] by dehydroiodination of $F_3CCH_2CH_2I$, is the original precursor of commercially available fluorinated silicone from Dow Corning, called Silastic (see Sect. 7.2). Such a polymer preserves interesting properties (inertness, fair surface tension, chain mobility) at low and rather high temperatures.

Table 18. Telomerisation of 1,1,1-trifluoropropene with various transfer agents (DBP, means dibenzoyl peroxide)

Telogen	Initiation	Structure of Telomers	Ref.
$H_3COCOC(CH_3)_2H$	Peroxides	$H_3COCOC(CH_3)_2[CH_2CH(CF_3)]_nH$	284
$(CH_3)_2C(OH)H$	γ rays (T < 90 °C)	$(CH_3)_2C(OH)[CH_2CH(CF_3)]_nH$ n = 1, 2	285
$C_6H_5CH_2$-Cl	$Fe(CO)_5$	$C_6H_5CHX\ [CH_2CH(CF_3)]_nY$ n = 1, 2 X = Cl, H Y = H, Cl	286
$C_5H_5CH_2Br$	$Fe(CO)_5$	as above with X = Br, H and Y = H, Br	287
Br_2CH-Br	$Fe(CO)_5$	$\{Br_2CH[CH_2CH(CF_3)]_nBr$ n = 1–3 $\{CF_3CBrHCH_2CBr_2[CH_2CH(CF_3)]_2H$	288
Br_2CH-Br	Peroxides	$XCBr_2[CH_2CH(CF_3)]_nY$ Y=X=H or Br n = 1, 2	288
$BrCH_2$-Br	$Fe(CO)_5$	$BrCH_2[CH_2CH(CF_3)]_nBr$ n = 1, 2	288
CBr_4	DBP	$Br_3C[CH_2CH(CF_3)]_nBr$ n = 1 – 3	289
CF_3I	UV/5 days	$CF_3[CH_2CH(CF_3)]_nI$ n = 1, 2	290
CF_3I	225 °C/36 h	$CF_3[CH_2CH(CF_3)]_nI$ n = 1–3	290
CF_3I	UV/variable T	Normal and reverse monoadducts few amounts of n = 2	283
Cl_3Si-H	UV	$Cl_3Si[CH_2CH(CF_3)]_nH$ n = 1, 2	110
$(C_2H_5O)_2P(O)H$	$(tBuO)_2$/130 °C	$(C_2H_5O)_2P(O)CH_2CH_2CF_3$ (39%)	148
CH_3SSCH_3	UV	$CH_3SCH_2CH(CF_3)SCH_3$	103
HBr	UV	$CF_3CH_2CH_2Br$	290

Interestingly, such a fluoroolefin is quite reactive in contrast to long perfluorinated chain-vinyl type monomers (e.g. $C_nF_{2n+1}CH{=}CH_2$ with $n > 1$).

But, several authors have preferred using such an olefin in telomerisation involving a rather wide variety of telogens (Table 18). This fluoroalkene gives satisfactory results in different initiation: redox, UV or γ radiations or in the presence of peroxides. In this field, most pertinent works were performed by Russian teams who chose esters [284], alcohols [285] or halogenated compounds [286–289]. In addition, Vasil'eva's team even proposed mechanisms of formation for unexpected telomers [288] and determined various chain transfer constants of telogens, showing that, in certain cases, the transfer to the telogen is more efficient than the radical chain growth [287, 289]. They have also observed that $C_6H_5CH_2Cl$ [286] and $C_6H_5CH_2Br$ [287] undergo both a C-H and a C-halogen cleavages since their reaction products are $C_6H_5CH_2(M)_n$-X and $C_6H_5CHX(M)_nH$, where X = Cl, Br and M represents the 1,1,1-trifluoropropene. In the same way, bromoform, in the presence of iron complex or peroxide (such as dibenzoyl peroxide or di-t-butyl peroxide) also leads to two series of telomers coming from its C-H and C-Br cleavages. Consequently, the formation of "false adduct" $CF_3CHBrCH_2CBr_2(CH_2CH(CF_3))_2H$ is not unexpected.

As for telomerisation of the above fluoroolefins, CF_3I is an effective telogen for that of 1,1,1-trifluoropropene, both thermally or photochemically. This latter

Table 19. Telomerisation of various minor fluoralkenes

Fluoroolefin	Telogen	Method of initiation	Structure of telomers (%)	Ref.
$F_2C{=}CCl_2$	Cl_3CBr	UV	$Cl_3CCF_2CCl_2Br$ (28)	277
	ICl	- 10 °C	$ClCF_2CCl_2I$ only	114
	ICl	Fe/20 °C	{$ClCF_2CCl_2I$ (70) {ICF_2CCl_3 (16) {$ClCF_2CCl_3$ (14)	114
	CF_3OF	- 7 °C to RT	{$CF_3OCF_2CCl_2F$ (major) {$CF_3O(C_2F_2Cl_2)_iOCF_3$ {CF_3CCl_2F	41, 47
	R_FOBr	- 83 °C to RT	{$R_FOCF_2CCl_2Br$ (75–90) {$R_FOCCl_2CF_2Br$ (25–10)	52
	RSH	Peroxide	$RSCF_2CCl_2H$	298
$F_2C{=}CHCF_3$	CF_3I	UV/30–100 °C	{$(CF_3)_2CHCF_2I$ (55–63) {$CF_3CHIC_2F_5$ (45–37)	299
	CF_3I	212 °C	{$(CF_3)_2CHCF_2I$ (17) {$CF_3CHIC_2F_5$ (83)	299
	HBr	UV	{$CF_3CH_2CF_2Br$ (28) {$CF_3CHBrCF_2H$ (12) {$BrCF_2CHBrCHF_2$ (60)	299
$F_2C{=}CFBr$	Cl_3CBr	UV	{$Cl_3CCF_2CFBr_2$ (97) {$Cl_3CCFBrCF_2Br$ (3)	277
	Cl_3CF	UV	$Cl_2CF(CF_2CFBr)_nCl$ $n=1–13$	300
	$Br_3CX(X=F, Br)$		$XCBr_2(CF_2CFBr)_nBr$ $n=1, 2$	301
	F_3CSSCF_3	UV	$F_3CS(CF_2CFBr)_nSCF_3$ $n=1–3$	102
$F_2C = CFOCH_3$	F_3CSH	UV	$F_3CSCF_2CFHOCH_3$ (71)	100
$F_2C = CFR_F$ R_F:C_6F_5;C_5F_{11}	CCl_4	radical	$Cl_3CCF_2CFClR_F$	302
$F_3CCF = CFCF_3$	F_3CSH	UV	$CF_3SCF(CF_3)CFHCF_3$ (low yield)	100
	F_3COF	- 50 °C	$CF_3OCF(CF_3)CF_2CF_3$	49

type of initiation has appeared more selective than the former [290]. Low et al. [271] investigated this reaction under UV radiations at various temperatures and obtained for the monoadduct a mixture of normal and reverse isomers. They observed that the higher the temperature, the higher the proportion of reverse derivative. In addition, they observed the presence of higher adducts from 175 °C.

Concerning other transfer agents, silanes, phosphonates, disulfides or hydrogen bromide are able to react with such an olefin, yielding the monoadduct mainly (Table 18).

4.2.3
Other Fluoroolefins

Table 19 lists non exhaustive results of telomerisation of various fluoroalkenes (classified by increasing number of fluorine atom or perfluorinated groups directly linked to ethylenic carbons) with different telogens, most of them being halogenated compounds.

First, it can be noted that several kinds of initiation are possible. Secondly, although many olefins have been investigated for the production of the monoadduct only (especially for kinetics or studies of regioselectivity), the obtaining of higher boiling telomers is not excluded since the authors used an excess of telogens. This is mentioned in most cases presented in Table 19.

More interesting features or reaction leading to telomeric distributions are described below. Concerning hydrosilylation reactions, a wide range of fluoroalkenes have been used and the results have been reported [291, 292] or very well reviewed by Marciniec [303] leading mainly to monoadducts, except for 1,1,1-trifluoropropene, TFE [122–124] and CTFE [122].

Even if the reactivity of monomers is described in Sect. 5, it is worth comparing those of $CF_3CH{=}CF_2$ and $CF_3\text{-}CF{=}CH_2$ about CF_3I. Only this latter olefin leads to telomers, useful for hydraulic fluids, lubricants and for heat transfer media [293]. Similarly, $C_2F_5CF{=}CF_2$ has been telomerised thermally by Paciorek et al. [294] who suggested a valuable mechanism.

Interestingly, 1,2-dichlorodifluoroethylene has been telomerised with various transfer agents using different initiation processes. Unfortunately fluorotrichloromethane, in the presence of $AlCl_3$ also leads to several by-products [78, 295], whereas several ethers, alcohols or esters are not quite efficient (8–30% monomer conversion, only) [296]. But, perhaloalkyl and trifluoromethyl hypofluorites gave the most encouraging results since telomers were produced [48].

The photochemically induced reaction of $F_2C{=}CHCl$ with CF_3I offered 92% of monoadduct but much lower yields were observed by Haszeldine's team [297] for thermal initiation from 225 °C. Besides, no reaction occurred at 190 °C for 100h. In successful reactions, both isomers were obtained (Table 19) whereas only one was produced from the photochemical addition of HBr to such an olefin. But, this team did not observe such a result for similar reaction of $F_2C{=}CHCF_3$ which formed the three following isomers [299]:

$$H{-}Br + F_2C{=}CHCF_3 \longrightarrow \begin{array}{ll} CF_3CH_2CF_2Br\ (24\%) & \underline{8} \\ CF_3CHBrCF_2H\ (9\%) & \underline{9} \\ BrCF_2CHBrCHF_2\ (50\%) & \underline{10} \end{array}$$

The authors assumed that dibromide *10* formed as major product may arise from the dehydrofluorination of monoadduct *9* followed by a radical addition of HBr to the intermediate olefin, according to the following scheme:

$$\underline{9} \xrightarrow{-HF} F_2C{=}CBrCHF_2 \xrightarrow{HBr} BrCF_2CHBrCHF_2$$

This would explain the poor amount of *9* isomer and the high proportion of *10*. Investigations performed on the same fluoroalkene with regards to the addition of trifluoroiodomethane have shown that the way of initiation has a drastic effect on the amount of both isomers [299]. For both photochemical and thermal initiations the proportion of isomers are inverse.

Similarly, 1,1-dichlorodifluoroethylene was also telomerised with various telogens from different initiations [41, 47, 52, 114, 277, 298]. Cl_3CBr and perfluoroalkyl hypohalites led to kinetic investigations. An interesting work concerns the addition of iodine monochloride to this olefin, producing $ClCF_2CCl_2I$ isomer exclusively at −10 °C. But, at higher temperatures and in the presence of iron as the catalyst, Cl_3CCF_2I isomer was also produced, and the formation of chlorinated product $ClCF_2CCl_3$ was observed under these conditions [114].

Bromotrifluoroethylene (BrTFE) telomerises well photochemically and it seems that no other way of initiation was mentioned in the literature. Even disulfides have provided telomers [102], and BrTFE telomers have found applications as gyroscope flotation oils [301].

Finally perfluorinated alkenes, prepared according to various routes (decarboxylation of perfluorinated acids or anhydrides, action of organolithium, magnesium reagents or metallic zinc-copper couple to perfluoroalkyl iodides, oligomerisation of TFE and of HFP, addition of perfluoroalkyl iodides to perfluoroallyl chloride [304], addition of KF to perhalogenated esters or alkanes [305]) are shown to be unreactive in telomerisation except radical addition of CCl_4 [302], photochemical addition of CF_3SH [155] or reaction of CF_3OF [49]. 1,3-Perfluorobutadiene (e.g., synthesised either by Narita [53] or Bargigia et al. [306] or Dedek and Chvatal [94]) can not be telomerised in a radical process but was efficiently polymerised anionically [53]. Similarly, perfluoro-1,5-hexadiene did not lead to any telomers [307].

Among all remaining monomers, those containing (per)fluorinated side chains such as fluorinated acrylates, vinyl ethers or esters, maleimides and styrenic monomers are also very interesting and have been studied in (co)telomerisation. Most of them have been previously reviewed [15]. However, they are not mentioned in this chapter.

4.3
Cotelomerisation of Fluoroalkenes

Interesting fluorinated copolymers, already commercially available, are known for their excellent properties, most of them have been synthesised to be used as elastomers [1–10]. The literature is abundant on the copolymerisation of two fluorinated olefins, or that of fluoroalkenes with non halogenated monomers, (e.g., vinyl ethers [308]) regarded as electron-withdrawing and electron-donating, respectively. However, few articles or patents outline the cotelomerisation of fluoromonomers. Brace [309] has provided an interesting summary on the radical addition of mercaptans to different kinds of monomers and yet other transfer agents (e.g., perfluoroalkyl iodides, brominated telogens, methanol) have already shown efficiency in cotelomerisation. This part deals with non exhaustive examples of that reaction for which two or three fluoroalkenes are considered, only.

Two alternatives are possible: either the direct cotelomerisation according to a kind of batch procedure or the step-by-step addition (i.e. the telomerisation of a second coalkene with a telomer produced from a first monomer, via a two step-process). This part is detailed in Sect. 6.3.

4.3.1
Cotelomerisation of Two Fluoroolefins

4.3.1.1
CTFE/VDF Cotelomers

Cotelomers produced by cotelomerisation of CTFE and VDF seem to have drawn much interest because they constitute interesting model of Kel F copolymers. Barnhart [118] used thionyl chloride as the telogen to produce Cl(VDF-CTFE)$_n$Cl ($n < 10$) while Hauptschein and Braid cotelomerised thermally these olefins with CF_3I [310] or $ClCF_2CFClI$ [311]. The obtained cotelomers exhibit the following random structure $ClCF_2CFCl[(CTFE)_n(VDF)_p]I$ where $n+p$ varies in the 3–20 range according to experimental conditions.

Furukawa [312] used perfluoroalkyl iodides and α,ω-diiodoperfluoroalkanes as the telogens for the synthesis of high molecular weight-cotelomers.
More recently, we used 2-hydroxyethyl mercaptan to cotelomerise both these co-olefins by radical way [101, 145], leading to $HOC_2H_4S(CTFE)_x(VDF)_yH$ with x and y close to 4. Interestingly, such cotelomers exhibit an almost alternating structure.

4.3.1.2
VDF/TFE Cotelomers

Few investigations were performed and the oldest one was conducted by Wolff who used isobutane [313] or diisopropyl ether [314] under radical initiation. Ono and Ukihashi [163] successfully carried out the radical cotelomerisation of VDF and TFE with perfluoroisopropyl iodide, but in similar conditions, that of TFE and ethylene failed.

4.3.1.3
VDF/HFP Cotelomers

Copolymers of VDF with HFP (and TFE) are well known Viton, Fluorel, Tecnoflon or Dai-El elastomers [2–4, 6]. Several investigations of cotelomerisation have been described in the literature, few of them are presented below.

Rice and Sandberg [29] first synthesised α,ω-dihydroxyl terminated VDF/HFP cotelomers from perfluoroadipic, perfluorosuccinic or perfluoroglutaric anhydrides under hydrogen peroxide initiation. Later, researchers from the Ausimont Company studied the peroxide-induced cotelomerisation of VDF and HFP with 1,2-dibromoperfluoroethane leading to fluorinated products having an average number molecular weight of 660 and a Tg of –94 °C [315].

In addition, a survey has been recently performed by Chambers et al. [154] from brominated telogens under γ ray or thermal initiations leading to various $\overline{DP_n}$ by monitoring initial molar ratio of reactants. Interestingly, the cotelomerisation seems more efficient than the compared telomerisations of separated fluoroalkenes. Tatemoto and Nakagawa [316] started from iC_3F_7I under peroxide initiation to cotelomerise VDF and HFP, useful for the preparation of block and graft copolymers. Several examples of the tertelomerisation of VDF with HFP and TFE are given in Sect. 4.3.2.

4.3.1.4
HFP/TFE Cotelomers

Hauptschein and Braid [317] studied the cotelomerisation of HFP and TFE with IC_3F_6Cl at 190–220 °C and obtained various molecular weight-cotelomers according to initial molar ratios of reactants.

4.3.2
Tertelomerisation of Three Fluoroalkenes

An interesting tertelomerisation of VDF with TFE and CTFE or VDF with TFE and HFP was performed by Bannai et al. [318] who used methanol as transfer agent under peroxide initiation. Excellent yields were reached and average number molecular weights were around 1,000.

These fluorinated cotelomers are novel precursors of the following original (meth)acrylic macromonomers ($R=H$ or CH_3):

$$H_2C{=}CRCO_2CH_2\left[(\mathrm{VDF})_{0\cdot5}\,(\mathrm{TFE\text{-}CTFE})_{0\cdot5}\right]_n\!-H$$

In similar radical conditions, perfluoro-α,ω-dibrominated telogens successfully cotelomerise VDF with HFP and TFE (and CTFE) [315].

Bromofluoroalkyl iodide [319] or α, ω-diiodoperfluoroalkanes [320] also allow the cotelomerisation of TFE with VDF and HFP.

Interestingly, even less activated CH_2I_2 telogen was efficient in similar cotelomerisations, leading to fluorinated cotelomers which were peroxide-vulcanised and moulded into a sheet showing good adhesion to metal plates through curable adhesives.

4.4
Conclusion

A large variety of telogens has been used in telomerisation of traditional fluoroolefins and numerous studies have been performed on the telomerisation of VDF, TFE, CTFE. Yet, less investigations were carried out on trifluoroethylene and hexafluoropropene. According to the telomers searched, more adequate ways of initiation have to be chosen. For instance, the peroxide induced telomerisation of VDF with CCl_4 leads to higher molecular weight-telomers than that catalysed by copper salts which appears more selective [92].

Beside most important and commercially available fluoroalkenes, a large variety of different fluoroolefins have been successfully telomerised. Although most of them led to monoadducts mainly, devoted to the perspective of basic research, a couple of them produced higher molecular weight telomers, and several of them have found interesting applications.

Vinyl fluoride and 1,1,1-trifluoropropene have shown a great efficiency in telomerisation with various telogens, but concerning the other ones, less interest has been focused on. This may be due to the lack of availability of monomers, and consequently studies on monoadduct were mostly investigated.

Bromotrifluoroethylene, even if less available than CTFE can easily be telomerised, and the bulkiness of bromine atom does not seem to hinder this reaction.

An interesting example is 1,1-dichlorodifluoroethylene which has been studied mainly for its monoadducts. But it can be imagined, because of steric hindrance of CCl_2-CCl_2 chaining, that propagation may occur from tail to head addition, thus producing much interest.

Hence, the synthesis of their telomers and, above all, their properties and their applications have not attracted research groups and it can be hoped that future surveys will be performed in this way.

The cotelomerisation appears as an increasing means to model copolymerisation and to explain the structure of complex copolymers. It can be expected that future surveys will be developed in order to design novel well-architectured-polymers.

5 Reactivity of Monomers and Telogens

5.1 Introduction and Theoretical Concepts

Examples in the literature about the reactivity of radicals to olefins are abundant. In Sect. 4, non exhaustive examples dealing with the reactivity of radicals with various olefins have already been mentioned.

According to Tedder and Walton [81–83], "none simple property can be used to determine the orientation of additions of free radicals" which depends upon a complex mixture of polar and steric parameters, and of bond-strengths.

As seen in Sect. 3, a telogen requires a weak X-Y cleavage to be efficient in telomerisation. Hence, the bond dissociation energy (BDE) of the telogen X-Y (yielding $X^{\cdot}$ and $Y^{\cdot}$ radicals) has to be taken into account as indicator of intrinsic reactivity factor. The BDE can be calculated according to the following equation, where $\Delta H_f^O i$ designates the enthalpy of formation of species i:

$$BDE_{XY} = \Delta H_f^O X^{\cdot} + \Delta H_f^O Y^{\cdot} - \Delta H_f^O XY$$

Tables 20 and 21 list the BDE of various bonds that may contain the telogen [322, 323]. Thus, according to such a BDE, the efficiency of different telogens can be compared. For instance, as it can be expected, the cleavage of C-F bond is the most difficult to perform among all the possible bonds, whereas CF_2-I one has better chance to be cleaved (Table 21). But the Table shows that the bond strength can not predict the reactivity, as noted in numerous examples.

Concerning the ability of the telogen to initiate the telomerisation, the reactivity and the obtained regioselectivity of the telogen radical $X^{\cdot}$ to the olefin can

Table 20. Mean bond dissociation energy (BDE) of various classic cleavable bonds of telogens [21, 322, 323]

	BDE (KJ mol^{-1} at 25 °C)					
X	(X-H)	(X-F)	(X-Cl)	(X-Br)	(X-I)	(X-C)
H	432	561	427	362	295	416
C	416	486	326	285	213	350
N	391	272	193	–	–	303
O	463	193	205	201	201	368
F	561	155	–	–	191	452
Cl	427	–	239	218	208	330
Br	362	–	218	193	–	285
I	295	191	208	–	151	213
Si	318	565	381	310	234	–
P	322	490	326	264	184	–
S	326	285	255	218	–	255

Table 21. Relative rate constants k of the photodissociation of R_FI (about that of CF_3I) and CF_2-I bond dissociation energy (BDE)

R_FI	k [341]	BDE ($kJmol^{-1}$) [342]
CF_3I	1	220
C_2F_5I	8	212
n-C_3F_7I	15	206
i-C_3F_7I	49	206
n-C_4F_9I	–	202

be monitored by several rules. The radical $X^{\cdot}$ produced can be either nucleophilic (e.g., $RS^{\cdot}$ or $CH_3^{\cdot}$ [324]) or electrophilic (e.g., $C_nF_{2n+1}^{\cdot}$). But, fluoroalkenes are regarded as poor electron-olefins and thus will react more easily with nucleophilic radicals.

However, in contrast to ionic addition to alkenes for which the regioselectivity is achieved according to a unique rule (Markovnikov's law), radical addition generally occurs at the less hindered carbon atom of the double bond. However, Tedder [81–83] suggested five rules which govern the radical additions linked mainly to polar and steric effects, these latter being more important and this was confirmed by Giese [325] in 1983 and by Rüchardt [326] and Beckwith [327]. Other interesting theoretical surveys and concepts were provided by Koutecky et al. [328] and Canadell et al. [329, 330] who suggested that the preferential attack of the radical occurs on the carbon which exhibits the highest linear combination of atomic orbitals coefficient in the highest occupied molecular orbital. In addition to this rule, Delbecq et al. [331] have considered a law of steric control (the preferential attack occurs on the less substituted carbon) and the thermodynamic control rule (the most exothermic reaction is the easiest). More recently, Wong et al. [332] have shown that the exothermicity of the reaction is the main parameter to govern the reactivity of the radical about the olefin.

In contrast to these concepts above obtained from numerous ab-initio molecular orbital calculations (usually from small radicals, e.g. $H^{\cdot}$, $F^{\cdot}$, $H_3C^{\cdot}$ or $HOCH_2^{\cdot}$ to fluoroalkenes), some semi-empirical calculation on more longer halogenated radicals were performed by Rozhkov et al. [333, 334] and Xu et al. [335]. The former team determined equilibrium geometries and electronic properties of perfluoroalkyl halogenides and showed that fluorine atoms in vinyl position strongly stabilise all the sigma molecular orbital.

The Chinese group [335] proposed the electronic structures of 17 perchlorofluoroolefins and perfluoroolefins via MNDO calculation. They have shown that the direction of the nucleophilic attack is governed not only by the perturbation energy of the ground state, but also by the stability of the anionic intermediate and the activation energy of the reaction. They suggested the following decreasing reactivity series: $F_2C{=}CFR_F > TFE > R_FCF{=}CFCl > R_FCF{=}CClR'_F > R_FCF{=}CFR'_F > R_FCF{=}CCl_2$

In the case of the addition of halogenated free radicals to fluoroalkenes, investigations were performed by Tedder and Walton. For example, this team has extensively studied the kinetics of addition of $H_3C^{\cdot}$ [144, 336], $Cl_3C^{\cdot}$ [337], $Br_3C^{\cdot}$ [93, 185], $Br_2CF^{\cdot}$ [93], $CF_2Br^{\cdot}$ [278, 338], $CH_2I^{\cdot}$ [143], $CFHI^{\cdot}$ [131], $F_3C^{\cdot}$ [151, 281], $C_2F_5^{\cdot}$ [281, 339], $nC_3F_7^{\cdot}$ [281], $iC_3F_7^{\cdot}$ [281, 340], $CF_3(CF_2)_n^{\cdot}$ n = 6,7 [281], $(CF_3)_3C^{\cdot}$ [269] to specific sides of unsymmetrical fluoroalkenes. They determined the rate constants of these radicals to both sites of these olefins and the Arrhenius parameters. The results were mainly correlated on the electrophilicities of radicals which add more easily to nucleophilic (or less electrophilic) alkene. For example, Table 22 illustrates the probability of addition of various halogenated radicals to VDF. It is noted that the more electrophilic the radicals (i.e., the more branched the perfluoroalkyl radical), the more selective the addition to the CH_2 site.

El Soueni et al. [282] suggested the following increasing electrophilicity series:

$$CF_3^{\cdot} < CF_3CF_2^{\cdot} < C_2F_5CF_2^{\cdot} < nC_3F_7CF_2^{\cdot} < nC_5F_{11}CF_2^{\cdot} < (CF_3)_2CF^{\cdot} < (CF_3)_3C^{\cdot}$$

This is in a good agreement with the relative rate constants of the photodissociation of perfluoroalkyl iodides into $R_F^{\cdot}$ and $I^{\cdot}$ [341] linked also to the bond dissociation energy [342] as shown in Table 21.

Table 22. Probabilities of addition of chloro and fluororadicals to both sites of VDF

Radicals	H_2C =	CF_2
$^{\cdot}CH_2F$	1	0.440
$^{\cdot}CHF_2$	1	0.150
$^{\cdot}CCl_3$	1	0.012
$^{\cdot}CF_3$	1	0.032
$^{\cdot}CF_2CF_3$	1	0.011
$^{\cdot}CF_2CF_2CF_3$	1	0.009
$^{\cdot}CF(CF_3)_2$	1	0.001

5.2 Reactivity of Telogens

The following two examples illustrate the above theories. First, if one considers such a concept to VDF, according to Hauptschein et al. [127, 128], the mono-adduct coming from the radical addition of $R_F^{\cdot}$ to VDF was composed of two isomers $R_FCH_2CF_2I$ (95%) and $R_FCF_2CH_2I$ (5%) and this was confirmed by Chambers [96, 129] from iC_3F_7I. But, from recent investigations, we have demonstrated by ^1H and ^{19}F NMR of the $R_F(C_2H_2F_2)I$ monoadducts and of $R_F(C_2H_2F_2)H$ (produced by reduction of these monoadducts), that $C_4F_9CH_2CF_2I$, $C_6F_{13}CH_2CF_2I$, $C_8F_{17}CH_2CF_2I$ [97] and $IC_nF_{2n}CH_2CF_2I$ n=2,4,6 [130] were the only isomers produced by thermal telomerisation of VDF from the corresponding transfer agents. This was mainly explained by the selective addition of the electrophilic $R_F^{\cdot}$ radical to CH_2 group regarded as the less electrophilic side of VDF [131].

However, for each case, the diadduct was composed of two isomers $R_FCH_2CF_2CH_2CF_2I$ (92%) and $R_FCH_2CF_2CF_2CH_2I$ (8%) [97] because $R_FCH_2CF_2^{\cdot}$ appears less electrophilic than $R_F^{\cdot}$ radical.

Furthermore, the higher the temperature, the higher the telogen conversion and the higher the $[R_FI]/[VDF]$ initial molar ratio, the more selective the telomerisation [97, 127, 128, 130]. Chambers et al. [129] suggested the following increasing reactivity series about VDF: $F_3C\text{-}I < C_2F_5\text{-}I < nC_3F_7\text{-}I < iC_3F_7\text{-}I$ mainly linked to the decrease of the strength of the C-I bond.

In the case of diiodides, it is worth mentioning that the longer the length of the $I(C_2F_4)_nI$ telogen, the higher the amount of the α, ω-diadduct (50% in the case of $ICF_2CH_2(C_2F_4)CH_2CF_2I$ and 95% for $ICF_2CH_2(C_6F_{12})CH_2CF_2I$ [130]).

Concerning the reactivity of free radicals provided from various telogens (especially perfluoroalkyl iodides) to trifluoroethylene, the electrophilic character of the perfluoroalkyl radical generated has a great influence on the proportion of reverse adduct: the higher its electrophilicity, the higher the amount of "normal" isomer. For instance, from $C_4F_9CH_2CF_2I$ the normal/reverse ratio is 1.5, whereas it is 13.3 from $C_5F_{11}CFICF_3$ [98].

5.3 Reactivity of Fluoroalkenes

Under the same conditions, two fluoroolefins are not supposed to react similarly with a given telogen. For instance, HFP requires much energy to react with a telogen compared to TFE or VDF and it homopolymerises with difficulties in contrast to other fluoroalkenes. Below are presented several examples of compared reactivities of fluoroolefins in telomerisation involving the same telogen.

Concerning methanol, in contrast to VDF which produces rather high molecular weights [134, 135] or trifluoroethylene [251] which gives telomers containing one to ten base units, TFE [87, 88], HFP [25] and CTFE [343] have led to monoadducts mainly. We have tried to explain such results by showing with semi-empirical calculations that in the case of VDF, the propagation rate is much higher than the transfer one [134].

On the contrary, TFE and VDF behave differently when they react with Cl_3CSO_2Br from peroxide initiation, yielding $Cl_3C(C_2F_4)_nBr$ (n = 1 – 10) or $Cl_3C(C_2H_2F_2)_pBr$ (p = 1,2), respectively.

Trifluoromethyl hypofluorite, known for achieving higher yields than those obtained from the corresponding hypochlorites, also behaves differently with respect to various fluoroalkenes. HFP leads to the monoadduct as expected [46], just like VDF [41, 42], 1,1-dichlorodifluoroethylene [47], 2-perfluorobutene [49] in contrast to the first three adducts or the first ten telomers obtained from 1,2-dichlorodifluoroethylene [48] and CTFE [43 – 45, 48], respectively.

Iodine monochloride has been studied with various fluoroolefins yielding the monoadduct in all cases, to produce original activated telogens containing CF_2I or, better, CFClI end group [114, 115, 156]. The addition of ICl to 1,1-dichlorodifluoroethylene was found to be bidirectional but to a lesser extent [114] than in the case of that to CTFE [115]; and the use of catalyst (e.g. iron) affects the yield and the amount of by-product. HFP requires heat contrary to other monomers [114].

The following decreasing reactivity series is suggested:

$$CTFE > VDF > CF_2{=}CCl_2 > HFP$$

Perfluoroalkyl iodides have been used as telogens with most fluoroalkenes. An extensive kinetic research on the synthesis of monoadducts was performed by Tedder and Walton. From a review of the literature [97 – 99, 172, 236], the following reactivity scale may be proposed:

$$VDF > TFE > F_2C{=}CFH > HFP > CTFE$$

and for less reactive monomers the more electrophilic the telogen radical the less easy the reaction.

5.4 Conclusion

The bond dissociation energy of cleavable bonds of telogens appears to be a key parameter to take into account regarding reactivity. This is followed by various factors controlling the reactivity (such as energetic, electronic, polar, steric, conformational effects and role of orbital interactions) of a radical and its orientation of addition.

However, three targets deserve to be investigated in telomerisation: regioselectivity, tacticity and the molecular weight of telomers. For the first, Tedder and Walton have extensively explored numerous additions of various radicals to different olefins, especially in determining the kinetics of formation of monoadducts. For achieving desired molecular weights, even if some trends have been given from the different results of telomerisation of fluoroalkenes depicted in the literature, it is still difficult to predict which olefin could be more easily telomerised than another one, even by a chosen initiation way.

6
Radical "Living" or Controlled Telomerisation and Polymerisation of Fluoromonomers

6.1
Introduction

To date, new techniques have been proposed and developed to control the reactivity of free radicals. Such a control may lead to give a "living" character to the radical polymerisation. While the first example of living polymerisation was introduced in 1956 [344] in anionic polymerisation, it is only in 1982 with the pioneering work of Otsu et al. [345, 346] that the possibility to have a "living" process in free radical polymerisation was demonstrated. Otsu, using thiuram disulphide compounds, developed the "iniferter concept" with a control of chain termination. Later on, Solomon et al. [347] and Georges et al. [348] have proposed the use of stable counter radicals (e.g., 2,2,6,6-tetramethyl piperidinyl-1-oxyl usually called Tempo) as a thermally labile capping agent for the growing polymeric chain. Such a radical does not allow initiation of polymerisation. This very promising approach is nowadays the subject of active research and allows in principle the preparation of polymers with narrow molecular weight distributions in one pot as with ionic methods but without high purity and vacuum requirements.

However, despite considerable progress, the truly living character is far from being attained and it seems preferable to use the term controlled process rather than living process. Recently, various methods to synthesise block copolymers by radical polymerisation or telomerisation were reviewed [349]. But, to our knowledge, the literature does not mention any investigation of controlled radical polymerisation of fluorinated polymers.

Two main types of counter radicals are proposed, purely organic ($^{\cdot}$CR) ones issued from nitroxyl [347, 350–354], alkoxyamines [347, 351], arylazooxyls [355] compounds or triphenyl methyl (called trityl) group [345, 356–358], CPh_2-G where G represents $OSi(CH_3)_3$, OPh or CN [359, 360], or sulphurated radical $^{\cdot}$SR with R=alkyl [84, 102, 103], aryl [164], C(O)R′ [356, 361] or C(*S*)R″ where R″ represents R′ [362], OR′, NR'_2 [363] (R′ being an alkyl group) and organometallic complexes (C) [364, 365] able to generate stable free radicals. Equilibrium may be written as follows:

$$I{-}M^{\cdot}_{n} + {}^{\cdot}CR \underset{\text{or UV}}{\overset{\Delta}{\rightleftharpoons}} I{-}M_n{-}CR$$

where I and M represent the initiator radical and the monomer, respectively.

The organometallic complexes are quite various: cobaltocene/bis(ethylacetoacetato) copper (II) [366], CuCl or CuBr/bipyridyl [367, 368], cobaltoxime complex [369], reduced nickel/halide system [370], organoborane [371], ruthenium complex/trialkoxyaluminum system [372].

Such above chemical equations show the essential difference between the traditional radical polymerisation and the controlled radical polymerisation in the presence of counter radicals or metallic complexes.

Difficulties arise from the various possible interactions between the counter radical, the initiator and the monomer. The truly living character is only demonstrated when some requirements are fulfilled. Molecular weight must increase in a linear fashion with conversion. Polydispersity must be narrow and lower than that in classical process (theoretical value, $\overline{M_{\omega}}/\overline{M_n}=1.1-1.5$) which supposes in particular, a rapid initiation step. Obtaining high and strictly controlled molecular weights must be possible.

According to the literature, the only examples of controlled radical telomerisation of fluoroolefins require adequate fluorinated telogens. The oldest method is based on the cleavage of the C-I bond that already led to industrial applications.

The cleavage of C-I bond can be achieved from various methods [373–375]. However, according to well chosen monomers, two main ways have been developed in order to control telomerisation from alkyl iodides: iodine transfer polymerisation (ITP) and degenerative transfer. ITP can be easily applied to fluorinated monomers whereas degenerative transfer concerns the controlled polymerisation of methyl methacrylate, butyl acrylate [376] or styrene [377] and will not be discussed in this chapter.

6.2 Iodine Transfer Polymerisation

Iodine transfer polymerisation is one of the radical living process being developed in the late seventies by Tatemoto [374, 375, 378–381]. Actually it requires (per)fluoroalkyl iodides because their highly electron withdrawing (per)-fluorinated group R_F allows the lowest level of the CF_2-I bond dissociation energy, such a C-I cleavage being not possible in $R_FCH_2CH_2I$. Various fluorinated monomers have been successfully used in ITP. Basic similarities in these living polymerisation systems are found in the stepwise growth of polymeric chains at each active species. The active living centre, generally located at the end-groups of the growing polymer, has the same reactivity at any time during polymerisation even when the reaction is stopped [374, 375, 378]. In the case of ITP of fluoro olefin, the terminal active bond is always the C-I bond originated from the initial iodine-containing chain transfer agent and monomer as follows:

$$C_nF_{2n+1}-I + (p+1)\,H_2C{=}CF_2 \xrightarrow{R^{\cdot}\text{ or }\Delta} C_nF_{2n+1}\,(C_2H_2F_2)_{\overline{p}}-CH_2CF_2-I$$

Usually molecular weights are not higher than 30 000 and yet polydispersity is narrow (1.2–1.3) [375, 380, 381].

Tatemoto et al. used peroxides as initiators of polymerisation. Improvement is also possible by using diiodide and polyiodide compounds [316, 382].

Several investigations have shown that iodine transfer polymerisation can occur by emulsion or radical initiation. When emulsion initiation is chosen, a perfluoroalkyl iodide is involved and limits the molecular weights [378, 379]. This is not described here but several articles and patents from Tatemoto are suggested [374, 375, 378], using ammonium persulphate as the initiator and involving tetrafluoroethylene (TFE), vinylidene fluoride (VDF) and hexafluoropropene (HFP) as the monomers. The fluoroelastomers produced by ITP can be peroxide-curable and lead to commercially available Dai-El [380] produced by Daikin. Such a polymer is stable up to 200 °C and finds many applications in high technology such as transportation and electronics.

The most remarkable use of "living" polymerisation is the preparation of block copolymers. The practical method for obtaining such copolymers is by direct cotelomerisation of two fluoromonomers with an efficient transfer agent. For instance, Tatemoto and Morita [378] used a mixture of VDF/HFP in 46/54% molar ratio and IC_4F_8I in the presence of trichloroperfluorohexanoyl peroxide as the initiator and obtained $I(HFP)_a(VDF)_bC_4F_8(VDF)_c(HFP)_dI$ that exhibited an average number molecular weight of 3,300 with a polydispersity of 1.27. It consists of a stepwise cotelomerisation as follows:

$$R_FI + nM_1 \longrightarrow R_F(M_1)_nI$$
$$R_F(M_1)_n I + mM_2 \longrightarrow R_F(M_1)_n (M_2)_mI.$$

M_1 and M_2 represent fluorinated monomers or a group of fluoroolefins which can be adequately chosen to bring softness and hardness, respectively, leading to a thermoplastic elastomer [380]. For instance, a hard segment sequence can be composed of E/HFP/TFE in 43/8/49 molar ratio (E represents ethylene) whereas the soft part can consist of TFE/HFP/VDF in 20/30/50 molar ratio [381]. A similar scheme can also be applied to α, ω-diiodides [316, 379]. Furthermore, the introduction of hard segments such as alternated copolymers of TFE and E, or PVDF, led to commercially available Dai-El Thermoplastic [379, 380]. They exhibit very interesting properties such as a high specific volume (1.90), a high melting point (160–220 °C), a high thermostability up to 380–400 °C, a refractive index of 1.357 and good surface properties ($\gamma_c \sim 19.6-20.5$ dynes.cm^{-1}). These characteristics offer excellent resistance against aggressive chemicals and strong acids, fuels, and oils. In addition, tensile modulus is close to that of cured fluoroelastomers.

This concept was also exploited in order to prepare "living" and well defined tetrafluoroethylene telomers in which the telomer produced acts as a further telogen as follows:

$$C_2F_5I \xrightarrow{C_2F_4} C_4F_9I \xrightarrow{C_2F_4} C_6F_{13}I \xrightarrow{C_2F_4} C_8F_{17}I \longrightarrow C_nF_{2n+1}I.$$

Such a living telomerisation can be initiated either thermally or in the presence of radical initiators or redox catalysts [383].

6.3
Stepwise Telomerisation

The application of such a process allows the step by step syntheses of block copolymers either from fluoroalkyl iodides or α,ω-diiodoperfluoroalkanes. Chambers' investigations or those performed in our laboratory have led to the extensive use of fluoroalkyl iodides with chlorotrifluoroethylene (CTFE) for the preparation of efficient transfer agents X-Y with X=Y=I [239] or Br [223, 224] and also X=I and Y=Cl [115] or F [235, 236]:

$$\text{X–Y} + \text{CF}_2\text{=CFCl} \longrightarrow \text{X(C}_2\text{F}_3\text{Cl)}_n\text{Y}$$

Chambers obtained difunctional oligomers especially for iodinated telogens [239]. It has been observed, however, that from ICl or (IF) generated *in situ* from iodine and iodine pentafluoride, monoadducts were obtained as the sole product [235, 236]. Addition of (IF) to CTFE yields CF_3CFClI which was successfully used as the telogen for the telomerisations of CTFE, hexafluoropropene (HFP) [235] or for the stepwise cotelomerisations of CTFE/HFP, CTFE/HFP/VDF, HFP/VDF and HFP/trifluoroethylene [384]. All these above cotelomers were successfully end capped by ethylene that allows further functionalisation [385]. CFClI end group shows a better reactivity than CF_2I [237].

Furthermore, stepwise cotelomerisations of various commercially available fluoromonomers lead also to interesting highly fluorinated derivatives as follows:

$$\text{C}_4\text{F}_9\text{I} + \text{CH}_2\text{=CF}_2 \longrightarrow \text{C}_4\text{F}_9(\text{C}_2\text{H}_2\text{F}_2)_n\,\text{I} \xrightarrow{\text{HFP}} \text{C}_4\text{F}_9\,(\text{VDF})_n\,(\text{HFP})\,\text{I} \quad [384]$$

$$\text{iC}_3\text{F}_7\text{I} + \text{CH}_2\text{=CF}_2 \longrightarrow \text{iC}_3\text{F}_7(\text{C}_2\text{H}_2\text{F}_2)_n\,\text{I} \xrightarrow{\text{HFP}} \text{iC}_3\text{F}_7(\text{C}_2\text{H}_2\text{F}_2)_n\text{CF}_2\text{CFICF}_3 \quad [96, 384]$$

$$\text{C}_n\text{F}_{2n+1}\,\text{I} + \text{HFP} \longrightarrow \text{C}_n\text{F}_{2n+1}\,(\text{HFP})\,\text{I} \xrightarrow{\text{VDF}} \text{C}_n\text{F}_{2n+1}\,(\text{HFP})\,(\text{VDF})_p\text{I} \quad [98, 384]$$

n = 4, 6, 8

$$\text{C}_n\text{F}_{2n+1}\,(\text{HFP})\,\text{I} \xrightarrow{\text{C}_2\text{F}_3\text{H}} \text{C}_n\text{F}_{2n+1}\,(\text{HFP})\,(\text{C}_2\text{F}_3\text{H})_x\text{I}$$

$$\text{C}_n\text{F}_{2n+1}\,(\text{HFP})\,(\text{VDF})_p\text{I} \xrightarrow{\text{C}_2\text{F}_3\text{H}} \text{F(TFE)}_n\,(\text{HFP})\,(\text{VDF})_p\,(\text{C}_2\text{F}_3\text{H})_m\text{I}$$

All these cotelomers were successfully end capped with ethylene that offered new ω-functionalised fluorocompounds [385]. In addition, α, ω-diiodofluoroalkanes also allow access to well-defined block cotelomers. Tortelli and Tonelli [117] and Baum and Malik [275, 386] performed the synthesis of such telechelic products by heating iodine crystal with tetrafluoroethylene. Such halogenated reactants were successfully involved in telomerisation of vinylidene fluoride (VDF) [130] or hexafluoropropene (HFP) [387] to form α, ω-diiodo VDF/TFE/VDF or α, ω-diiodo HFP/TFE/HFP triblock cotelomers. These products are potential starting materials for Viton type multiblock cotelomers [387] as shown in the following examples:

$$\text{I}\,(\text{VDF})_p(\text{TFE})_n(\text{VDF})_q\text{I} + \text{HFP} \longrightarrow \text{I(HFP)}_X(\text{VDF})_p(\text{TFE})_n(\text{VDF})_q(\text{HFP})_y\,\text{I}$$

$$\text{I}\,(\text{HFP})(\text{TFE})_n(\text{HFP})_t\text{I} + \text{VDF} \longrightarrow \text{I(VDF)}_Z(\text{HFP})\,(\text{TFE})_n(\text{HFP})_t(\text{VDF})_\omega\,\text{I}$$

These novel α, ω-diiodides underwent functionalisations, especially for the preparation of original fluorinated nonconjugated dienes [388] utilised in the preparation of hybrid fluorosilicones [389] which exhibit excellent properties at low and high temperatures.

6.4
Conclusion

Even if in 1955 Haszeldine [237] or in 1957 Hauptschein et al. [114, 273] started to show a certain livingness of the radical telomerisation of CTFE with $ClCF_2CFClI$, the up to date research has attracted many academic or industrial chemists towards such a fascinating area. In addition, Dear and Gilbert [102] or Sharp et al. [84, 103] performed the telomerisation of various fluoroolefins with disulfides but they did not examine the living character of this reaction.

However, the step-wise cotelomerisation of fluoroalkenes has already shown original livingness, and the polymerisation via organometallic systems which is quite successful for hydrogenated monomers could be an encouraging route for these halogenated olefins.

7
Applications of Fluorinated Telomers

The applications of fluorinated telomers or fluoro oligomers are linked to their exceptional properties [1]: excellent thermostability, remarkable surface properties, low refractive indexes, low dielectric constants and very high chemical stabilities.

The only application of the oligomers, obtained directly, are as lubricant oils, for example Kel F oils. In general however, oligomers need to undergo chemical transformation before an application is found. In this latter case, the number of functional groups (generally one or two) and the nature of the linkage between the functional group and the fluorinated chain plays an important role in the kind of application.

Hereafter are given several examples of non functional, monofunctional and telechelic telomers.

7.1
From Nonfunctional Telomers Mainly Used as Lubricants

Three series of products are used as lubricants: silicone-oils or polytrifluoropropyl methyl siloxane (Silastic 15–53 produced by the Dow Corning Company), perfluorinated polyethers (Fomblin, Montedison now Ausimont Company) and Kel F oils (3 M Company). Only these latter ones are obtained directly by telomerisation.

Initially, the telomerisation of CTFE with SO_2Cl_2 was used by the Kellogg Company to prepare $Cl\text{-}(CF_2\text{-}CFCl)_n\text{-}Cl$ thermostable oligomers [246]. A few years later, inspired by the redox telomerisation of CTFE with CCl_4 [90], the Occidental Chemical Company [390] prepared $CCl_3\text{-}(CF_2\text{-}CFCl)_n\text{-}Cl$ telomers subsequently fluorinated by ClF_3 in order to increase thermal stability [391].

Recently, other new methods have been investigated by Marraccini [43, 51] using CF_3OF as the telogen, for the production of original $CF_3O(C_2F_3Cl)_nF$ telomers (see Sect. 4.1.3).

However, the research programs in this area are difficult to develop because of several chemical problems: e.g. inversion of addition in CTFE oligomerisation which leads to a decrease of thermal stability brought by -CFCl-CFCl- linkages and, above all nowadays, the availability of CTFE in the world because the production of chlorofluorinated ethane $CF_2Cl\text{-}CFCl_2$ is to be interrupted.

7.2 From Monofunctional Telomers

The monofunctional oligomers can be synthesised either by chemical change of the end-groups [85] or by direct telomerisation with a functional telogen (e.g. $Cl_3CCO_2CH_3$ [92], HOC_2H_4SH [101], CH_3OH [134, 135], $HP(O)(OEt)_2$ [148, 149]). Another method is the electrochemical fluorination of carboxylic or sulphonic hydrogenated acid [392].

In any case, applications are mainly devoted to surfactants and numerous products are commercially available (anionic, cationic, non ionic but also mono- or polydispersed compounds).

One well known application concerns the fire fighting foam-agents (A3F), as well as the use of fluorosurfactants in emulsions (for fluoromonomers) and microemulsions.

In addition, these fluorinated monofunctional telomers can be interesting precursors of original macromonomers leading to novel polymeric optical fibers [393] or grafted copolymers [134].

The monofunctional oligomers are also used as precursors of numerous monomers. One of the most famous ones is the series of fluoroacrylates for which the synthesis, the properties and the applications have been described in detail [15]. First of all, their use in the protection of textiles has challenged several companies: Scotchguard (3 M), Foraperle (Atochem) and Lezanova (Daikin). Nowadays, these products are efficient for the protection of leathers, stones and in the coating with high weatherability. The second important application of fluoroacrylates concerns the materials for optics: optical fibers (core and cladding), optical components (guides) and optical glasses. The third one is the coating of amorphous silicon in order to prepare positive photoresists for electronics. Other minor applications concern fluoroelastomers, solid propellants, dental materials.

Two other families of polymers with fluorinated side groups have been investigated in the field of elastomers: polysiloxanes and polyphosphazenes [394, 395]. These latter macromolecules, called NPF, have been obtained by chemical change of poly(dichlorophosphazene) as for classic phosphazenes:

$$-\!\left(N=PCl_2\right)_n\!- + NaOR_F \longrightarrow \left(-N=\underset{\underset{OR_F}{|}}{\overset{\overset{OR_F}{|}}{P}}-\right)_n$$

The most known and commercialised by Firestone is the following:

$$\left(N = P \underset{\underset{OCH_2(CF_2)_4H}{|}}{\overset{\overset{OCH_2CF_3}{|}}{}} \right)_n$$

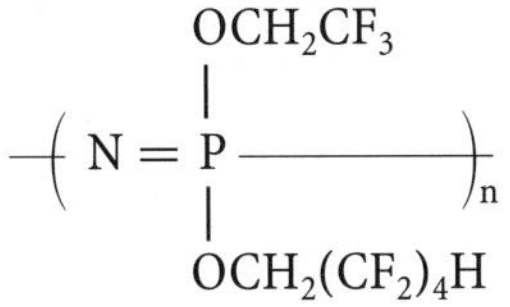

It is obtained with a higher molecular weight ($\overline{M}_n > 10^6$) and its glass transition temperature is close to $-68\,°C$ [396]. Interestingly, this fluoropolyphosphazene preserves its good properties up to $175\,°C$ and exhibits excellent chemical inertness. However, the synthesis of such a polymer is not easy (mainly because of corrosion) and the purification of the trimer precursor is difficult to perform.

Concerning fluorinated polysiloxanes, the most common and commercially available (by Dow Corning) has the following formulae:

$$\left(\underset{\underset{C_2H_4CF_3}{|}}{\overset{\overset{CH_3}{|}}{Si}} - O \right)_n$$

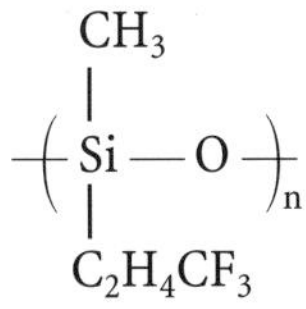

It is synthesised from the corresponding D_3 cyclosiloxane and exhibits a Tg of $-68\,°C$. Such a polymer can easily be crosslinked by various processes of silicone chemistry (e.g., peroxides, SiH or Si-vinyl derivatives). Fluorosilicones can be used for their resistance to swelling in non polar medium (e.g., Skydrol fluids) even at $100\,°C$; but their main application is to replace elastomers of fluorinated monomers at low temperatures.

Furthermore, important and interesting investigations have been performed by Kobayashi and Owen [397] who prepared fluorosilicones with perfluorinated lateral group:

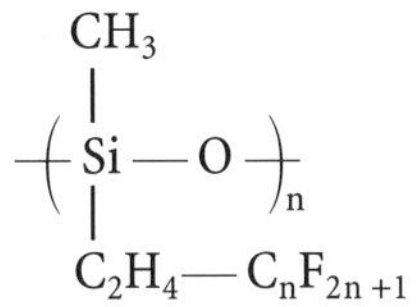

$$\left(\underset{\underset{C_2H_4 - C_nF_{2n+1}}{|}}{\overset{\overset{CH_3}{|}}{Si}} - O \right)_n \qquad \text{with } n = 1 \text{ to } 8.$$

At the same time, the fluorinated chains have found applications in various thermostable polymers such as polyquinazolones [398] and polyimides [399]. In this latter case, the presence of the fluorinated group decreases both the water absorption and the dielectric constant.

7.3
From Telechelic Oligomers

Telechelic (or α, ω-difunctional) oligomers exhibit functional groups at both ends of the oligomeric backbone. These compounds are quite useful precursors for well defined architectured polymers (e.g. polycondensates, polyadducts and other fluoropolymers previously reviewed [400, 401]). As mentioned above, a large effort has been done for these last ten years, in order to prepare new telechelic fluorinated telomers. These investigations have provided other compounds than polyethers (Krytox, Fomblin and Demnum).

Thus, starting from fluorinated α, ω-diiodinated telomers, various α, ω-dienes [388], α,ω-diols [275, 386] and diacids have been prepared. Interestingly, fluorinated polyurethanes were synthesised from diols which exhibit fluorinated groups either in the backbone or in the side chain. The former series [386, 402] has led to fluoropolymers with interesting thermally improved properties whereas the second family of diols has favoured lower surface tensions [70].

The synthesis of hybrid polysiloxanes is depicted in the following scheme [389]:

$$
CH_2{=}CH{-}R_F{-}CH{=}CH_2 + H{-}\underset{\underset{R_{F'}}{|}}{\overset{\overset{CH_3}{|}}{Si}}{-}Cl \xrightarrow[3°)copolycondensation]{\overset{1°)\,H_2PtCl_6}{\underset{2°)H_2O/\overset{\ominus}{OH}}{}}}
$$

$$
\left(\underset{\underset{R_{F'}}{|}}{\overset{\overset{CH_3}{|}}{Si}}{-}C_2H_4{-}R_F{-}C_2H_4{-}\underset{\underset{R_{F'}}{|}}{\overset{\overset{CH_3}{|}}{Si}}{-}O\right)_x\left(\underset{\underset{R_{F''}}{|}}{\overset{\overset{Me}{|}}{Si}}{-}O\right)_y
$$

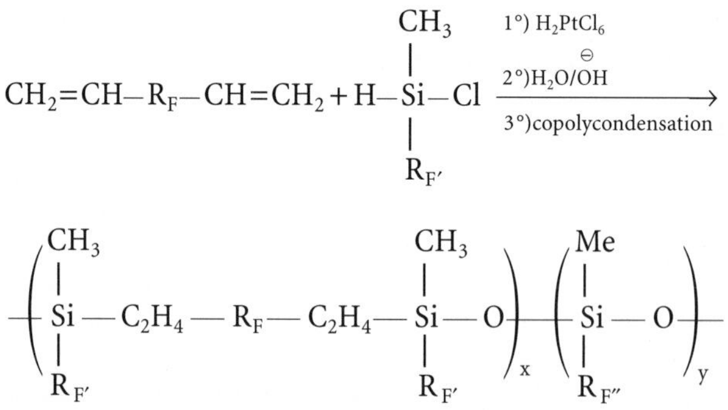

In the field of soft sequences in thermostable polymers, it is interesting to note the real progress concerning the following reaction, previously studied by McLoughlin and Thrower [403] and optimised by Chen et al. [404]:

$$
I{-}R_F{-}I + I{-}\langle\bigcirc\rangle{-}\text{Ⓖ} \xrightarrow[DMSO/biPyr.]{copper} \text{Ⓖ}{-}\langle\bigcirc\rangle{-}R_F{-}\langle\bigcirc\rangle{-}\text{Ⓖ}
$$

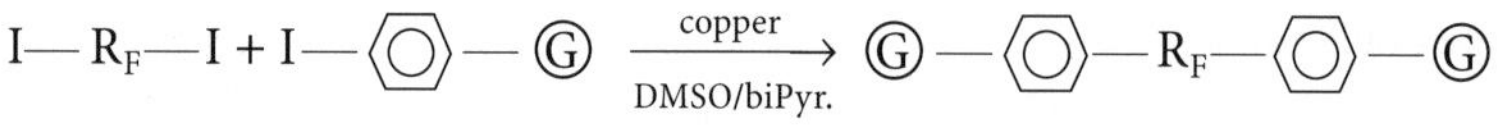

The important increase of the yield allows to forecast new generation of thermostable polymers.

Thus, fluorinated telomers are effective intermediates which can be utilised in further reactions, producing high value added materials with relevant properties. Two main series of properties are obtained: the polymers containing the perfluorinated group in the backbone exhibit higher thermostability whereas those having such a group as a side chain are expected to show very good surface properties.

8
Conclusion

This article outlines the important number of investigations performed on the telomerisation of well-known fluorinated alkenes. These fluorotelomers allow to synthesise novel fluoropolymers regarded as high value added materials with usually exceptional properties.

The telomerisation reaction appears a powerful tool to prepare well-defined molecules and their well-characterised end-groups allow further key-reactions. However, an adequate way of initiation has to be chosen and even if traditional initiations are still up to date, recent processes appear promising (e.g. in supercritical fluids).

However, it is known that some defects in chaining may enhance dramatic issues (e.g. in polychlorofluorinated oils, two vicinal chlorine atoms induce weak points which affect the thermostability).

Fortunately, much work has still to be developed, and especially a better knowledge of the synthesis of tailor-made polymers with well-defined architecture in order to improve the properties by a better control of the regioselectivity and of the tacticity. Even if much work on the reactivity has been extensively carried out by Tedder and Walton, methods are searched for a better orientation of the sense of addition, since most fluoroalkenes are unsymmetrical.

In our opinion, two main processes should bring original solutions. The first one concerns the step-by-step addition of fluoroalkenes (i.e., homostepwise telomerisation as described by Tatemoto) in order to favour the generated radical to react on the same site of the olefin.

The second opportunity deals with the use of organometallic systems which allow a very good control of both the regioselectivity and the tacticity. This kind of polymerisation has been used successfully on hydrogenated monomers, but, has not yet been investigated on fluoroalkenes.

Hence, such systems should induce interesting consequences on the structures and on the properties of materials thus obtained. For example, in the case of fluoroacrylic monomers, it is well-known that the crystallisation of perfluorinated chains insures improvement of surface properties. In addition, by introducing well-defined fluorotelomers in the backbone or in the side chains, building block silicones can be synthesised with monitored thermal properties especially at low temperatures.

Such novel targets are real challenges and should attract the interest of many academic and industrial researchers.

9
References

1. Yamabe M (1992) Makromol Chem Macromol Symp 64:11
2. Banks RE, Willoughby BG (1994) Fluoropolymers. In: Bloor D, Brook RJ, Flemings MC, Mahajan S (eds) The Encyclopedia of Advanced Materials, Pergamon, Oxford 862
3. Smith S (1982) Fluoroelastomers. In: Banks RE (ed) Preparation, Properties and Industrial Applications of Organofluorine Compounds, Ellis Horwood Chichester 235
4. Wall LA (1972) Fluoropolymers. Wiley, New-York 381
5. Uschold RE (1985) Polym J 17:253
6. Lynn MM, Worm AT (1987) Ency Polym Sci Eng 7:257
7. Logothetis AL (1989) Prog Polym Sci 14:251
8. Cook D, Lynn MM (1995) Rapra Review Reports Shrewsbury vol 3 report 32
9. Feiring AE (1994) J Macromol Sci Pure Appl Chem A31(11):1657
10. Johns K (1994) Fluorocoatings: an overview, Fluorine in Coating Symposium, Salford, UK, September 28–30

11. Vanoye D, Ballot E, Legros R, Loubet O, Boutevin B, Améduri B (1994) French Patent 94,14367 (30-11-1994) to Elf Atochem
12. Bonardi C (1991) Eur Patent Appl EP 426,530 (to Elf Atochem) (Chem Abst 155:161338z)
13. Head RA, Johnson S (1987) Eur Patent Appl 260,842 (to ICI)
14. Barraud JY, Gervat S, Ratovelomanana V, Boutevin B, Parisi JP, Cahuzac A, Octeur RJ (1992) French Patent 9,204,222 (to Alcatel) (07-04-1992)
15. Boutevin B, Pietrasanta Y (1988) Les Acrylates et Polyacrylates Fluorés: Dérivés et Applications, Erec, Puteaux (Fr)
16. Kippling B (1982) ACS Symp Ser 180:477
17. Grot WG (1994) Macromol Symp 82:161
18. Riess JG, Chang TMS (1993) Blood substitutes and oxygen carriers, Marcel Dekker, New-York
19. Hanford WE, Joyce RM (1942) US patent 2,440,800 (10-04-1942) (to Dupont)
20. Freidlina R Kh, Terent'ev AB, Kharlina M Ya, Aminov SN (1966) Zh Vses Khim Obshchestva im DI Mendeleva 11:211 (Chem Abstr (1966) 65:8692)
21. Starks CM (1974) Free Radical Telomerization 1st edn, Academic Press, New-York
22. Gordon R, Loftus RD (1989) Telomerization In: Kirk RE, Othmer DF (eds) Ency Poly Sci Tech, Wiley 16:533
23. Boutevin B, Pietrasanta Y (1989) Telomerization In: Allen G, Bevington JC, Eastmond AL, Russo A (eds) Comprehensive Polymer Sc, Pergamon, Oxford 3:185
24. Améduri B, Boutevin B (1994) Telomerization. In: Bloor D, Brook RJ, Flemings RD, Mahajan S (eds) Encyclopedia Adv Materials, Pergamon, Oxford 2767
25. Haszeldine RN, Rowland R, Sheppard RP, Tipping AE (1985) J Fluorine Chem 28:291
26. Tedder JM, Walton JC (1967) Progr React Kinet 4:37
27. Masson JC (1989) Decomposition rates of organic free radical initiators. In: Brandrup J, Immergut EH (eds) Polymer Handbook, John Wiley, New-York 1:1
28. Bauduin G, Boutevin B, Mistral JP, Sarraf L (1985) Makromol Chem 186:1445
29. Rice DE, Sandberg CL (1971) Polym Prepr Amer Chem Soc Div Polym Chem 12:396 and (1969) US Patent 3,457,245 (to 3 M) 22-07-1969
30. Sawada H (1993) J Fluorine Chem 61:253
31. David C, Gosselain PA (1962) Tetrahedron 18:639
32. Boutevin B, Pietrasanta Y (1976) Eur Polym J 12:219
33. Tsuchida E, Kitamura K, Shinohara J (1972) J Polym Sci 10:3639
34. Boutevin B, Parisi JP, Vaneeckoutte P (1991) Eur Polym J 27:159 and 27:1029
35. Lustig M, Shreeve JM (1973) Adv Fluorine Chem 7:175
36. Shreeve JM (1983) Adv Inorg Radiochem 26:119
37. Mukhametshin FM (1980) Russ Chem Rev 49:668
38. Schack CJ, Christe KO (1978) Isr J Chem 17:20
39. Storzer W, DesMarteau DD (1991) Inorg Chem 30:4821
40. Kellog KB, Cady GH (1948) J Amer Chem Soc 70:3986
41. Johri KK, DesMarteau DD (1983) J Org Chem 48:242
42. Sekiya A, Ueda K (1990) Chem Lett 609
43. Marraccini A, Pasquale A, Vincenti M (1990) Eur Patent Appl 379,070 (11-01-90) to Ausimont (Chem Abst 113:232722)
44. Randolph BB, DesMarteau DD (1993) J Fluorine Chem 64:129
45. Campbell D, Fifolt M, Saran M (1985) German Patent 3,438,934 (to Occidental Chem Co) 09-05-1985 (Chem Abst 104:5565)
46. Dos Santos Afonso M, Schumacher HJ (1984) Int J Chem Kinet 16:103
47. Dos Santos Afonso M, Czarnowski J (1988) Z Phys Chem 158:25
48. Marraccini A, Pasquale A, Vincenti M (1990) Eur Patent Appl EP 348,980 (03-01-1990) to Ausimont (Chem Abst 112:199361)
49. Dos Santos Afonso M, Schumacher HJ (1988) Z Phys Chem 158:15
50. Di Loreto HD, Czarnowski J (1994) J Fluorine Chem 66:1
51. Marraccini A, Perego G, Guastalla G (1989) Eur Patent Appl EP 321,990 (28-06-1989) to Ausimont (Chem Abst 111:215077)

52. Anderson JDO, DesMarteau DD (1996) J Fluorine Chem 77:147
53. Narita T (1994) Macromol Symp 82:185
54. Warnell JL (1967) US Patent 3,311,658 (to Du Pont de Nemours) 23-03-1967 (Chem Abst 67:43423)
55. Moore LO (1966) US Patent 3,250,808 (02-05-1966)
56. Ito TI, Kaufman J, Kratzer RH, Nakahara JH, Paciorek KJL (1979) J Fluorine Chem 14:93
57. Kvicala J, Paleta O, Dedek V (1990) J Fluorine Chem 47:441
58. Flynn RM (1988) US Patent 4,749,526 (to 3 M) 07-06-1988
59. Hill JT (1974) J Macromol Sci Chem 8:499
60. Eleuterio HS (1972) J Macromol Sci Chem 6:1027
61. Sokolov SV (1975) Zh Win Org 11:303
62. Yohnosuke O, Takashi T, Shoji T (1983) Eur Patent Appl 148,482 (to Daikin) 26-12-1983 (Chem Abst 104:69315)
63. Fielding HC (1968) British Patent 1,130,822 (to ICI) 16-10-1968 (Chem Abst 70:11364t)
64. Fielding HC, Deem WR, Houghton LE (1973) German Patent 2,215,385 (to ICI) (Chem Abst 78:147560)
65. Caporiccio G, Viola G, Corti C (1983) Eur Patent 89,820 (to Montedison)
66. Caporiccio G (1986) J Fluorine Chem 33:314
67. Tanesi D, Pasetti A, Corti C (1969) US Patent 3,442,942 (to Montefluos)
68. Sianesi D, Pasetti A, Belardinelu G (1984) US Patent 4,451,646 (to Montefluos)
69. Trischler FD, Hollander J (1967) J Polym Sc Part A1 5:2343
70. Hirn B, Cathebras N, Collet A, Commeyras A, Viguier M (1995) Eur Symp in Fluorine Chem (Bled) Proceedings p104
71. Karam L, Améduri B, Boutevin B (1993) J Fluorine Chem 65:43
72. Gong M, Fuss W, Kompa KL (1990) J Phys Chem 94:6332
73. Zhang L, Zhang J (1995) Wuli Huaxue Xuebao 11:308 (Chem Abst 123:8893)
74. Krespan CG, Petrov VA (1995) International Patent WO9532936 (to Du Pont de Nemours) 07-12-1995 (Chem Abst 124:177222)
75. Krespan CG, Petrov VA (1995) US Patent 5,459,212 (to DuPont de Nemours) 17-10-1995 (Chem Abst 124:30631)
76. Yoshida Y, Ishizaki K, Horiuchi T, Matsushige K (1993) Kogaku Shuho Kyushu Daigaku 66:199 (Chem Abst 120:271981)
77. Paleta O (1977) Fluorine Chem Rev 8:39
78. Paleta O, Posta A, Liska F (1978) Sb Vys Sk Chem Techno Prag Org Chem Tech C25:105 (Chem Abst 92 (1980) 6026)
79. Ikeya T, Kitazume T (1987) 12th Symp Fluorine Chem Hakuta, Japan, Abst 3R17
80. Combes JR, Guan Z, De Simone JM (1994) Macromolecules 24:865
81. Tedder JM (1982) Angew Chem Int Ed Engl 21:401
82. Tedder JM (1978) Directive Effects in Gas-Phase Radical Addition Reactions. In: Gold V, Bethell D (eds) Adv Phys Org Chem, Academic Press, New-York 16:51
83. Tedder JM, Walton JC (1980) Tetrahedron 36:701
84. Sharp DWA, Miguel HT (1978) Isr J Chem 17:144
85. Toyoda Y, Sakauchi N, Nobuo C (1967) Japanese patent 77319 (to Kureha Chem Ind Co Ltd) 02-12-1967
86. Chambers RD, Grievson BG, Drakesmith FG, Powell RL (1985) J Fluorine Chem 29:323
87. Paleta O, Dedek V, Reutschek H, Timpe HJ (1989) J Fluorine Chem 42:345
88. Joyce RM Jr (1951) US Patent 2,559,628 (Chem Abst 1952, 46:3063)
89. Posta A, Paleta O (1966) Collect Czech Chem Commun 31:2389
90. Boutevin B, Pietrasanta Y (1976) Eur Polym J 12:231
91. Kotora M, Hajek M (1993) J Fluorine Chem 64:101
92. Boutevin B, Furet Y, Lemanach L, Vial-Reveillon F (1990) J Fluorine Chem 47:95
93. Sloan JP, Tedder JM, Walton JC (1973) J Chem Soc Faraday Trans 1 69:1143
94. Dedek V, Chvatal Z (1986) J Fluorine Chem 31:363
95. Hauptschein M, Braid M, Fainberg AH (1958) J Amer Chem Soc 80:851
96. Apsey GC, Chambers RD, Salisbury MJ, Moggi G (1988) J Fluorine Chem 40:261

97. Balagué J, Améduri B, Boutevin B, Caporiccio G (1995) J Fluorine Chem 70:215
98. Balagué J, Améduri B, Boutevin B, Caporiccio G (1995) J Fluorine Chem 73:237
99. Balagué J, Améduri B, Boutevin B, Caporiccio G (1995) J Fluorine Chem 74:49
100. Harris JF, Stacey FW (1963) J Amer Chem Soc 85:749
101. Boutevin B, Furet Y, Hervaud Y, Rigal G (1994) J Fluorine Chem 69:11
102. Dear REA, Gilbert EE (1974) J Fluorine Chem 4:107
103. Haran G, Sharp DWA (1972) J Chem Soc Perkin Trans 4:34
104. Grigor'ev NA, German LS, Freidlina RK (1979) Bull Acad Sc USSR 863 and (1979) Izv Akad Nauk SSSR Ser Khim 918 (Chem Abst 91:19833)
105. Burch GM, Goldwhite H, Haszeldine RN (1963) J Chem Soc 1083
106. Brace NO (1961) J Org Chem 3:3197
107. Roberts HL (1967) Ger Patent 1,233,375 (to ICI) 02-02-1967 (Chem Abst 66:85452)
108. Barnhart WS (1957) US Patent 2,786,827 (to Kellog Co)
109. Boutevin B, Pietrasanta Y, Youssef B (1986) J Fluorine Chem 31:57
110. Bell TN, Haszeldine RN, Newlands MJ, Plum JB (1965) J Chem Soc 2107
111. Haszeldine RN, Steele BR (1954) J Chem Soc London 3747
112. Shaw GC, Bissel ER (1962) J Chem Soc 27:1482
113. Boutevin B, Gornowicz G, Caporiccio G (1992) US Patent 3671 (to Dow Corning) 05-08-1992
114. Haupstchein M, Braid M, Fainberg AH (1961) J Amer Chem Soc 83:2495
115. Améduri B, Boutevin B, Kostov GK, Petrova P (1995) J Fluorine Chem 74:261
116. Haszeldine RN (1952) J Chem Soc 4423
117. Tortelli V, Tonelli C (1990) J Fluorine Chem 47:199
118. Barnhart WS (1959) US Patent 2,898,382 (to 3 M)
119. Dannels BF, Fifolt MJ, Tang DY, Amherst E (1989) US Patent 4,808,760 (to Occidental Chem Co) (28-02-1989) Chem Abst 111:78806
120. Boutevin B, Pietrasanta Y, Rousseau A, Bosc D (1987) J Fluorine Chem 37:151
121. Tezuka Y, Imai K (1984) Makromol Chem Rapid Commun 5:559
122. Ponomarenko VA, Cherkaev VG, Petrov AD, Zadorozhnyi NA (1958) Isvest Akad Nauk SSSR Otdel Khim Nauk 247 (Chem Abst 52:12751)
123. Geyer AM, Haszeldine RN (1957) J Chem Soc London 3925
124. Geyer AM, Haszeldine RN (1957) J Chem Soc London 1038
125. Zimin AV, Matyuk VM, Yankelevich AZ, Shapet'ko NN (1976) Dokl Akad Nauk SSSR Ser Khim 231:870
126. Boutevin B, Guida-Pietrasanta F, Ratsimihety A, Caporiccio G, Gornowicz G (1993) J Fluorine Chem 60:211
127. Hauptschein M, Braid M, Lawlor FE (1958) J Amer Chem Soc 80:846
128. Hauptschein M, Oesterling RE (1960) J Amer Chem Soc 82:2868
129. Chambers RD, Hutchinson J, Mobbs RH, Musgrave WKR (1964) Tetrahedron 20:497
130. Manséri A, Améduri B, Boutevin B, Chambers RD, Caporiccio G, Wright AP (1995) J Fluorine Chem 74:59
131. Sloan JP, Tedder JM, Walton JC (1975) J Chem Soc Perkin II 1846
132. Chen QY, Ma ZZ, Jiang XK, Zhang YF, Jia SM (1980) Hua Hsueh Hsueh Pao 38:175 (Chem Abst (1981) 94:174184)
133. Duc M, Améduri B, Boutevin B, Sage JM to be submitted to J Fluorine Chem
134. Oku J, Chan RJH, Hall HK, Hughes OR (1986) Polym Bull 16:481
135. Duc M, Kharroubi M, Améduri B, Boutevin B, Sage JM submitted to Macromolecules
136. Tarrant P, Lovelace AM (1955) J Amer Chem Soc 77:768
137. Tarrant P, Lovelace AM, Lilyquist MR (1955) J Amer Chem Soc 77:2783
138. Modena S, Pianca M, Tato M, Moggi G, Russo S (1985) J Fluorine Chem 29:154
139. Modena S, Pianca M, Tato M, Moggi G, Russo S (1989) J Fluorine Chem 43:15
140. Tarrant P, Lilyquist MR (1954) J Chem Soc 77:3640
141. Ashton DS, Sand DJ, Tedder JM, Walton JC (1975) J Chem Soc Perkin II 320
142. Moggi G, Modena S, Marchionni G (1990) J Fluorine Chem 49:141
143. Mac Murray N, Tedder JM, Vertommen LLT, Walton JC (1976) J Chem Soc Perkin II, 63

144. Low HC, Tedder JM, Walton JC (1976) J Chem Soc Faraday Trans 1 72:1707
145. Boutevin B, Furet Y, Hervaud Y, Rigal G (1995) J Fluorine Chem 74:37
146. Tiers GVD (1958) US Patent 2,846,472 (to 3 M)
147. Zhu MY, Sun SSX, Zhang YH, Jiang XK (1993) Ch Chem Lett 4:583
148. Haszeldine RN, Hobson DL, Taylor DR (1976) J Fluorine Chem 8:115
149. Duc M, Améduri, B, Boutevin B, Sage JM to be submitted to J Fluorine Chem
150. Haszeldine RN, Steele BR (1954) J Chem Soc 923
151. Cape JN, Greig AC, Tedder JM, Walton JC (1975) J Chem Soc Farad 1 592
152. Tedder JM, Walton JC (1966) J Chem Soc Farad 1 62:1859
153. Tedder JM, Walton JC (1974) J Chem Soc Farad 1 70:308
154. Chambers RD, Proctor LD, Caporiccio G (1995) J Fluorine Chem 70:241
155. Harris JF, Stacey FW (1961) J Amer Chem Soc 83:840
156. Manséri A, Améduri B, Boutevin B, unpublished results
157. Rondestvedt CS Jr (1977) J Org Chem 42:1985
158. Kiryukin DP, Nevel'Skaya TI, Kim IP, Barkalov IM (1982) Vys Soed Ser A 24:307 (Chem Abst 96:200211)
159. Jeffrey GC (1966) US Patent 3,235,611 (to Dow Chemical Co) 15-02-1966 (Chem Abst 64:17421)
160. Fearn JE (1971) J Research Nat Bur Stand 75A:41
161. Haszeldine RN (1949) J Chem Soc 2856
162. Ashton DS, Tedder JM, Walton JC (1974) J Chem Soc Faraday Trans 1 70:299
163. Ono Y, Ukihashi H (1970) Asahi Garasu Henkyn Hokoku 20:55 (Chem Abst 74:99234)
164. Zaitseva EL, Rozantseva TV, Chicherina II, Yakubovich A Yu (1971) Zh Org Khim 7:2548 (English transl) Chem Abst 76:72156
165. Roberts HL (1964) Deutch Appl 6,402,992 (to ICI) Chem Abst 62:13043
166. Roberts HL British patents (1961) 877,961 and (1962) 898,309 to ICI
167. ICI (1974) German Patent 2,416,261
168. Terjeson RJ, Gard GL (1987) J Fluorine Chem 35:653
169. Zhang Y, Guo C, Zhou X, Chen Q (1982) Huaxue Xuebao 40:331 (Chem Abst 97:126982)
170. Pennsalt Chemicals (1962) Belgian Patent 612,520 and (1967) Neth Appl 6,609,483 (13-01-1967) (Chem Abst 67:12621)
171. Pennsalt Chemicals (1965) US patent 3,219,712
172. Bertocchio R, Lacote G, Vergé C (1993) European Patent 552,076 (to Elf Atochem) 21-07-1993
173. Hutchinson J (1973) J Fluorine Chem 3:429
174. Moore LO (1983) Macromolecules 16:357
175. Peavy RE (1994) US Patent 5,310,870 (to DuPont de Nemours and Co) 10-05-1994 (Chem Abst 121:206293)
176. Afanas'ev IB, Safronenko ED, Beer AA (1967) Vysokomol Soedin Ser B 9:802 (Chem Abst 68:30099)
177. Asahi Glass Co (1975) Japanese Patent 47,908
178. Satokawa T, Fujii T, Ohmori A, Fujita Y (1982) US Patent 4,346,250 (to Daikin Kogyo Co) 24-08-1982 (Chem Abst 97:197835) and (1979) Jap Patent 154,707 (to Daikin Kogyo Co) 06-12-1969 (Chem Abst 93:25890)
179. No VB, Bakhmutov Yu L, Pospelova NB (1976) Zh Org Khim 12:1825 (Chem Abst 85:159134)
180. Daikin Kogyo Co (1981) Jap patent 43,225 (21-04-1981) Chem Abst 95:186628
181. Huang WY, Liang WX, Lin WD, Chen JH, Zhou XY (1980) Hua Hsueh Hsueh Pao 38:283 (Chem Abst 94:30115)
182. Chkubianishvili NG (1969) Soobshch Akad Nauk SSSR 56:329 (Chem Abst 72:79598)
183. Moore LO (1971) J Phys Chem 75:2075
184. Du Pont de Nemours and Co (1967) French Patent 1,471,692
185. Ashton DS, Shand DJ, Tedder JM, Walton JC (1975) J Chem Soc Perkin Trans II 4:320
186. Rudolph W, Massonne J (1973) German Patent 2,162,368 (to Kali-Chemie AG) 20-06-1973 (Chem Abst 79:79519)
187. Asahi Glass (1973) Jap Patent 48–42852

188. Rebsdat S, Schuierer E, Hahn H (1970) German Patent 1,915,395 (to Hoechst AG) 02-07-1970 (Chem Abst 73:76644)
189. Rudolph W, Massonne J (1972) French Patent 2,163,444 (to Kali-Chemie AG) 31-08-1973 (Chem Abst 80:3066)
190. Felix B (1973) German patent 2,164,567 (to Hoechst AG) 05-07-1973 (Chem Abst 79:78076)
191. Hoescht AG (1976) German patent 2,542,496
192. Nagai M, Shinkai H, Kato T, Asaoka M, Nakatsu T, Fujii Y (1974) Jap patents 61,103 and 61,104 (to Daikin Kogyo Co) 13-06-1974 (Chem Abst 81:119910 and 119911)
193. EI Du Pont de Nemours (1965) US Patent 3,226,449
194. Iserson H, Magazzu JJ, Osborn SW (1972) German Patent 2,130,378 (to Thiokol Chem Co) 27-01-1972 (Chem Abst 76:99090)
195. Daikin Kogyo (1981) Eur Pat Appl 0,045,070
196. Tittle B, Platt AE (1965) British Patent 1,007,542 (to ICI) 13-10-1965 (Chem Abst 64:1956)
197. Battais A, Boutevin B, Pietrasanta Y, Bertocchio R, Lantz A (1989) J Fluorine Chem 42:215
198. Chen Q, Su D, Zhu R (1987) J Fluorine Chem 36:483
199. EI Du Pont de Nemours (1965) French patent 1,385,682
200. Jaeger H (1969) South African Patent 06,187 (to Ciba Ltd) 17-03-1969 (Chem Abst 71:101282)
201. Jaeger H (1970) German Patent 2,001,140 (to Ciba Ltd) 30-07-1970 (Chem Abst 73:109259)
202. Rudolph W, Massonne J, Jaeger HH (1972) German Patent 2,034,472 (to Kali-Chemie AG) 10-02-1972 (Chem Abst 77:6148)
203. ICI (1968) British Patent 1,127,045
204. Hong L, Hu C, Ying S, Mao S (1987) Gaofenzi Xuebao 1:30 (Chem Abst 107:116038)
205. Tittle B (1971) British Patent 1,242,712 (to ICI) 11-08-1971 (Chem Abst 75:129321)
206. Caporiccio G, Bargigia G, Guidetti G (1978) German Patent 2,735,210 (to Montedison) 09-02-1978 (Chem Abst 88:153252)
207. Shanghai Guangming Electroplating Factory (1977) Hua Hsueh Hsueh Pao 35:209 (Chem Abst 89:196929)
208. Boutevin B, Doheim M, Pietrasanta Y, Rigal G (1979) J Fluorine Chem 13:29
209. Boutevin B, Pietrasanta Y (1988) French patent 01,882 (to Atochem) 17-02-1988
210. Boutevin B, Hervaud Y, Rolland L (1983) French patent 01,128 (to Gaz de France)
211. Dannels B (1985) Eur Pat Appl 140,385 (to Occidental Chem)
212. Boutevin B, Maubert C, Mebkhout A, Pietrasanta Y (1981) J Polym Sci Part A Polym Chem 19:499 and 511
213. Boutevin B, Doheim M, Pietrasanta Y, Rigal G (1979) J Polym Sci J Polym Sci Part A Polym Chem 14:29
214. Mohan S (1983) Eur Pat Appl 93,580 (to Occidental Chem) 24-04-1983
215. Boutevin B, Pietrasanta Y, Sideris A (1975) Compte Rendu Acad Sci (Ser C) 281:405
216. Henne AL, Kraus DW (1951) J Amer Chem Soc 73:15303
217. Pietrasanta Y, Rabat JP, Vernet JL (1974) Eur Polym J 10:633
218. Dannels B, Olsen D (1989) German Patent 3,837,394 (to Occidental Chem) 18-05-89 (Chem Abst 112:21425)
219. Battais A, Boutevin B, Pietrasanta Y (1979) J Fluorine Chem 14:467
220. Boutevin B, Maliszewicz M (1983) Makromol Chem 184:977
221. Boutevin B, Hervaud Y, Pietrasanta Y (1984) Phosphorous Sulfur 20:189
222. Park JD, Lacher JP (1973) Daehan Hwahak Hwoejee 17:70 (Chem Abst 73:158818)
223. Paleta O, Posta A, Frantisek C (1970) Coll Czech Chem Comm 35:1302 (Chem Abst 72:131976)
224. Rogozinski M, Shorr LM, Hasman U, Ader-Barcas D (1968) J org Chem 33:3859
225. Ehrenfeld RL (1957) US Patent 2,788,375 09-04-1957 (Chem Abst 51:11759)
226. Dannels BF, Tang DY (1989) German patent 3,820,934 (to Occidental Chem) 18-05-1989 (Chem Abst 111:177752)

227. Barnhart WS, Cranford NJ (1959) US Patent 2,875,253 (to 3 M) 24-02-1959 (Chem Abst 53:12751)
228. Henne AL, Kraus DW (1954) J Amer Chem Soc 76:1175
229. Yung KK (1968) French Patent 1,533,794
230. Boutevin B, Cals J, Pietrasanta Y (1976) Eur Poly J 12:225
231. Boutevin B, Cals J, Pietrasanta Y (1976) Eur Poly J 12:231
232. Yung KK (1967) J Org Chem 32:3673
233. Hauptschein M, Braid M (1965) US Patent 3,219,712 (to Pennsalt Chem Co) 23-11-1965 (Chem Abst 64:8031)
234. Gumprecht WH, Dettre RH (1975) J Fluorine Chem 5:245
235. Amiry MP, Chambers RD, Greenhall MP, Améduri B, Boutevin B, Caporiccio G, Gornowicz GA, Wright AP (1993) Polym Prepr Amer Chem Soc Div Polym Chem 34:411 and to be submitted in J Fluorine Chem
236. Chambers RD, Musgrave WKR, Savory J (1961) J Chem Soc 3779 and (1961) Proc Chem Soc 113
237. Haszeldine RN (1955) J Chem Soc 4291
238. Kremlev MM, Cherednichenko PG, Moklyachuk LI, Yagupolskii LM (1989) Zh Org Khim 25:2582 (Engl Transl) Chem Abst 113:39872
239. Chambers RD, Greenhall MP, Wright AP, Caporiccio G (1995) J Fluorine Chem 73:87
240. Haszeldine RN, Steele BR (1953) J Chem Soc 1592
241. Henne RN, Kraus DW (1951) J Amer Chem Soc 73:1791
242. Caporiccio G (1991) US Patent 5,041,588 (to Dow Corning Co) 20-08-1991 (Chem Abst 115:281856)
243. Matsuo H, Kawakami S, Ito K (1985) Jap patent 60,184,032 (to Asahi Glass Co Ltd) 19-09-1985 (Chem Abst 104:109007)
244. Kiseleva LN, Cherstkov JE, Sterlin SA, Mysov EI, Velichko FK, German LS (1989) Izv Akad Nauk SSSR Ser Khim 5:1130 (Chem Abst 112:76318)
245. Barnhart WS (1969) German patent 1,296,383 (to 3M) 29-05-1969 (Chem Abst 71: 39817)
246. Barnhart WS (1958) US Patent 2,820,772 (to 3M) 21-02-1958 (Chem Abst 52:6845)
247. Liska F, Nemec M, Dedek V (1974) Coll Czech Chem Com 39:580 and 689
248. Chutny B (1980) Czech Patent 182999 15-04-1980 (Chem Abst 94:140412)
249. Chutny B, Liska F, Dedek V (1969) Large Radiat Sources Ind Process ; Proc Symp Util Large Radiat Sources Accel Ind Process 31 (Chem Abst 73:44804) and Ustav Jad Vyzk N° 2237-ch (Chem Abst 74:17968)
250. Emrick DD (1964) US Patent 3,136,808 (to Standard Oil Co) 09-06-1964 (Chem Abst 61:6921)
251. Powell RL, Chambers RD (1996) British Patent 2292151 (to ICI) 14-02-1996 (Chem Abst 125:59403)
252. Fleming GL, Haszeldine RN, Tipping AE (1973) J Chem Soc Perkin Trans 1 574
253. Haszeldine RN, Mir I, Tipping AE, Wilson AG (1976) J Chem Soc Perkin Trans 1 1170
254. Haszeldine RN, Steele BR (1957) J Chem Soc 2800
255. Haszeldine RN, Keen DW, Tipping AE (1970) J Chem Soc C 414
256. Bissel ER (1964) J Chem Soc 252
257. Améduri B, Boutevin B, Kostov G, Petrova P to be submitted to J Fluorine Chem
258. Cosca AT (1961) J Chem Soc 2995
259. Anhudinov AK, Ryazanova RM, Sokolov SV (1974) J Org Chem USSR 10:2520
260. Kremlev MM, Moklyachuk LI, Fialkov Yu A, Yagupol'skii LM (1984) Zh Org Khim 20:1162
261. Haszeldine RN, Osborne JC (1956) J Chem Soc 61
262. Putnam RE (1989) Polymerization of Fluoro Monomers. In: Allen G, Bevington JC, Eastmond AL, Russo A (eds) Comprehensive Polymer Sc, Pergamon, Oxford 3:321
263. Eleuterio HS (1960) US Patent 2,958,685 Chem Abst (1960) 60:20875.
264. Jaeger H (1978) US Patent 4,067,916 (to Ciba Ltd) Chem Abst (1978) 89:120598.
265. Lin WT, Chen CH, Hung HC, Chou HY, Hsun TF, Huang WY (1964) Ko Fen Tzu T'ung Hsun 6:363 (Chem Abst 63:16147)

266. Chambers RD, Fuss RW, Jones M, Sartori P, Swales AP, Herkelmann R (1990) J Fluorine Chem 49:409
267. Haszeldine RN, Rowland R, Tipping AE, Tyrrell G (1982) J Fluorine Chem 21:253
268. Haszeldine RN (1953) J Chem Soc 3559
269. Tedder JM, Walton JC, Vertommen LLT (1979) J Chem Soc Farad Trans 1 75:1040
270. Pollet T (1991) Thése de Doctorat CNAM, University of Montpellier
271. Low HC, Tedder JM, Walton JC (1976) J Chem Soc Faraday Trans 1 72:1300
272. Kirchenbaum AD, Streng AG, Hauptschein M (1953) J Chem Soc 75:3141
273. Hauptschein M, Braid M, Lawlor FE (1957) J Amer Chem Soc 79:2549
274. Deev LE, Nazarenko TI, Pashkevich KI (1988) Izv Akad Nauk SSSR Ser Khim 402 (Chem Abst 109:169809)
275. Baum K, Malik AA (1994) J Org Chem 59:6804
276. Dyatkin BL, Molchalina EP, Knunyants IL (1961) Dokl Akad Nauk SSSR (english translation) 139:106
277. Johari DP, Tedder JM, Walton JC (1971) J Chem Soc 95
278. Tedder JM, Walton JC (1974) J Chem Soc Faraday I 70:308
279. Haszeldine RN, Steele BR (1953) J Chem Soc 1199
280. Rondestvedt CS Jr (1967) French Patent 1,521,775 (to Du Pont) 19-04-1968 (Chem Abst 71:2955)
281. Ashton DS, Mackay AF, Tedder JM, Tipney DC, Walton JC (1973) J Chem Soc Chem Comm 14:496
282. El Soueni AMR, Tedder JM, Vertommen LLT, Walton JC (1977) Coll Int CNRS 278:411
283. Haszeldine RN (1951) J Chem Soc 2495
284. Ikonnikov NS, Lamova NI, Terent'ev AB (1988) Izv Akad Nauk SSSR Ser Khim 117 (Chem Abst 109:169831)
285. Shostenko AG, Dobrov IV, Chertorizhskii AV (1983) Khim Prom-st 339 (Chem Abst 99:123027)
286. Terent'ev AB, Vasil'eva TT (1995) Ind Chem Library 7:180
287. Vasil'eva TT, Fokina IA, Vitt SV (1991) Izv Akad Nauk SSSR Ser Khim 1384 (Chem Abst 116:40981) and Gasanov RG, Vasil'eva TT, Gapusenko SI (1991) Kinet Katal 32:1466 (Chem Abst 116:105413)
288. Vasil'eva TT, Kochetkava VA, Dostovalova VI, Nelyubin BV, Freidlina RK (1989) Izv Akad Nauk SSSR Ser Khim 2558 (Chem Abst 113:5666)
289. Vasil'eva TT, Kochetkova VA, Nelyubin BV, Dostovalova VI, Freidlina RK (1987) Izv Akad Nauk SSSR Ser Khim 808 (Chem Abst 108:111753)
290. Haszeldine RN (1952) J Chem Soc (London) 2504
291. Boutevin B, Pietrasanta Y, Youssef B (1987) J Polym Sci Part A Polym Chem 25:3025
292. Boutevin B, Pietrasanta Y (1985) Prog Org Coat 13:297
293. Hauptschein M, Braid M, Lawlor FE (1966) US Patent 3,240,825 (to Pennsalt Chem Co) 15-03-1966 (Chem Abst 64:17422)
294. Paciorek KL, Merkl BA, Lenk CT (1962) J Org Chem 27:1015
295. Paleta O, Kvicala J, Gunter J, Dedek V (1986) Bull Soc Chim Fr 6:920
296. Modena S, Fontana A, Moggi G (1985) J Fluorine Chem 30:109
297. Gregory R, Haszeldine RN, Tipping AE (1968) J Chem Soc (C) 3020
298. Beaune O, Bessiére JM, Boutevin B, El Bachiri A (1995) J Fluorine Chem 73:27
299. Gregory R, Haszeldine RN, Tipping AE (1969) J Chem Soc (C) 991
300. Nema SK, Francis AU, Narendranath PK, Rao KVC (1979) Ind Polym Radiat Proc Symp 89 (Chem Abst 94:122008)
301. Dittman AL (1972) US Patent 3,668,262 (to Halocarbon Pdts Co) (06-06-1972)
302. Chen LS (1990) J Fluorine Chem 47:261
303. Marciniec B (1992) Comprehensive Handbook on Hydrosilylation, Pergamon, Oxford, p220
304. Cirkva V, Paleta O, Améduri B, Boutevin B (1995) J Fluorine Chem 75:87
305. Battais A, Boutevin B, Cot L, Granier W, Pietrasanta Y (1979) J Fluorine Chem 13:531 and 14:467 and ref therein

306. Bargigia G, Tortelli V, Tonelli C, Modena S (1988) Eur Pat Appl 270,956 (to Ausimont) 15-06-1988
307. Fearn JE, Brown DW, Wall LA (1966) J Polym Sci Part A1 4:131
308. Améduri B, Boutevin B (1994) Macromol Symp 82:1
309. Brace NO (1993) J Fluorine Chem 62:217
310. Hauptschein M, Braid M (1963) US Patent 3,089,911 (to Pennsalt Chem Co) 14-05-1963
311. Hauptschein M, Braid M (1963) US Patent 3,091,648 (to Pennsalt Chem Co) 28-05-1963
312. Furukawa Y (1981) Eur Pat Appl 45,070 (to Daikin Kogyo Co Ltd)
313. Wolff NE (1959) US Patent 2,907,795 (to Dupont)
314. Wolff NE (1958) US Patent 2,856,440 (to Dupont)
315. Modena S, Fontana A, Moggi G, Bargigia G (1988) Eur Pat Appl 251,284 (to Ausimont) 07-01-1988 (Chem Abst 109:7844).
316. Tatemoto M, Nakagawa T (1978) German Patent 2,729,671 (12-01-1978) to Daikin Kogyo Co ltd (Chem Abst 88:137374)
317. Hauptschein M, Braid M (1961) US Patent 3,002,031 (to Pennsalt Chem Co)
318. Bannai N, Yasumi H, Hirayama S (1986) US Patent 4,580,981 (to Kureha) 08-04-1986
319. Tatsu H, Okabe J, Naraki A, Abe M, Ebina Y (1987) German Patent 3,710,818 (to Nippon Mecktron Co Ltd) 08-10-1987 (Chem Abst 108:76908)
320. Caporiccio G (1995) US Patent 5,350,878 (to Dow Corning) 24-09-1994 (Chem Abst 123:113164)
321. Kasahara M, Kai M (1994) Jap Patent 6,331,040 (to Asahi Chem Ind) 29-11-1994 (Chem Abst 122:190156)
322. Fossey J, Lefort D, Sorba J (1995) Free Radicals in Organic Chemistry, Wiley, New-York
323. March J (1992) Advanced Organic Chemistry, 4th edn. Wiley, New York
324. Arnaud R, Vidal S (1992) New J Chem 16:471 and (1986) J Chem Soc Perkin Trans II 1517
325. Giese B (1983) Angew Chem Int Ed Engl 22:753
326. Rüchardt C (1980) Top Curr Chem 88:1
327. Beckwith ACJ (1981) Tetrahedron 37:3073
328. Koutecky VB, Koutecky J, Salem L (1977) J Amer Chem Soc 99:842
329. Canadell E, Eisenstein O, Ohanessian G, Poblet JM (1985) J Phys Chem 89:4856
330. Poblet JM, Canadell E, Sordo T (1983) Can J Chem 61:2068
331. Delbecq F, Ilavsky D, Anh NT, Lefour JM (1985) J Amer Chem Soc 107:1623
332. Wong MW, Pross A, Radom L (1994) J Amer Chem Soc 116:11938
333. Rozhkov IN, Borisov Yu A (1989) Izv Akad Nauk SSSR Ser Khim 8:1801
334. Rempel GD, Borisov Yu A, Raevskii Ni, Igumnov SM, Rozhkov IN (1990) Izv Akad Nauk SSSR Ser Khim 5:1059
335. Xu ZQ, Li SS, Hu CM (1991) Ch J Chem 9:335
336. Tedder JM, Walton JC, Winton KDR (1971) J Chem Soc 1046
337. Sidebottom HW, Tedder JM, Walton JC (1972) Int J Chem Kinet 4:249
338. Tedder JM, Walton JC (1980) Adv Free-Radical Chem 6:155
339. El Soueni A, Tedder JM, Walton JC (1978) J Fluorine Chem 11:407
340. Vertommen LLT, Tedder JM, Walton JC (1977) J Chem Res Part S (Synop) 18
341. Klabunde KJ (1970) J Amer Chem Soc 92:2427
342. Okafo EN, Whittle E (1975) Inter J Chem Kinet 7:287
343. Liska F, Simek S (1970) Collect Czech Chem Commun 35:1752
344. Szwarc M, Levy M, Milkovich M (1956) J Am Chem Soc 78:2656
345. Otsu T, Yoshida M (1982) Makromol Chem Rapid Commun 3:127
346. Otsu T, Yoshida M, Tazaki T (1982) Makromol Chem Rapid Commun 3:133
347. Solomon DH, Rizzardo E, Cacioli P (1984) Eur Pat Appl 0,135,280
348. Georges MK, Veregin RPN, Kazmaier PM, Hamer GK (1993) Macromolecules 26:2987
349. Améduri B, Boutevin B, Gramain P (1997) Adv Polym Sci 127:87
350. Vegerin RPN, Georges MK, Hamer GK, Kazmaier PM (1995) Macromolecules 28:4391
351. Matyjaszewski K, Gaynor S, Greszta D, Mardare D, Shigemoto T (1995) Macromol Symp 98:73
352. Solomon DH, Rizzardo E, Cacioli P (1986) US Patent 4,581,429

353. Bertin D, Boutevin B (1996) Polym Bull 37:337; Bertin D, Boutevin B, Nicol P (1996) French Patent 96,05909 (to ElfAtochem SA) 13-05-1996
354. Hawker CJ, Carter KR, Hedrick JL, Voeksen W (1995) Polym Prepr 36:10
355. Druliner JD (1991) Macromolecules 24:6079
356. Otsu T, Yoshida M (1986) Polym Bull 16:277
357. Otsu T, Matsumoto A (1994) "Macroiniferters: controlled synthesis and design through living radical polymerization". In: Mishra MK (ed) Macromolecular Design: Concept and Practice, Polymer Frontier International, Hopewell Jct (USA) Chapter 12:471
358. Demircioglu P, Acar MH, Yagci Y (1992) J Appl Poly Sci 46:1639
359. Braun D, Skrzek T, Steinhauer-Beisser S, Tretner H, Lindner HJ (1995) Macromol Chem Phys 196:573
360. Braun D, Skrzek T (1995) Makromol Chem Phys 196:4039
361. Otsu T (1956) J Polym Sci 21:559
362. Otsu T, Nayatani K, Muto I, Ima M (1958) Makromol Chem 27:142
363. Reghunadhan Nair CP, Chaumont P, Clouet G (1994) Application of thermal iniferters in free radical polymerization: a new trend in macromolecular design. In: Mishra MK (ed), Macromolecular Design: Concepts and Practice. Polymer Frontiers International, Hopewell Jct, 11:433
364. Teyssié Ph (1994) Macromol Chem Macromol Symp 88:1
365. Wayland BB, Poszmik G, Mukerjee SL, Fryd M (1994) J Amer Chem Soc 116:7943
366. Mun Y, Sato T, Otsu T (1984) Makromol Chem 185:1507
367. Percec V, Barboin B (1995) Macromolecules 28:7970
368. Wang JS, Matyjaszewski K (1995) J Amer Chem Soc 117:5614
369. Arvanitopoulos LD, Greuel MP, Harwood HJ (1994) ACS Polym Prepr 35(2):549
370. Otsu T, Tazaki T, Yoshioka M (1990) Chemistry Express 5:801
371. Chung TC, Jiang GJ (1992) Macromolecules 25:4816
372. Kato M, Kamigaito M, Sawamoto M, Higashimura T (1995) Macromolecules 28:1721
373. Kotora M, Kvicala J, Améduri B, Hajek M, Boutevin B (1993) J Fluorine Chem 64:259
374. Tatemoto M (1990) Kagaku Kogyo 41(1):78 (Chem Abst 114:8081v) and (1992) Kobunshi Ronbunshu 49(10):765 (Chem Abst 118:22655z)
375. Tatemoto M, Tomoda M, Ueta Y (1980) Ger Patent DE 29,401,35 (to Daikin Kogyo Co Ltd Jap) (Chem Abst 93:27580)
376. Matyjaszewski K, Gaynor SG, Greszta D, Mardare D, Shigemoto T, Wang JS (1955) Macromol Symp 95:217
377. Gaynor SC, Wang JS, Matyjaszewski K (1995) Polym Preprints 36:467
378. Tatemoto M, Morita S (1981) Eur Pat Appl EP 27,721 (to Daikin Kogyo Co Ltd) (Chem Abst 95:170754)
379. Tatemoto M, Furukawa Y, Tomoda M, Oka M, Morita S (1980) Eur Pat Appl EP 14,930 (to Daikin Kogyo Co Ltd Jap) (Chem Abst 94:48603)
380. Tatemoto M (1985) Int Poly Sc Tech 12(4):85 translated in english from (1984) Nippon Gomu Kyokaishi 57(11):761
381. Oka M, Tatemoto M (1984) Contemp Topics Polym Sci 4:763
382. Tatemoto M, (1990) Eur Patent Appl 399,543 (to Daikin Kogyo Co Ltd Jap) (Chem Abst 114:166150)
383. Vergé C (1991) Thése de l'Université de Montpellier II, France
384. Balagué J, Améduri B, Boutevin B, Caporiccio G submitted to J Fluorine Chem
385. Balagué J, Améduri B, Boutevin B, Caporiccio G submitted to J Fluorine Chem
386. Baum K, Archibald TG, Malik AA (1993) US Patent 5,204,441 (to Fluorochem Inc)
387. Manséri A, Améduri B, Boutevin B, Chambers RD, Caporiccio G, Wright AP (1996) J Fluorine Chem 78:145
388. Manséri A, Améduri B, Boutevin B, Caporiccio G (1997) J Fluorine Chem 81:103
389. Améduri B, Boutevin B, Caporiccio G, Guida-Pietrasanta F, Ratsimihéty A. Use of fluorinated telomers in the preparation of hybrid silicones. In: Hougham G, Davidson T, Cassidy P (Eds) Fluorinated Polymers: Synthesis and Applications, Plenum, New York (in press)

390. Dannels BF (1984) Eur Pat Appl 140,385 A3 (to Occidental Chem Co) (31-10-1984)
391. Saran NS (1983) Eur Pat Appl 93,579 A2 (to Occidental Chem Co) (27-04-1983) (Chem Abst 100:68969)
392. Abe T, Nagase S (1982) Electrochemical Fluorination as a Route to Perfluorinated Organic Compounds of Industrial Interest. In: Banks RE (ed) Preparation, Properties and Industrial Applications of Organofluorine Compounds. Ellis Horwood, Chichester 1:19
393. Mitsubishi Rayon Co (1983) Jap patent 58007602 A2 (17-01-1983) Chem Abst 99:72093
394. Kuharcik SE (1996) PhD Dissertation, Pennsylvania State University (Chem Abst 124:318077)
395. Ref 3 p278 and Valaitis JK, Kyker GS (1979) J Appl Polymer Sci 23:765
396. Tate DP (1974) J Polym Sci Part C Phys Chem 48:33
397. Kobayashi H, Owen MJ (1995) Trends Polym Sci 3:330
398. Boutevin B, Ranjalahy Rasoloarijao L, Rousseau A, Garapon J, Sillion B (1992) Makromol Chem 193:1995
399. Takekoshi T (1990) Adv Polym Sci 94:1
400. Boutevin B, Robin JJ (1992) Adv Polym Sci 102:105
401. Améduri B, Boutevin B (1992) Adv Polym Sci 102:133
402. Trombetta T, Tonelli C, Castiglioni G, Ajroldi G (1992) 10th European Symp in Fluorine Chem, Padova, Proceedings p210
403. McLoughlin VCR, Thrower J (1969) Tetrahedron Lett 25:5921
404. Chen GJ, Chen LS, Eapen KC (1993) J Fluorine Chem 63:113

Author Index Volumes 151–192

Author Index Vols. 26–50 see Vol. 50
Author Index Vols. 51–100 see Vol. 100
Author Index Vols. 101–150 see Vol. 150

The volume numbers are printed in italics

Springer and the environment

Printing: Saladruck, Berlin
Binding: Buchbinderei Lüderitz & Bauer, Berlin